W0230105

Isaac Ishaaya (Ed.)

Biochemical Sites of Insecticide Action and Resistance

Springer

Berlin
Heidelberg
New York
Barcelona
Hong Kong
London
Mailand
Paris
Singapore
Tokyo

Isaac Ishaaya (Ed.)

Biochemical Sites of Insecticide Action and Resistance

With 46 Figures

 Springer

Professor Dr. Isaac Ishaaya
Department of Entomology
Agricultural Research Organization
The Volcani Center
Bet Dagan 50250
Isarael

ISBN 3-540-67625-2 Springer-Verlag Berlin Heidelberg New York

Library of Congress Cataloging-in-Publication Data

Biochemical sites of insecticide action and resistance / Isaac Ishaaya (ed.)
    p.   cm.
    Includes bibliographical references.
    ISBN 3540676252
    1. Insecticides – Mechanism of action.   2. Insecticide resistance.   I. Ishaaya, I.

  SB951.5 .B56   2000
  632′.9517 – dc21   00-063580

This work is subject to copyright. All rights reserved, whether the whole or part of the material is concerned, specifically the rights of translation, reprinting, reuse of illustrations, recitation, broadcasting, reproduction on microfilm or in any other way, and storage in data banks. Duplication of this publication or parts thereof is permitted only under the provisions of the German Copyright Law of September 9, 1965, in its current version, and permission for use must always be obtained from Springer-Verlag. Violations are liable for prosecution under the German Copyright Law.

Springer-Verlag Berlin Heidelberg New York
a member of Bertelsmannspringer Science + Business Media GmbH

© Springer-Verlag Berlin Heidelberg 2001
Printed in Germany

The use of general descriptive names, registered names, trademarks, etc. in this publication does not imply, even in the absence of a specific statement, that such names are exempt from the relevant protective laws and regulations and therefore free for general use.

Cover design: Design & Production GmbH, Heidelberg
Typesetting: Best-set Typesetter Ltd., Hong Kong
SPIN 10731629   31/3130 – 5 4 3 2 1 0 – Printed on acid-free paper

Preface

In recent years many of the conventional methods of insect control by broad-spectrum synthetic chemicals have come under scrutiny because of their undesirable effects on human health and the environment. In addition, some classes of pesticide chemistry, which generated resistance problems and severely affected the environment, are no longer used. It is against this background that the authors of this book present up-to-date findings relating to biochemical sites that can serve as targets for developing insecticides with selective properties, and as the basis for the elucidation of resistance mechanisms and countermeasures.

The book consists of eight chapters relating to biochemical targets for insecticide action and seven chapters relating to biochemical modes of resistance and countermeasures. The authors of the chapters are world leaders in pesticide chemistry, biochemical modes of action and mechanisms of resistance. Biochemical sites such as chitin formation, juvenile hormone and ecdysone receptors, acetylcholine and GABA receptors, ion channels, and neuropeptides are potential targets for insecticide action. The progress made in recent years in molecular biology (presented in depth in this volume) has led to the identification of genes that confer mechanisms of resistance, such as increased detoxification, decreased penetration and insensitive target sites. A combination of factors can lead to potentiation of the resistance level. Classifications of these mechanisms are termed gene amplification, changes in structural genes, and modification of gene expression.

This book is intended to serve as a text for researchers, university professors and graduate students involved in developing new groups of insecticides suited to integrated pest management (IPM) and insecticide resistance management (IRM) programs. The data presented are essential for the establishment of new technologies for developing compounds that will impact our future agricultural practices.

In the preparation of the manuscript, the editor and the authors are indebted to the reviewers of the various chapters for valuable suggestions and criticism: J.R. Bloomquist (USA), D. Dean (USA), L.M. Field (UK), K. van Frankenhuyzen (Canada), R.V. Gunning (Australia), I. Ishaaya (Israel), A. Koch (USA), R. Laprade (Canada), H. Oberlander (USA), J.-L. Schwartz (Canada), J.G. Scott (USA), B.D. Siegfried (USA), S. Sohi (Canada), C.-N. Sun (Taiwan),

V. Vachon (Canada), P. Weintraub (Israel), H. Wieczorek (Germany), M.S. Williamson (UK), and M. Wolfersberger (USA).

I thank Mrs. Svetlana Kontsedalov for her patience in typing and organizing the various sections. I am indebted to my wife Mrs. Eulalie Ishaaya-van Hoye for her support and assistance throughout the preparation of the book.

Isaac Ishaaya

Contents

Insecticides Affecting Voltage-Gated Ion Channels
E. ZLOTKIN

Acetylcholine Receptors as Sites for Developing Neonicotinoid Insecticides
R. NAUEN, U. EBBINGHAUS-KINTSCHER, A. ELBERT, P. JESCHKE, and K. TIETJEN

Ecdysteroid and Juvenile Hormone Receptors: Properties and Importance in Developing Novel Insecticides
S.R. PALLI and A. RETNAKARAN

Imaginal Discs and Tissue Cultures as Targets for Insecticide Action
H. OBERLANDER and G. SMAGGHE

Insect Neuropeptide Antagonists: a Novel Approach for Insect Control
M. ALTSTEIN and C. GILON

Ion Balance in the Lepidopteran Midgut and Insecticidal Action
of *Bacillus thuringiensis*
J.L. GRINGORTEN

Evolution of Amplified Esterase Genes as a Mode of Insecticide Resistance In Aphids
L.M. FIELD, R.L. BLACKMAN, and A.L. DEVONSHIRE

Insensitive Acetylcholinesterase as Sites for Resistance to Organophosphates and Carbamates in Insects: Insensitive Acetylcholinesterase Confers Resistance in Lepidoptera
R.V. GUNNING and G.D. MOORES

Glutathione *S*-Transferases and Insect Resistance to Insecticides
C.-N. SUN, H.-Y. HUANG, N.-T. HU, and W.-Y. CHUNG

Cytochrome P450 Monooxygenases and Insecticide Resistance: Lessons from *CYP6D1*

J.G. SCOTT

Mechanisms of Organophosphate Resistance in Insects

B.D. SIEGFRIED and M.E. SCHARF

Insect Midgut as a Site for Insecticide Detoxification and Resistance
G. SMAGGHE and L. TIRRY

**Impact of Insecticide Resistance Mechanisms
on Management Strategies**
A.R. HOROWITZ and I. DENHOLM

Contributors

M. ALTSTEIN (e-mail: vinnie2@netvision.net.il)
Department of Entomology, The Volcani Center, ARO, Bet Dagan,
50250 Israel

R.L. BLACKMAN (e-mail: r.blackman@nhm.ac.uk)
Natural History Museum, Cromwell Road, London, SW7 5BD, UK

J.R. BLOOMQUIST (e-mail: jbquist@vt.edu)
Department of Entomology, Virginia Polytechnic Institute and State
University, Blacksburg, VA 24061, USA

W.-Y. CHUNG (e-mail: cnsun@hotmail.com)
Department of Entomology, National Chung-Hsing University, Taichung,
Taiwan 40227, Republic of China

I. DENHOLM (e-mail: ian.denholm@bbsrc.ac.uk)
Department of Biological and Ecological Chemistry, IACR-Rothamsted,
Harpenden, Herts AL5 2JQ, UK

A.L. DEVONSHIRE (e-mail: alan.devonshire@bbsrc.ac.uk)
IACR-Rothamsted, Harpenden, Herts AL5 2JQ, UK

U. EBBINGHAUS-KINTSCHER
(e-mail: ulrich.ebbinghaus-kintscher.ue@bayer-ag.de)
Bayer AG, Agrochemicals Division, Research Insecticides,
51368 Leverkusen, Germany

A. ELBERT (e-mail: alfred.elbert.ae@bayer-ag.de)
Bayer AG, Agrochemicals Division, Research Insecticides,
51368 Leverkusen, Germany

L.M. FIELD (e-mail: lin.field@bbsrc.ac.uk)
IACR-Rothamsted, Harpenden, Herts AL5 2JQ, UK

C. GILON (e-mail: gilon@vms.huji.ac.il)
Institute of Organic Chemistry, The Hebrew University, 91904 Jerusalem,
Israel

J.L. GRINGORTEN (e-mail: lgringo@nrcan.gc.ca)
Canadian Forest Service, Great Lakes Forestry Centre, 1219 Queen Street
East, Sault Ste. Marie, Ontario P6A 5M7, Canada

R.V. GUNNING (e-mail: rgunning@enternet.com.au)
NSW Agriculture, Tamworth Centre for Crop Improvement,
RMB 944 Calala Lane, Tamworth, NSW, Australia 2340

A.R. HOROWITZ (e-mail: hrami@netvision.net.il)
Department of Entomology, Agricultural Research Organization,
The Volcani Center, Bet Dagan 50250, Israel

N.-T. HU (e-mail: nthu@dragon.nchu.edu.tw)
Graduate Institute of Biochemistry, National Chung-Hsing University,
Taichung, Taiwan 40227, Republic of China

H.-Y. HUANG (e-mail: cnsun@hotmail.com)
Department of Entomology, National Chung-Hsing University, Taichung,
Taiwan 40227, Republic of China

I. ISHAAYA (e-mail: vpisha@netvision.net.il)
Department of Entomology, Agricultural Research Organization,
The Volcani Center, Bet Dagan 50250, Israel

P. JESCHKE (e-mail: peter.jeschke.pj@bayer-ag.de)
Bayer AG, Agrochemicals Division, Research Insecticides,
51368 Leverkusen, Germany

G.D. MOORES (e-mail: graham.moores@bbsrc.ac.uk)
Department of Biological and Ecological Chemistry, IACR-Rothamsted,
Harpenden, Herts AL5 2JQ, UK

R. NAUEN (e-mail: ralf.nauen.rn@bayer-ag.de)
Bayer AG, Agrochemicals Division, Research Insecticides,
51368 Leverkusen, Germany

H. OBERLANDER (e-mail: hoberlander@gainesville.usda.ufl.edu)
Center for Medical, Agricultural and Veterinary Entomology, Agricultural
Research Service, U.S. Department of Agriculture, P.O. Box 14565,
Gainesville, FL 32604, USA

S.R. PALLI (e-mail: rahsub@rohmhaas.com)
Rohm and Haas Research Laboratories, 727 Norristown Road,
Spring House, PA 19477, USA

A. RETNAKARAN (e-mail: aretnak@nrcan.gc.ca)
Canadian Forest Service, Great Lakes Forestry Centre, 1219 Queen Street
East, Sault Ste. Marie, Ontario, Canada P6A 5M7

M.E. SCHARF (e-mail: mscharf1@unl.edu)
Department of Entomology, 202 Plant Industry Building, University of
Nebraska, Lincoln, NE 68583-0816, USA

J.G. SCOTT (e-mail: Jgs@cornell.edu)
Department of Entomology, Comstock Hall, Cornell University, Ithaca,
NY 14853-0901, USA

B.D. SIEGFRIED (e-mail: bsiegfried1@unl.edu)
Department of Entomology, 202 Plant Industry Building, University of
Nebraska, Lincoln, NE 68583-0816, USA

G. SMAGGHE (e-mail: guy.smagghe@rug.ac.be)
Laboratory of Agrozoology, Faculty of Agricultural and Applied Biological
Sciences, Ghent University, Coupure Links 563, 9000 Ghent, Belgium

C.-N. SUN (e-mail: cnsun@hotmail.com)
Department of Entomology, National Chung-Hsing University, Taichung,
Taiwan 40227, Republic of China

K. TIETJEN (e-mail: klaus.tietjen.kt@bayer-ag.de)
Bayer AG, Agrochemicals Division, Research Insecticides, 51368
Leverkusen, Germany

L. TIRRY (e-mail: luc.tirry@rug.ac.be)
Laboratory of Agrozoology, Faculty of Agricultural and Applied Biological
Sciences, Ghent University, Coupure Links 653, 9000 Ghent, Belgium

E. ZLOTKIN (e-mail: zlotkin@vms.huji.ac.il)
Department of Cell and Animal Biology, Institute of Life Sciences,
The Hebrew University, 91904 Jerusalem, Israel

Biochemical Processes Related to Insecticide Action: an Overview

Isaac Ishaaya[1]

1
Introduction

Throughout modern history, man has devised various methods to combat insect pests such as the use of sulfur, chalk, wood ash and plant extracts. Further progress came with the introduction of botanical compounds such as pyrethrum, deris, quassia, and others. The inventory of insecticides used in the 19th century includes sulfur, arsenic, fluorides, soaps, kerosene and various botanicals such as nicotine, rotenone, pyrethrum, sabadilla and quassia (for more details, see Retnakaran et al. 1985; Perry et al. 1998).

During the 20th century, significant progress in the synthesis of new chemicals and standardization of techniques and bioassays has resulted in an evaluation of the structures and biological activity of various compounds. Synthetic insecticides, such as chlorinated hydrocarbons, organophosphates and carbamates, have been developed and used to control insect pests over the past five decades, minimizing losses in agricultural yield and improving human health. Unfortunately, many of these chemicals are harmful to man and beneficial organisms. In some cases they are too persistent in the environment and cause ecological disturbances.

Efforts have been made during the past three decades to develop novel insecticides with selective properties to act on biochemical sites or physiological processes present in a special insect group but differ from others in their properties. This approach has led to the formation of compounds which affect the hormonal regulation of molting and developmental processes in insects such as ecdysone agonists (Wing 1988; Dhadialla et al. 1998), juvenile hormone mimics (Ishaaya and Horowitz 1992, 1995; Ishaaya et al. 1994) and chitin synthesis inhibitors (Cohen 1987; Ishaaya 1990; Oberlander and Silhacek 1998). In addition, compounds which inhibit or enhance the activity of biochemical sites, such as respiration (diafenthiuron) (Ishaaya et al. 1993), or interact with nicotine acetylcholine receptors (imidacloprid, acetamiprid and thiamethoxam) have been introduced for the control of

[1] Department of Entomology, Agricultural Research Organization, The Volcani Center, Bet Dagan 50250, Israel

Isaac Ishaaya (Ed.): Biochemical Sites of Insecticide Action and Resistance
© Springer-Verlag Berlin, Heidelberg 2001

aphids and whiteflies (Elbert et al. 1998; Ishaaya and Horowitz 1998). Compounds originating from natural products (abamectin, emamectin, milbemectin and spinosad) and which act on specific biochemical sites, such as GABA and glutamate receptors and chloride channels, have been developed and used successfully to combat agricultural pests (Jansson and Dybas 1998; Bloomquist, this Vol.). *Bacillus thuringiensis* δ-endotoxin, which affects the midgut ion exchange, has been developed as an insecticide to control lepidopterans or introduced as a component in transgenic field crops aimed at suppressing important agricultural insect pests (Whalon and McGaughey 1998; Gringorten, this Vol.).

The aim of this report is to present general information on biochemical sites related to insecticide action, while the various chapters of the book discuss up-to-date details and uses of biochemical and physiological sites as targets for developing novel insecticides and of processes relating to resistance mode of action and management.

2
Chitin Synthesis Inhibition

Over the past three decades, two groups of compounds, the benzoylphenyl ureas and buprofezin, have been developed and used as commercial compounds for controlling agricultural pests. Benzoylphenyl ureas act on insects of various orders by inhibiting chitin formation (Ishaaya and Casida 1974; Post et al. 1974) thereby causing abnormal endocuticular deposition and abortive molting (Mulder and Gijswijt 1973). Studies with diflubenzuron [1-(4-chlorophenyl)-3-(2,6-diflubenzoyl)urea], the first commercial compound and the most investigated one in this group, revealed that the compound alters cuticle composition – especially that of chitin – thereby affecting the elasticity and firmness of the endocuticle (Grosscurt 1978; Grosscurt and Anderson 1980). The reduced level of chitin in the cuticle seems to result from inhibition of biochemical processes leading to chitin formation (Post et al. 1974; Hajjar and Casida 1979; Van Eck 1979). Chitin synthetase is not the primary biochemical site for the reduced level of chitin since, in some studies, benzoylphenyl ureas do not inhibit its activity in cell-free systems (Cohen and Casida 1980; Mayer et al. 1981; Cohen 1985). Some of the reports indicate the possibility that benzoylphenyl ureas might: affect the insect hormonal sites, thereby resulting in physiological disturbances such as inhibition of DNA synthesis (Mitlin et al. 1977; DeLoach et al. 1981; Soltani et al. 1984); alter carbohydrase and phenoloxidase activities (Ishaaya and Casida 1974; Ishaaya and Ascher 1977) or suppress microsomal oxidase activity (Van Eck 1979). Recent studies, using imaginal discs and cell-free systems, indicate that benzoylphenyl ureas inhibit 20E-dependent GlcNAc incorporation into chitin (Mikolajczyk et al. 1994; Oberlander and Silhacek 1998). These findings suggest that benzoylphenyl ureas affect ecdysone-dependent biochemical sites which lead to chitin inhibition.

The search for potent acylureas has led to the development of new compounds such as chlorfluazuron (Haga et al. 1982), teflubenzuron (Becher et al. 1983) and hexaflumuron (Sbragia et al. 1983), which are far more potent than diflubenzuron on various agricultural pests (Ishaaya 1990). One of the recent benzoylphenyl ureas, which is in the process of commercialization, is novaluron {Rimon EC-10, 1-[chloro-4-(1,1,2-trifluoromethoxyethoxy)phenyl]-3-(2,6-difluorobenzoyl)urea}. It acts by both ingestion and contact. It is a powerful suppressor of lepidopteran larvae such as *Spodoptera littoralis, S. exigua, S. frugiperda, Helicoverpa armigera* and *Tuta absoluta*, species known to attack cotton, corn and vegetables. It also efficiently controls the whiteflies *Bemisia tabaci* and *Trialeurodes vaporariorum* and the leaf miners *Liriomyza huidobrensis* and *Perileucoptera cofeella*. Our studies indicated that the LC-50 value of Rimon on third-instar *S. littoralis* fed on treated leaves is ~0.1 mg a.i./l. This value resembles that of chlorfluazuron and is tenfold lower than that of teflubenzuron. Novaluron affects larvae of *B. tabaci* to a much greater extent than chlorfluazuron and teflubenzuron. Total larval mortality was obtained at a concentration of 1 mg a.i./l (Ishaaya et al. 1996, 1998). Artificial rain at a rate of 40 mm/h applied 5 and 24 h after treatment in a cotton field had no appreciable effect on the potency of novaluron on *S. littoralis* larvae (unpubl. results). Hence, novaluron can be used in tropical areas and in rainy seasons.

In general, benzoylphenyl ureas have no effect on parasitoids and are considered to have a mild effect on other natural enemies (Ishaaya 1990). As such, they are considered important additions in integrated pest management (IPM) programs.

Buprofezin (Applaud, 2-*tert*-butylimino-3-isopropyl-5-phenyl-1,3,5-thiadiazinan-4-one), a chitin synthesis inhibitor which acts specifically on sucking pests such as plant hoppers and whiteflies, has been developed by Nihon Nohyaku, Japan (Kanno et al. 1981). Its mode of action resembles that of benzoylphenyl ureas, although its structure is not analogous. The compound inhibits incorporation of ^{3}H-glucose and N-acetyl-D-^{3}H-glucosamine into chitin (Izawa et al. 1985; Uchida et al. 1985). The characteristic symptoms in the greenhouse whitefly *Trialeurodes vaporariorum* resembled those obtained with benzoylphenyl ureas (De Cock and Degheele 1991). As a result of chitin deficiency, the procuticle of the whitefly nymphs loses its elasticity and the insect is unable to molt (De Cock and Degheele 1998). Buprofezin is a powerful suppressor of larval metamorphosis and embryogenesis in the rice plant hopper *Nilaparvata lugens* and in the whiteflies *B. tabaci* and *T. vaporariorum* (Nagata 1986; Yasui et al. 1987; Ishaaya et al. 1988).

The chitin synthesis inhibition site has proved to be important for developing control agents which act selectively on important groups of insect pests.

3
Ecdysone and Juvenile Hormone Receptors

Physiological and biochemical processes that govern growth, development, and reproduction in insects are regulated by juvenile and molting hormones. During the past two decades, many investigations have been directed towards elucidating the possible use of ecdysteroids and juvenile hormone (JH) receptors as target sites for developing novel insecticides (Bergamasco and Horn 1980; Horn et al. 1981; Riddiford 1985; Riddiford et al. 1987; Dhadialla et al. 1998). Several substituted dibenzoyl hydrazines which act as ecdysone agonists have been synthesized by Rohm and Haas Co. (Spring House, Pennsylvania, USA). Two compounds, tebufenozide and methoxyfenozide, have been commercialized and used to control lepidopteran pests (Dhadialla et al. 1998; Smagghe and Degheele 1998). These compounds bind to the ecdysteroid receptors thereby initiating the molting process (Wing 1988; Wing et al. 1988; Retnakaran et al. 1995; Palli et al. 1996). Ecdysone agonists are powerful toxicants which act specifically on lepidopteran pests such as *Manduca sexta* (Wing et al. 1988), *Plodia interpunctella* (Silhacek et al. 1990), *Spodoptera frugiperda* (Monthéan and Potter 1992), and *S. littoralis* (Smagghe and Degheele 1992; Ishaaya et al. 1995). The potency of tebufenozide against agricultural pests has been demonstrated against several lepidopteran pests such as the codling moth *Carpocapsa* (*Cydia*) *pomonella* in apple orchards (Heller et al. 1992), the fall armyworm *S. frugiperda* in maize, the bollworm *Heliothis zea* (Chandler et al. 1992) and the armyworms *S. exigua, S. littoralis* and *S. exempta* in cotton, cereal, rice and vegetables (Chandler et al. 1992; Smagghe and Degheele 1994; Ishaaya et al. 1995). Methoxyfenozide, a recently developed ecdysone agonist, was five- to tenfold more potent than tebufenozide, and both have no appreciable cross-resistance with conventional insecticides such as pyrethroids and organophosphates (Ishaaya et al. 1995). These compounds are considered highly selective with no harm to parasitoids and predators (Dhadialla et al. 1998; Smagghe and Degheele 1998) and fit well in IPM and insecticide resistance management (IRM) programs.

Among the JH mimics, fenoxycarb and pyriproxyfen exhibit reasonable field stability and high potency on agricultural pests. Fenoxycarb, ethyl [2-(4-phenoxyphenoxy)ethyl] carbamate, was the first commercial compound to be marketed for the control of agricultural pests (Dorn et al. 1981; Masner et al. 1987; Peleg 1988). Pyriproxyfen, 2-[1-methyl-2-(4-phenoxyphenoxy)ethoxy] pyridine, is a fenoxycarb derivative in which a part of the aliphatic chain has been replaced by pyridyl oxyethylene. The compound is a potent JH mimic affecting the hormonal balance in insects resulting thereby in strong suppression of embryogenesis, metamorphosis and adult formation (Itaya 1987; Kawada 1988; Langley 1990; Koehler and Patterson 1991). Pyriproxyfen is considered a leading compound for controlling whiteflies (Ishaaya and Horowitz

1992, 1995; Ishaaya et al. 1994) and scale insects (Peleg 1988), and it is one of the most important components in IRM strategy in cotton fields (Horowitz and Ishaaya 1994; Dennehy and Williams 1997; Horowitz et al. 1999).

Compounds which act on insect hormonal receptors are considered important selective insecticides for controlling insect pests. Extensive research is being undertaken by several chemical companies aimed at synthesizing more potent and selective ecdysone agonists and JH mimics to be used in our agricultural systems.

4
Acetylcholine Receptors

Efforts have been made to develop nicotinyl insecticides with high affinity to insect nicotinic acetylcholine receptors (nAChR), resulting in the development of a new group of neonicotinoid insecticides (Abbink 1991; Tomizawa and Yamamoto 1992; Liu and Casida 1993; Elbert et al. 1998). Neonicotinoids of potential use in agriculture are imidacloprid, acetamiprid and thiamethoxam. These compounds interact with nAChR in a structure-activity relationship (Tomizawa et al. 1995a,b), resulting in excitation and paralysis followed by death. Their selectivity results from a higher affinity to the insect nAChR as compared to that of vertebrates, in contrast to the original nicotine compound (Tomizawa et al. 1995b). Hence it has been suggested that imidacloprid and related compounds be called neonicotinoids (Yamamoto et al. 1995).

Imidacloprid displaced radiolabeled α-bungarotoxin, a specific ligand of nAChR, in the ganglia of the American cockroach, indicating its direct interaction with the receptor (Elbert et al. 1998). Electrophysiological studies with imidacloprid on the cholinergic motor neuron of *Periplaneta americana* revealed a depolarization of the cell membrane similar to that of acetylcholine (Bai et al. 1991). High affinity of radiolabeled imidacloprid to the binding site was also observed in housefly head membrane (Liu and Casida 1993) and the green peach aphid (Nauen et al. 1996). One of the important features of these compounds is their selectivity towards insects; their affinity to interact with rat nAChR was 1000 times weaker than that with the insect receptor (Methfessel 1992; Zwart et al. 1994).

Imidacloprid, 1-(6-chloro-3-pyridylmethyl)-N-nitro-2-imidazolidinimine, and acetamiprid, (E)-N^1-[(6-chloro-3-pyridyl)methyl]-N^2-cyano-N^1-methylacetamidine, are the first neonicotinoids to be commercially used to control agricultural pests. They are also called chloronicotinyl insecticides, indicating the biological importance of the *chlorine* moiety in their chemical structure (Leicht 1993). Imidacloprid is a relatively polar material with good xylem mobility suitable for seed treatment and soil application (Elbert et al. 1998). Comparative assays carried out in standardized growth chambers indicate that in soil application imidacloprid gave superior performance for controlling the whitefly *B. tabaci* as compared with acetamiprid, while acetamiprid was much more potent than imidacloprid in foliar application (Horowitz et al. 1998). Hence acetamiprid has been introduced in Israel as a component in the IRM program to control *B. tabaci* foliarly in cotton fields, while imidacloprid is used systemically through the soil to control whiteflies and aphids in vegetable and ornamentals. Thimethoxam is a neonicotinoid compound developed recently by Novartis and is considered a potent compound for the control of aphids and whiteflies. A combination of neonicotinoid overuse, coupled with a risk of cross resistance between the various members of the group, threatens the effectiveness of the group as a whole (Cahill and Denholm 1999).

The neonicotinoids act specifically on sucking pests and have mild or no effect on parasitoids and predators, and as such fit well in various IPM programs. Hence nAChR proved to be an important site for the development of a new group of insecticides able to control homopteran pests in various agricultural systems.

5
GABA and Glutamate Receptors and Ion Channels

Voltage-dependent sodium channels and γ-aminobutyric acid (GABA)-gated chloride channels are primary sites of action of a number of established insecticides. Examples include pyrethroids and DDT which affect sodium channels (Zlotkin 1999), polychlorocycloalkane fipronyl which blocks the chloride ion channels (Cole et al. 1993; Bloomquist 1994; Scharf and Siegfried 1999) and the avermectins which activate the GABA receptor (Deng and Casida 1992; Arena et al. 1995; Jansson and Dybas 1998). Excellent reviews of GABA and glutamate receptors and of ion channels are presented by Bloomquist and Zlotkin (this Vol.). Hence, I will concentrate on the agricultural importance of some novel insecticides which act on these biochemical sites.

The avermectins are a group of macrocyclic lactones isolated from fermentation of the soil microorganism *Streptomyces avermitilis*. They bind with high affinity to sites in the head and muscle neuronal membranes of various insect species (Deng and Casida 1992; Rohrer et al. 1995) acting thereby as agonists for GABA-gated chloride channels (Mellin et al. 1983; Albrecht and Sherman 1987). Abamectin (Vertimec), developed for agricultural use by Merck Sharp & Dohme (New Jersey, USA), consists of about 80% avermectin B1a and 20% avermectin B1b (Fischer and Mrozik 1989). The compound is specifically toxic to phytophagous mites and to a select panel of insect species, but it is markedly less potent against some lepidopteran and homopteran species (Dybas 1989; Lasota and Dybas 1991). It is considerably less toxic to beneficial insects such as honey bees, parasitoids and predators (Hoy and Cave 1985; Dybas 1989; Zhang and Sanderson 1990) and may be considered a selective insecticide. Despite its rapid decomposition, abamectin provides residual activity in the field due to its translaminar activity (Wright et al. 1985; Dybas 1989). Addition of ultrafine oil to the spray solution enhances considerably the potency of abamectin on the sweetpotato whitefly *Bemisia tabaci* as a result of higher translaminar activity (Horowitz et al. 1997). New avermectin derivatives, emamectin benzoate (Proclaim) developed by Merck (New Jersey, USA) and milbemectin (Milbeknock) by Sankyo (Tokyo, Japan), act on a diversity of insect species. Emamectin acts specifically on lepidopteran pests (Cox et al. 1995). Studies carried out in our laboratory indicate that emamectin is a powerful toxicant against *Spodoptera littoralis, Helicoverpa armigera,* and the western flower thrips *Frankliniella occidentalis* (unpubl. results). Milbemectin acts on a wide range of mite pests such as the two-spotted spider mites *Tetranychus urticae,* the spider mite *T. cinnabarinus,* and the citrus red mite *Panonychus citri* (Sankyo 1997). The product is a mixture of milbemycin A3 and milbemycin A4, both of which are metabolites of *Streptomyces hygroscopicus* (Barrett et al. 1985). It exhibits a considerable toxicity against the whitefly *B. tabaci* especially against young stages; the LC-90 value for 1st-instar is 0.06 mg a.i./l (Pluschkell et al. 1999).

Spinosad is a naturally occurring mixture of two major components – spinosyn A and spinosyn D – produced by the soil actinomycete *Saccharopolyspora spinosa* (Kirst et al. 1992). Spinosyns initially cause involuntary muscle contractions and tremors by exciting neurones in the central nervous system (Salgado 1998). They excite the nervous system directly when applied to insect ganglia at nanomole concentrations (Salgado 1998). These effects are consistent with the activation of nicotinic acetylcholine receptors; they affect also GABA function, which may contribute further to their potency (Thompson and Hutchins 1999). Spinosad is a potent compound for controlling lepidopterans and thrips and also has activity on various insect orders (Thompson and Hutchins 1999). The compound has a translaminar activity, affecting hidden insect pests not targeted by the spray solution. The compound is among the most powerful toxicants for controlling the western flower thrips *Frankliniella occidentalis* (unpubl. results). It has a mild or no toxicity to most of the natural enemies and as such it may fit well in various IPM programs.

Phenylpyrazoles comprise a new class of pesticides which exhibit herbicidal and insecticidal characteristics (Yanase and Andoh 1989; Klis et al. 1991). Fipronil, 5-amino-1-[2,6-dichloro-4-(trifluoromethyl)phenyl-3,4-cyano-4-trifluoromethylsulphinylpyrazole, a member of the relatively new phenylpyrazole class, is active at the neuro-inhibitory GABA-gated chloride channels (Scharf and Siegfried 1999). It exhibits a broad activity against several insect pests (Colliot et al. 1992) and it is highly toxic to cockroaches (Scott and Wen 1997; Silverman and Liang 1999), house flies (Cole et al. 1993; Scott and Wen 1997) and fruit flies (Bloomquist 1994). Competitive bindings have demonstrated that phenylpyrazoles have greater affinity toward GABA-activated chloride channels of insects than of mammals, suggesting that phenylpyrazoles can be selectively toxic towards insects (Cole et al. 1993).

It can be concluded that GABA and glutamate receptors, and ion channels, are important sites for developing novel selective insecticides.

6
Other Biochemical Sites

Several novel insecticides, which act on biochemical sites other than those described above, have been developed during the past decade. Pymetrozine, 4,5-dihydro-6-methyl-4-[(3-pyridinylmethylene)amino]-1,2,4-triazin-3(2H)one, is a new insecticide with a novel mode of action. It affects the nerve which controls the salivary pump of some sucking pests causing irreversible cessation of feeding, followed by starvation and death (Schwinger et al. 1994). Further studies are required to elucidate the target mechanisms of pymetrozine at a molecular or enzymatic level. The compound is a powerful toxicant against aphids, whiteflies and plant hoppers (Fuog et al. 1998). The compound has a systemic and translaminar activity and can be used in soil or foliar applications (Flückiger et al. 1992a,b). In some cases, host plants have a major effect on the potency of pymetrozine, due to its rate of penetration and translocation in various plant crops. It was a more powerful toxicant to white-flies when applied on Bulgarian beans as compared with cotton plants (Ishaaya and Horowitz 1998). Pymetrozine has no appreciable effect on natural enemies and the environment and as such it is considered a potential component in IPM programs for controlling aphids and whiteflies and for suppressing virus transmission in various vegetable crops (Harrewijn and Piron 1994; Harrewijn and Kayser 1997; Fuog et al. 1998).

Diafenthiuron, 3-(2,6-diisopropyl-4-phenoxyphenyl)-1-tert-butylthiourea, is a new type of thiourea derivative which acts specifically on sucking pests such as mites, whiteflies and aphids (Streibert et al. 1988; Anonymous 1989; Kadir and Knowles 1991; Ishaaya et al. 1993). It is photochemically converted within a few hours in sunlight to its carbodiimide derivative, which is a more powerful acaricide/insecticide than diafenthiuron (Steinemann et al. 1990). Feeding tests revealed a decreased toxicity of diafenthiuron in the presence of piperonyl butoxide which is known to be an inhibitor of cytochrome P-450. This indicates the importance of cytochrome P-450 in metabolizing diafenthiuron after ingestion to the toxic carbodiimide product (Ruder et al. 1992). The carbodiimide is a potent inhibitor of mitochondrial ATP synthesis and reacts covalently with the proteolipid subunit of the ATPase and porin which forms the outer membrane of the mitochondria (Ruder et al. 1992). The compound inhibits ATPase activities in preparations from the bulb mite *Rhizoglyphus echinopus,* the two-spotted spider mite *Tetranychus urticae,* and the bluegill *Lepomis macrochirus Rafinesque* (Kadir and Knowles 1991). It has a favorable acute mammalian toxicity coupled with relatively low toxicity to beneficial insects and predatory mites (Streibert et al. 1998; Anonymous 1989). As such it is an important component in IPM and IRM programs for controlling white-flies in cotton fields (Horowitz and Ishaaya 1995).

Azadirachtin is one of the major constituents of the neem extract, exhibiting about 90% of the total activity. It interacts with the corpus cardiacum, thereby blocking the activity of the molting hormone. As such the compound acts as an insect growth regulator, suppressing fecundity, molting, pupation and adult formation. Other constituents which exhibit high potency on insects and are structurally related to azadirachtin are salannin, meliantrol, and nimbin (Schmutterer 1995). Neem extract is a potential insecticide, acting mainly by ingestion and affecting a diversity of insects (Jacobson 1988; Ascher 1993; Schmutterer 1995). The extract seems to have no harmful effect on beneficials and is considered a potential component in IPM programs.

Bacillus thuringiensis (*Bt*) δ-endotoxins are large proteins which act specifically in the midgut of lepidopterans. These molecules are proteolytically converted in the midgut into small toxic polypeptides (Bulla et al. 1981; Andrews et al. 1985; Hofte and Whiteley 1989). These toxins bind with high affinity to receptors on the midgut epithelium (Knowles and Ellar 1986; Van Rie et al. 1989, 1990). Studies with Cry IV mosquito-specific toxins suggest that a combination of phospholipid, glycoprotein or perhaps other components are involved in the bindings (Ward and Ellar 1986; Chilcott et al. 1990). The bound toxins generate pore production in the cell membrane thereby disrupting osmotic balance (Knowles and Ellar 1987; Hofte and Whiteley 1989). Formulations containing *Bt* δ-endotoxins are used as selective insecticides for controlling lepidopteran pests. Most *Bt* insecticides are based on the species *kurstaki* HD-1 (Navon et al. 1990; Navon 1993) such as Dipel, Thuricide, Biobit and Javelin. In addition, several transgenic plants (cotton, potatoes, tomatoes) which produce *Bt* δ-endotoxins are resistant or tolerant to major lepidopteran pests such as *Helicoverpa, Pectinophora, Earias* and *Plusia* (Perlak et al. 1990; Ely 1993). In-depth details on the biochemical mode of action are presented by Gringorten (this Vol.).

Bt insecticides are considered safe to the environment with little or no harm to natural enemies. Hence midgut epithelium can serve as an important site for developing selective insecticides that are similar in their potency to δ-endotoxins.

7
Conclusions

Novel insect control agents, which act on biochemical sites present in insects but not in mammals, have been developed in the past two decades and introduced for the selective control of insect pests. Chitin synthesis inhibitors such as diflubenzuron, chlorfluazuron, hexaflumuron, teflubenzuron, lufenuron, novaluron and buprofezin have been commercialized and used to control insect pests in various agricultural systems. Compounds which interact with juvenile hormone and ecdysone receptors such as fenoxycarb and pyriproxyfen (juvenile hormone mimics), and tebufenozide and methoxyfenozide (ecdysone agonists), are considered selective insecticides for controlling scale insects, whiteflies and lepidopteran pests.

In addition, compounds which interact more efficiently with biochemical sites present in insects as compared to mammals have been introduced for the selective control of important groups of insect pests. The neonicotinoids, imidacloprid, acetamiprid and thiamethoxam, which act on acetylcholine receptors, are considered an important group of insecticides acting on aphids and whiteflies. Abamectin, emamectin, milbemectin and spinosad, which act on GABA receptors and ion channels, have been developed to control mites and other agricultural pests.

Novel approaches for developing insecticides which can act specifically or more efficiently on insect biochemical sites as compared to mammals are essential in the future progress in developing safe and efficient insect control agents which can serve as important components in integrated pest management programs.

Acknowledgements. The author thanks Phylis Weintraub for her valuable comments and Svetlana Kontsedalov for assisting in the preparation of the manuscript.

References

Abbink J (1991) The biochemistry of imidacloprid. Pflanzenschutz-Nachr Bayer 44:183–194

Albrecht CP, Sherman M (1987) Lethal and sublethal effect of avermectin B_1 on three fruit fly species (Diptera: Tephritidae). J Econ Entomol 80:344–347

Andrews RE Jr, Bibilos MM, Bulla LA Jr (1985) Protease activiation of the entomocidal protoxin of *Bacillus thuringiensis* subsp. *kurstaki*. Appl Environ Microbiol 50:737–742

Anonymous (1989) Polo (diafenthiuron, CGA 106630), Technical Data Sheet. Ciba Geigy, Basel, pp 1–18

Arena JP, Liu KK, Paress PS, Frezier EG, Cully DF, Mrozik H, Schaeffer J (1995) The mechanism of action of avermectin in *Caenorhabditis elegans*: correlation between activation of glutamate-sensitive chloride current, membrane binding and biological activity. J Parasitol 81: 286–294

Ascher KRS (1993) Non-conventional insecticidal effects of pesticide available from the neem tree, *Azadirachta indica*. Arch Insect Biochem Physiol 22:433–449

Bai D, Lummis SCR, Leicht W, Breer H, Satelle DB (1991) Actions of imidacloprid and related nitromethylene on cholinergic receptor of an identified insect motor neurone. Pestic Sci 33:197–204

Barrett AGM, Curr RA, Attwood SV, Finch MAW, Richardson G (1985) The application of novel carbanion chemistry in milbemycin-avermectin synthesis. In: James NF (ed) Recent advances in the chemistry of insect control. R Soc Chem, London, pp 257–271

Becher HM, Becker P, Prokic-Immel R, Wirtz W (1983) CME, a new chitin synthesis inhibiting insecticide. Brighton Crop Prot Conf 1:408–415

Bergamasco R, Horn DHS (1980) The biological activities of ecdysteroids and ecdysteroid analogues. In: Hoffman JA (ed) Progress in ecdysone research. Elsevier, Amsterdam, pp 299–324

Bloomquist JR (1994) Cyclodiene resistance at the insect GABA receptor/chloride channel
complex confers broad cross-resistance to convulsants and experimental phenylpyrazole
insecticides. Arch Insect Biochem Physiol 26:69–79

Bulla LA, Jr, Kramer KJ, Cox DJ, Jones BL, Davidson LI, Lookhart GL (1981) Purification and characterization of the entomocidal protoxin of *Bacillus thuringiensis*. J Biol Chem 256:3000–3004

Cahill M, Denholm I (1999) Managing resistance to chloronicotinyl insecticides: rhetoric or
reality? In: Yamamoto I, Casida JE (eds) Nicotinoid insecticides and the nicotinic acetylcholine
receptor. Springer, Berlin Heidelberg New York, pp 253–270

Chandler LD, Pair SD, Harrison WE (1992) RH-5992, a new insect growth regulator active against
corn earworm and fall armyworm (Lepidoptera: Noctuidae). J Econ Entomol 85:1099–1103

Chilcott CN, Knowles BH, Ellar DJ, Brobniewski FA (1990) Mechanism of action of *Bacillus
thuringiensis israelensis* paraproposal body. In: deBarjac H, Southerland DJ (eds) Bacterial
control of mosquitoes and black flies. Rutgers Univ Press, New Brunswick, NJ, pp 45–65

Cohen E (1985) Chitin synthetase activity and inhibition in different insect microsomal preparations. Experientia 41:470–472

Cohen E (1987) Chitin biochemistry: synthesis and inhibition. Annu Rev Entomol 322:71–93

Cohen E, Casida JE (1980) Inhibition of *Tribolium* gut synthetase. Pestic Biochem Physiol 13:
129–136

Cole L, Nicholson R, Casida JE (1993) Action of phenylpyrazole insecticides at the GABA-gated
chloride channel. Pestic Biochem Physiol 46:47–54

Colliot F, Kukorowski KA, Hawkins DW, Robert DA (1992) Fipronil: a new soil and foliar broad
spectrum insecticide. Brighton Crop Protection Conference: pests and diseases. British Crop
Protection Council, Farnham, UK, pp 29–34

Cox DL, Knight AL, Biddinger DG, Lasota JA, Pikounis B, Hull LA, Dybas RA (1995) Toxicity and
field efficacy of avermectins against codling moth (Lepidoptera: Tortricidae) on apples. J Econ
Entomol 88:708–715

De Cock A, Degheele D (1991) Effects of buprofezin on the ultrastructure of the third instar
cuticle of *Trialeurodes vaporariorum*. Tissue Cell 23:755–762

De Cock A, Degheele D (1998) Buprofezin: a novel chitin synthesis inhibitor affecting specifically
planthoppers, whiteflies and scale insects. In: Ishaaya I, Degheele D (eds) Insecticides with
novel modes of action: mechanism and application. Springer, Berlin Heidelberg New York,
pp 74–91

DeLoach JR, Meola SM, Mayer RT, Thompson JM (1981) Inhibition of DNA synthesis by diflubenzuron in pupae of the stable fly *Stomoxys calcitrans* (L.). Pestic Biochem Physiol 15:172–180

Deng Y, Casida JE (1992) Housefly head GABA-gated chloride channel: toxicological relevant
binding site for avermectins coupled to site for ethynyl-bicycloortho benzoate. Pestic Biochem
Physiol 43:116–122

Dennehy TJ, Williams L (1997) Management of resistance in *Bemisia* in Arizona cotton. Pestic Sci
51:398–406

Dhadialla TS, Carlson GR, Le DP (1998) New insecticides with ecdysteroidal and juvenile
hormone activity. Annu Rev Entomol 45:545–569

Dorn J, Frischknecht ML, Martinez V, Zurflüh R, Fischer U (1981) A novel non-neurotoxic insecticide with a broad activity. Z Pflanzenkr Pflanzenschutz 88:269–275

Dybas RA (1989) Abamectin use in crop protection. In: Campbell WC (ed) Ivermectin and
abamectin. Springer, Berlin Heidelberg New York, pp 287–310

Elbert A, Nauen R, Leicht W (1998) Imidacloprid, a novel chloronicotinyl insecticide: biological
activity and agricultural importance. In: Ishaaya I, Degheele D (eds) Insecticides with novel
modes of action: mechanism and application. Springer, Berlin Heidelberg New York, pp 50–73

Ely J (1993) The engineering of plants to express *Bacillus thuringiensis* δ-endotoxins. In: Entwistle
PF, Cory JS, Bailey MJ, Higgs S (eds) *Bacillus thuringiensis*, an environmental biopesticide:
theory and practice. Wiley, Chichester, pp 105–124

Fischer MH, Mrozik H (1989) Chemistry. In: Campbell WC (ed) Ivermectin and abamectin.
Springer, Berlin Heidelberg New York, pp 1–23

Flückiger CR, Kristinsson H, Senn R, Rindlisbacher A, Buholzer H, Voss G (1992a) CGA 215'944 – a novel agent to control aphids and whiteflies. Brighton Crop Prot Conf (Pests and Diseases) 1:43–50

Flückiger CR, Senn R, Buholzer H (1992b) CGA 215'944 – opportunities for use in vegetables. Brighton Crop Prot Conf (Pests and Diseases) 3:1187–1192

Fuog D, Fergusson SJ, Flückiger C (1998) Pymetrozine: a novel insecticide affecting aphids and whiteflies. In: Ishaaya I, Degheele D (eds) Insecticides with novel modes of action: mechanism and application. Springer, Berlin Heidelberg New York, pp 40–49

Grosscurt AC (1978) Effect of diflubenzuron on mechanical penetrability, chitin formation, and structure of the elytra of *Leptinotarsa decemlineata*. J Insect Physiol 24:827–831

Grosscurt AC, Anderson SO (1980) Effect of diflubenzuron on some chemical and mechanical properties of the elytra of *Leptinotarsa decemlineata*. Proc K Ned Akad Wet 83C:143–150

Haga T, Tobi T, Koyanagi R (1982) Structure activity relationship of series of benzoyl-pyridyloxyphenyl-urea derivatives. Abstr 5th Int Congr Pestic Chem (IUPAC), August 1982, Kyoto, p IId-7

Hajjar NP, Casida JE (1979) Structure activity relationship of benzoylphenyl ureas as toxicants and chitin synthesis inhibitors in *Oncopeltus fasciatus*. Pestic Biochem Physiol 11:33–45

Harrewijn P, Piron PGM (1994) Pymetrozine, a novel agent for reducing virus transmission by *Myzus persicae*. Brighton Crop Protection Conference (Pests and Diseases) 2:923–928

Harrewijn P, Kayser H (1997) Pymetrozine, a fast acting and selective inhibitor of aphid feeding. *In situ* studies with electronic monitoring of feeding behaviour. Pestic Sci 49:130–140

Heller JJ, Mattioda H, Klein E, Sagenmüller A (1992) Field evaluation of RH-5992 on lepidopteran pests in Europe. Brighton Crop Prot Conf (Pests and Diseases) 1:59–66

Hofte H, Whiteley HR (1989) Insecticidal crystal of *Bacillus thuringiensis*. Microbiol Rev 53:242–255

Horn DHS, Galbraith MN, Kelly BA, Kinnear JF, Martin MD, Middleton EJ, Virgonia CTF (1981) Moulting hormones LIII. The synthesis and biological activity of some ecdysone analogues. Aust J Chem 34:2607–2618

Horowitz AR, Ishaaya I (1994) Monitoring resistance to IGRs in the sweetpotato whitefly (Homoptera: Aleyrodidae). J Econ Entomol 87:866–871

Horowitz AR, Ishaaya I (1995) Chemical control of *Bemisia* – management and application. In: Gerling D, Mayer RT (eds) Bemisia: 1995 – taxonomy, biology, damage, control and management. Intercept, Andover, UK, pp 537–556

Horowitz AR, Klein M, Yablonski S, Ishaaya I (1992) Evaluation of benzoylphenyl ureas for controlling the spiny bollworm, *Earias insulana* (Baisd.), in cotton. Crop Prot 11:465–469

Horowitz AR, Mendelson Z, Ishaaya I (1997) Effect of abamectin mixed with mineral oil on the sweet potato whitefly (Homoptera:Aleyrodidae). J Econ Entomol 90:349–353

Horowitz AR, Mendelson Z, Weintraub PG, Ishaaya I (1998) Comparative toxicity of foliar and systemic application of acetamiprid and imidacloprid against the cotton whitefly, *Bemisia tabaci* (Hemiptera: Aleyrodidae). Bull Entomol Res 88:437–442

Horowitz AR, Mendelson Z, Cahill M, Denholm I, Ishaaya I (1999) Managing resistance to the insect growth regulator pyriproxyfen in *Bemisia tabaci*. Pestic Sci 55:272–276

Hoy MA, Cave FE (1985) Laboratory evaluation of avermectin as a selective acaricide for use with *Metasciulus occidentalis* (Nesbitt) (Acarina: Phytoseiidae). Exp Appl Acarol 1:139–152

Ishaaya I (1990) Benzoylphenyl ureas and other selective control agents – mechanism and application. In: Casida JE (ed) Pesticides and alternatives. Elsevier, Amsterdam, pp 365–376

Ishaaya I, Ascher KRS (1977) Effect of diflubenzuron on growth and carbohydrate hydrolases of *Tribolium castaneum* Phytoparasitica 5:149–158

Ishaaya I, Casida JE (1974) Dietary TH-6040 alters cuticle composition and enzyme activity of housefly larval cuticle. Pestic Biochem Physiol 4:484–490

Ishaaya I, Horowitz AR (1992) Novel phenoxy juvenile hormone analog (pyriproxyfen) suppresses embryogenesis and adult emergence of the sweet potato whitefly (Homoptera: Aleyrodidae). J Econ Entomol 85:2113–2117

Ishaaya I, Horowitz AR (1995) Pyriproxyfen, a novel insect growth regulator for controlling whiteflies – mechanism and resistance management. Pestic Sci 43:227–232

Ishaaya I, Horowitz AR (1998) Insecticides with novel modes of action: an overview. In: Ishaaya I, Degheele D (eds) Insecticides with novel modes of action: mechanism and application. Springer, Berlin Heidelberg New York, pp 1–24

Ishaaya I, Mendelson Z, Melamed-Madjar V (1988) Effect of buprofezin on embryogenesis and progeny formation of sweetpotato whitefly (Homoptera: Aleyrodidae). J Econ Entomol 81: 781–784

Ishaaya I, Mendelson Z, Horowitz AR (1993) Toxicity and growth suppression exerted by diafen-thiuron in the sweetpotato whitefly Bemisia tabaci. Phytoparasitica 21:199–204

Ishaaya I, De Cock A, Degheele D (1994) Pyriproxyfen, a potent suppressor of egg hatch and adult formation of the greenhouse whitefly (Homoptera: Aleyrodidae). J Econ Entomol 87: 1185–1189

Ishaaya I, Yablonski S, Horowitz AR (1995) Comparative toxicity of two ecdysteroid agonists, RH-2485 and RH-5992, on susceptible and pyriproxyfen-resistant strains of the Egyptian cotton leafworm, Spodoptera littoralis. Phytoparasitica 23:139–145

Ishaaya I, Yablonski S, Mendelson Z, Mansour Y, Horowitz AR (1996) Novaluron (MCW-275), a novel benzoylphenyl urea, suppressing developing stages of lepidopteran, whitefly and leafminer pests. Brighton Crop Prot Conf (Pests and Disease), pp 1013–1020

Ishaaya I, Danme N, Tirry L (1998) Novaluron, optimization and use for the control of the beet armyworm and the greenhouse whitefly. Brighton Crop Prot Conf (Pests and Diseases), pp 49–56

Itaya W (1987) Insect juvenile hormone analogue as an insect growth regulator. Sumitomo Pyrethroid World 8:2–4

Izawa Y, Uchida M, Sugimoto T, Asai T (1985) Inhibition of chitin biosynthesis by buprofezin analogs in relation to their activity controlling Nilaparvata lugens Stål. Pestic Biochem Physiol 24:343–347

Jacobson M (ed) (1988) Focus on phytochemical pesticides: the neem tree, vol 1. CRC Press, Boca Raton

Jansson RK, Dybas RA (1998) Avermectins: biochemical mode of action, biological activity and agricultural importance. In: Ishaaya I, Degheele D (eds) Insecticides with novel modes of action: mechanism and application. Springer, Berlin Heidelberg New York, pp 152–170

Kadir HA, Knowles CO (1991) Toxicological studies of the thiourea diafenthiuron in diamond-back moth (Lepidoptera: Yponomeutidae), two-spotted spider mites (Acari: Tetranychidae), and bulb mite (Acari: Acaridae). J Econ Entomol 84:780–784

Kanno H, Ikeda K, Asai T, Maekawa S (1981) 2-tert-Butylimino-3-isopropyl-5-phenylperhydro-1,3,5-thiodiazin-4-one (NNI 750), a new insecticide. Brighton Crop Protection 1:56–69

Kawada H (1988) An insect growth regulator against cockroaches. Sumitomo Pyrethroid World 11:2–4

Kirst HA, Michel KH, Mynderse JS, Chao EH, Yao RC, Nakatsukasa WM, Boeck LD, Occlowitz J, Paschel JW, Deeter JB, Thompson GD (1992) Discovery, isolation and structure elucidation of a family of structurally unique fermentation-derived tetracyclic macrolides. In: Baker DR, Fenyes JG, Steffens JJ (eds) Synthesis and chemistry of agrochemicals III. Am Chem Soc, Washington, DC, pp 214–225

Klis SFL, Vijverberg HPM, van den Berken J (1991) Phenylpyrazoles, a new class of pesticides: electrophysiological investigation into basis effects. Pestic Biochem Physiol 39:210–218

Knowles BH, Ellar DJ (1986) Characterization and partial purification of a plasma membrane receptor for Bacillus thuringiensis var. krustaki lepidopteran-specific δ-endotoxin. J Cell Sci 83:89–101

Knowles BH, Ellar DJ (1987) Colloid-osmotic lysis is a general feature of the mechanism of action of Bacillus thuringiensis delta endotoxin with different insect specificity. Biochem Biophys Acta 924:509–518

Koehler PG, Patterson RJ (1991) Incorporation of pyriproxyfen in German cockroach (Dictyoptera: Blattellidae) management program. J Econ Entomol 84:917–921

Langley P (1990) Control of the tsetse fly using a juvenile hormone mimic, pyriproxyfen. Sumitomo Pyrethroid World 15:2–5

Lasota JA, Dybas RA (1991) Avermectin, a novel class of compounds: implications for use in arthropod pest control. Annu Rev Entomol 36:96–117

Leicht W (1993) Imidacloprid – a chloronicotinyl insecticide. Pestic Outlook 4:17–21

Liu M-Y, Casida JE (1993) High affinity binding of [^{3}H]-imidacloprid in the insect acetylcholine receptor. Pestic Biochem Physiol 46:40–46

Masner P, Angst M, Dorn S (1987) Fenoxycarb, an insect growth regulator with juvenile hormone activity: a candidate for *Heliothis virescens* (F.) control on cotton. Pestic Sci 18:89–94

Mayer RT, Chen AC, DeLoach JR (1981) Chitin synthesis inhibiting insect growth regulators do not inhibit chitin synthase. Experientia 37:337–338

Mellin TN, Busch RD, Wang CC (1983) Postsynaptic inhibitions of invertebrate neuromuscular transmission by avermectin B1a. Neuropharmacology 22:89–96

Methfessel C (1992) Action of imidacloprid on the nicotinic acetylcholine receptor in rat muscle. Pflanzenschutz-Nachr Bayer 45:369–380

Mikolajczyk P, Oberlander H, Silhacek DL, Ishaaya I, Shaaya E (1994) Chitin synthesis in *Spodoptera frugiperda* wing imaginal discs. I. Chlorfluazuron, diflubenzuron, and teflubenzuron inhibit incorporation but not uptake of [^{14}C]-N-acetyl-D-glucosamine. Arch Insect Biochem Physiol 25:245–258

Mitlin N, Wiygul G, Haynes JW (1977) Inhibition of DNA synthesis in boll weevils (*Anthonomus grandis* Boheman) sterilized by Dimilin. Pestic Biochem Physiol 7:559–563

Monthéan C, Potter DE (1992) Effects of RH-5849, a novel insect growth regulator, on Japanese beetle (Coleoptera: Scarabaeidae) and fall armyworm (Lepidoptera: Noctuidae) in turfgrass. J Econ Entomol 85:507–513

Mulder R, Gijswijk MT (1973) The laboratory evaluation of two promising new insecticides which interfere with cuticle deposition. Pestic Sci 4:737–745

Nagata T (1986) Timing of buprofezin application for control of the brown planthopper, *Nilaparvata lugens* Stål. (Homoptera: Delphacidae). Appl Entomol Zool 21:357–362

Nauen R, Strobel J, Otsu K, Tietjen K, Erdelen C, Elbert A (1996) Aphicidal activity of imidacloprid against a carbamate and organophospate resistant Japanese strain of the tobacco feeding *Myzus persicae* (Homoptera: Aphididae) closely related to *Myzus nicotianae*. Bull Entomol Res 86:165–171

Navon A (1993) Control of lepidopteran pests with *Bacillus thuringiensis*. In: Entwistle PF, Cory JS, Bailey MJ, Hidds S (eds) *Bacillus thuringiensis*, an environmental biopesticide: theory and practice. Wiley, Chichester, pp 125–146

Navon A, Klein M, Braun S (1990) *Bacillus thuringiensis* potency bioassays against *Heliothis armigera*, *Earias insulana* and *Spodoptera littoralis* larvae based on standardized diets. J Invertebr Pathol 55:387–393

Oberlander H, Silhacek DL (1998) New perspectives on the mode of action of benzoylphenyl urea insecticides. In: Ishaaya I, Degheele D (eds) Insecticides with novel modes of action: mechanism and application. Springer, Berlin Heidelberg New York, pp 92–105

Palli SR, Ladd TR, Sohi SS, Cook BJ, Retnakaran A (1996) Cloning and developmental expression of Choristoneura hormone receptor 3, an ecdysone-inducible gene and a member of the steroid hormone receptor superfamily. Insect Biochem Mol Biol 26:485–499

Peleg BA (1988) Effect of new phenoxy juvenile hormone analog on California red scale (Homoptera: Diaspididae), Florida wax scale (Homoptera: Coccidae) and the ectoparasite *Aphytis holoxanthus* DeBache (Hymenoptera: Aphelinidae). J Econ Entomol 81:88–92

Perlak FJ, Deaton RW, Armstrong TA, Fuchs RL, Sims SR, Greenplate JT, Fischhoff (1990) Insect resistant cotton plants. Biotechnology 8:939–942

Perry AS, Yamamoto I, Ishaaya I, Perry RY (1998) Insecticides in agriculture and environment: retrospects and prospects. Springer, Berlin Heidelberg New York, pp 1–3

Pluschkell U, Horowitz AR, Ishaaya I (1999) Effect of milbemectin on the sweetpotato whitefly *Bemisia tabaci*. Phytoparasitica 27:183–191

Post LC, de Jong BJ, Vincent WR (1974) 1-(2,6-Disubstituted benzoyl)-3-phenylurea insecticides: inhibitors of chitin synthesis. Pestic Biochem Physiol 4:473–483

Retnakaran A, Granett J, Ennis T (1985) Insect growth regulators. In: Kerkut GA, Gilbert LI (eds) Comprehensive insect physiology, biochemistry and pharmacology, vol 12. Pergamon Press, Oxford, pp 529–601

Retnakaran A, Hiruma K, Palli SR, Riddiford LM (1995) Molecular analysis of the mode of action of RH-5992, a lepidopteran-specific, non-steroidal ecdysteroid agonist. Insect Biochem Mol Biol 25:109–117

Riddiford LM (1985) Hormone action at the cellular level. In: Kerkut GA, Gilbert LI (eds) Comprehensive insect physiology, biochemistry and pharmacology, vol 18. Pergamon Press, Oxford, pp 37–84

Riddiford LM, Osir EO, Fikinghoff CM, Green JM (1987) Juvenile hormone analog binding in *Manduca epidermis*. Insect Biochem 17:1039–1043

Rohrer SP, Birzin ET, Costa SD, Arena JP, Hayes EC, Schaeffer JH (1995) Identification of neuron-specific ivermectin binding sites in *Drosophila melanogaster* and *Schistocerca americana*. Insect Biochem Mol Biol 25:11–17

Ruder FJ, Benson JA, Kayser H (1992) The mode of action of the insecticide/acaricide diafenthiuron. In: Otto D, Weber B (eds) Insecticides: mechanism of action and resistance. Intercept, Andover, UK, pp 263–276

Salgado VL (1998) Studies on the mode of action of Spinosad: insect symptoms and physiological correlates. Pestic Biochem Physiol 60:91–92

Sankyo Co (1997) Technical datasheet on milbemectin. Sankyo Co, Tokyo, Japan

Sbragia R, Bisabri-Ershadi, Rigterink RH (1983) XRD-473, a new acylurea insecticide effective against *Heliothis*. Brighton Crop Prot Conf 1:417–424

Scharf ME, Siegfried BD (1999) Toxicity and neurophysiological effects of fipronil and fipronil sulfone on the western corn rootworm (Coleoptera: Chrysomelidae). Arch Insect Biochem Physiol 40:150–156

Schmutterer H (ed) (1995) Neem tree – source of unique natural products for integrated pest management, medicine industry and other purposes. VCH, Weinheim, 696 pp

Schwinger M, Harrewiju P, Kayser H (1994) Effect of pymetrozine (CGA 215'944), a novel aphicide, on feeding behavior of aphids. Proc 8th IUPAC Int Cong Pestic Chem, Washington, DC 1:230

Scott JG, Wen Z (1997) Toxicity of fipronil to susceptible and resistant strain of German cockroaches (Dictyoptera: Blattelidae) and houseflies (Diptera: Muscidae). J Econ Entomol 90:1152–1156

Silhacek DL, Oberlander H, Procheron P (1990) Action of RH-5849, a non-steroidal ecdysteroid mimic, on *Plodia interpunctella* (Hübner) in vivo and in vitro. Arch Insect Biochem Physiol 15:201–212

Silverman J, Liang D (1999) Effect of fipronil on bait formulation-based aversion in the German cockroach (Dictyoptera: Blattelidae). J Econ Entomol 92:886–889

Smagghe G, Degheele D (1992) Effect of RH-5849, the first nonsteroidal ecdysteroid agonist, on larvae of *Spodoptera littoralis* (Boisd.) (Lepidoptera: Noctuidae). Arch Insect Biochem Physiol 21:119–128

Smagghe G, Degheele D (1994) Action of a novel nonsteroidal ecdysteroid mimic, tebufenozide (RH-5992), on insects of different orders. Pestic Sci 42:85–92

Smagghe G, Degheele D (1998) Ecdysone agonists: mechanisms and biological activity. In: Ishaaya I, Degheele D (eds) Insecticides with novel modes of action: mechanism and application. Springer, Berlin Heidelberg New York, pp 25–39

Soltani N, Besson MT, Delachambre J (1984) Effect of diflubenzuron on the pupal-adult development of *Tenebrio molitor* L. (Coleoptera: Tenebrionidae): growth and development, cuticle secretion, epidermal cell density and DNA synthesis. Pestic Biochem Physiol 21:256–264

Steinmann A, Stamm E, Frei B (1990) Chemodynamics in research and development of new plant protection agents. Pestic Outlook 1(3):3–7

Streibert HP, Drabek J, Rindlisbacher A (1988) CGA 106630 – a new type of acaricide/insecticide for the control of the sucking pest complex in cotton and other crops. Brighton Crop Prot Conf (Pests and Diseases) 1:25–33

Tomizawa M, Yamamoto I (1992) Binding of nicotinoids and the related compounds to the insect nicotinic acetylcholine receptor. J Pestic Sci 17:231

Tomizawa M, Otsuka H, Miyamoto T, Eldefrawi ME, Yamamoto I (1995a) Pharmacological characteristics of insect nicotinic acetylcholine receptor with its ion channel and the comparison of the effect of nicotinoids and neonicotinoids. J Pestic Sci 20:57–64

Tomizawa M, Otsuka H, Miyamoto T, Yamamoto I (1995b) Pharmacological effects of imidacloprid and its related compounds on the nicotinic acetylcholine receptor with its ion channel from the *Torpedo* electric organ. J Pestic Sci 20:49–56

Tompson G, Hutchins S (1999) Spinosad. Pestic Outlook 10:78–81

Uchida M, Asai T, Sugimoto T (1985) Inhibition of cuticle deposition and chitin biosynthesis by a new insect growth regulator buprofezin in *Nilaparvata lugens* Stål. Agric Biol Chem 49:1233–1234

Van Eck WH (1979) Mode of action of two benzoylphenyl ureas as inhibitors of chitin synthesis in insects. Insect Biochem 9:295–300

Van Rie J, Jansens S, Hofte H, Degheele D, Van Mellaert H (1989) Specificity of *Bacillus thuringiensis* δ-endotoxin: importance of specific receptors on the brush border membrane of the midgut of target insects. Eur J Biochem 186:239–247

Van Rie J, Jansens S, Hofte H, Degheele D, Van Mellaert H (1990) Receptors on the brush border membrane of the insect mid-gut as determinants of the specificity of *Bacillus thuringiensis* delta-endotoxins. Appl Environ Microbiol 56:1378–1385

Ward ES, Ellar DJ (1986) *Bacillus thuringiensis* var. *israelensis* delta endotoxin: nucleotide sequence and characterization of the transcripts in *Bacillus thuringiensis* and *Escherichia coli*. J Mol Biol 191:1–11

Whalon ME, McGaughey WH (1998) *Bacillus thuringiensis*: use and resistance management. In: Ishaaya I, Degheele D (eds) Insecticides with novel modes of action: mechanism and application. Springer, Berlin Heidelberg New York, pp 106–137

Wing KD (1988) RH-5849, a nonsteroidal ecdysone agonist: effect on *Drosophila* cell line. Science 241:467–469

Wing KD, Slawecki RA, Carlson GR (1988) RH-5849, a nonsteroidal ecdysone agonist: effect on larval lepidoptera. Science 241:470–472

Wright DJ, Loy A, Green ASJ, Dybas RA (1985) The translaminar activity of abamectin (MK-936) against mites and aphids. Meded Fac Landbouwwet Rijksuniv Gent 50:633–637

Yamamoto I, Yabuta G, Tomizawa M, Saito T, Miyamoto T, Kagabu S (1995) Molecular mechanism of selective toxicity of nicotinoids and neonicotinoids. J Pestic Sci 20:33–40

Yanase D, Andoh A (1989) Porphyrin synthesis involvement in diphenyl ether-like mode of action of TNPP-ethyl, a novel phenylpyrazole herbicide. Pestic Biochem Physiol 35:70–79

Yasui M, Fukada M, Mackawa S (1987) Effect of buprofezin on reproduction of the greenhouse whitefly, *Trialeurodes vaporariorum* (Westwood) (Homoptera: Aleyrodidae). Appl Entomol Zool 22:266–271

Zhang Z, Sanderson JP (1990) Relative toxicity of abamectin to the predatory mite *Phytoseiulus persimilis* (Acari: Phytoseiidae) and the two-spotted spider mite (Acari: Tetranychidae). J Econ Entomol 83:1783–1790

Zlotkin E (1999) The insect voltage-gated sodium channel target of insecticides. Annu Rev Entomol 44:429–455

Zwaart R, Oortigiesen M, Vijverberg HPM (1994) Nitromethylene heterocycles: selective agonists of nicotinic receptors in locust neurons compared to mouse N1E-115 and BC3H1 cells. Pestic Biochem Physiol 48:202

GABA and Glutamate Receptors as Biochemical Sites for Insecticide Action

Jeffrey R. Bloomquist[1]

1
Introduction

The γ-aminobutyric acid (GABA) receptor/chloride ionophore complex has been the focus of intense interest by industrial and academic scientists as a site of insecticide action (Bloomquist 1993). This receptor was initially exploited as a site of action for commercial insecticides over 40 years ago by the polychlorocycloalkane compounds (e.g., cyclodienes, toxaphene, and lindane), although the suggestion that they might be disrupting GABA receptor function was not advanced until 1982 (Ghiasuddin and Matsumura 1982). This hypothesis was roughly contemporaneous with the discovery of the avermectins, which were originally hypothesized to affect GABA-gated chloride channels, but now are thought to mainly affect the glutamate-gated chloride channel of invertebrate muscle (Rohrer and Arena 1995).

The present review is focused on the role of GABA and glutamate receptors as biochemical sites of action of insecticides, as well as their role in resistance. A number of reviews on the chemistry, toxicology, and biochemical action of insecticides which act on the GABA receptor have been published, although the glutamate receptor has received relatively less attention. Whenever possible, I will rely heavily on previous reviews to cite well-established findings. The first section of the review offers a limited survey of similarities and differences between mammalian and insect GABA receptors in order to understand their role in insecticide action and resistance. This section includes information on the molecular structure of the GABA receptor, its tissue distribution, its role in mediating inhibition in electrically excitable cells, and pharmacology. The next section summarizes a large body of work on the action of convulsants on the GABA receptor and includes studies of the cyclodienes, picrotoxinin (PTX), and the trioxabicyclooctanes (TBPS, TBOB, and EBOB). Also included is a summary of findings on the role played by $GABA_a$ receptors in the action of avermectins in mammals. In the subsections that follow, I will discuss more recent results on the role of GABA receptors as a site of action of channel-blocking convulsants, new chemistry acting on this site, and altered GABA

[1] Department of Entomology, Virginia Polytechnic Institute and State University, Blacksburg, VA 24061, USA

Isaac Ishaaya (Ed.): Biochemical Sites of
Insecticide Action and Resistance
© Springer-Verlag Berlin, Heidelberg 2001

receptors in target site resistance. The final section of this chapter will be devoted to what is known about the invertebrate glutamate-gated chloride channel and its role in the mode of action of the avermectins.

2
GABA Receptors in Mammals and Insects

2.1
Classification of GABA Receptors

Vertebrate GABA receptors are classified as $GABA_a$, $GABA_b$, or $GABA_c$, but insect receptors do not fall into identical categories (Lummis 1990; Rauh et al. 1990; Hosie et al. 1997). The $GABA_a$ receptor is broadly distributed within the mammalian central nervous system (CNS) and spinal cord, and is coupled to an intrinsic chloride ion channel. These receptors are thought to be pentameric, and contain α, β, and γ or δ subunits, where each subunit is composed of an external dicysteine loop on the N-terminus and four transmembrane helices (M_1-M_4), with the M_2 transmembrane regions of each subunit forming the ion channel. A similar type of receptor/chloride ionophore complex is the $GABA_c$ receptor, which is a homomultimer made up of ρ subunits, and is mostly found in the vertebrate retina. The $GABA_b$ receptor is coupled to second messengers via G-proteins and, although found in insects (Hue 1991; Bai and Sattelle 1995), is not a known target of insecticides.

2.2
Structure and Physiological Role of Insect GABA Receptors

The insect $GABA_a$ receptor homologue is probably also a heteromultimer, although its component subunits remain to be defined (Hosie et al. 1997). To date, only a few types of ligand-gated chloride channel subunits have been identified from insects, and include Rdl (resistance to dieldrin) homologues from *Drosophila melanogaster* and *Heliothis virescens* (Wolff and Wingate 1998), Grd (GABA and glycine-like receptor of *D. melanogaster*), and LCCh3 (ligand-gated chloride channel homologue 3), which is homologous to vertebrate β subunits (Knipple et al. 1995). The insect GABA receptor/chloride ionophore complex (hereafter simply referred to as the GABA receptor) is found in the CNS, and also at peripheral neuromuscular sites (Lummis 1990). In central and peripheral nerves, it mediates inhibition to allow for proper integration of neuronal activity, and on muscles it promotes relaxation. Another proposed location and function of GABA receptors is on chemosensory cells, where it is hypothesized that they mediate gustatory responses to certain plant compounds and perhaps insecticides (Mullin et al. 1994).

2.3
Pharmacology of GABA Receptors

When activated by stimulation of inhibitory nerve fibers or by application of GABA, the binding of two molecules of GABA to the receptor opens the intrinsic chloride ion channel (Lummis 1990; Rauh et al. 1990; Hosie et al. 1997). This action causes a large increase in chloride ion conductance in the cell membrane and typically an influx of chloride ions that results in membrane hyperpolarization. Insect GABA receptors are generally insensitive to bicuculline, a GABA antagonist acting at the GABA binding site. They share this characteristic with GABA$_c$ receptors, but differ in this property from mammalian GABA$_a$ receptors, which are sensitive to bicuculline. GABA receptors of cultured *D. melanogaster* neurons are blocked by PTX, but insensitive to bicuculline, as are homomultimers of susceptible wild-type Rdl subunits (Zhang et al. 1995). Co-expression of Rdl with the vertebrate β homologue LCCh3 forms channels that are sensitive to bicuculline, but insensitive to PTX, suggesting to Zhang et al. (1995) that LCCh3 expression may determine bicuculline sensitivity in vivo. The resistance to PTX was taken to mean that Rdl and LCCh3 do not co-assemble in vivo and the pattern of antibody labeling to Rdl and LCCh3 in *D. melanogaster* whole mounts showed little temporal or spatial overlap. However, it is interesting to note that there are certain GABA receptor preparations in crayfish muscle (Albert et al. 1986) and cockroach neurons (Corronc and Hue 1999) that are naturally resistant to picrotoxinin in otherwise sensitive strains of animals. For the purposes of this review, the most important pharmacological properties of insect GABA receptors are their sensitivity to convulsants that block the chloride channel or to avermectins which activate the chloride channel. Both classes of toxicant bind to sites on the receptor distinct from that occupied by GABA.

3
Summary of Effects of Convulsants and Avermectins on the GABA Receptor

The signs of intoxication caused by convulsants are indicative of an action on central GABA receptors, and this idea is supported by a wealth of information on their electrophysiological and biochemical actions (Casida et al. 1988; Bloomquist 1993, 1998; Casida 1993; Deng 1995). In insects and mammals, hyperactivity, hyperexcitability, and convulsions are correlated with increased spontaneous nerve activity and the generation of prolonged, high frequency discharges following nerve stimulation. A number of studies using extracellular or intracellular recordings from neurons have shown that micromolar concentrations of lindane, picrotoxinin, dieldrin, endrin, or TBPS strongly antagonized the depression of firing, membrane hyperpolarization, or inward chloride current caused by application of GABA. TBPS, PTX, and cyclodienes showed a mixed type of GABA antagonism having competitive and noncompetitive components. In part, the blocking action of PTX and other convulsants can be overcome, in vitro, by prolonged exposure to GABA (Bloomquist et al. 1991). The mixed inhibition observed with TBPS was explained by a stabilization of a closed, GABA-bound form of the channel. Similar conclusions were reached by other workers studying polychlorocycloalkanes, as well as Newland and Cull-Candy (1992) with PTX. ffrench-Constant et al. (1995) suggested that blockers preferentially stabilized the desensitized form of the receptor and identified the alanine 302 residue on the M_2 helix as critical for convulsant action.

GABA-stimulated ^{36}Cl uptake and radioligand binding experiments using mouse and rat brain vesicle preparations confirmed convulsant block of the GABA receptor and facilitated a quantitative structure-activity analysis. Cyclodienes and other convulsants showed inhibition of GABA-stimulated chloride uptake in mouse and rat brain vesicles that displayed noncompetitive kinetics, consistent with their neurophysiological effects. Good correlations were obtained between the potency for blocking chloride uptake, displacement of $[^{35}S]TBPS$ binding from the convulsant site, and acute lethality for a range of polychlorocycloalkanes and trioxabicyclooctanes. However, all the structural classes of chloride channel blockers could not be grouped together, because they fell on different regression lines when correlated with toxicity or binding activity, suggesting subtle differences in action or binding sites in mammals. The utility of $[^{35}S]TBPS$ as a radioligand did not extend to insects, but $[^3H]EBOB$ proved to be an excellent label for the convulsant site in the housefly, and there was a robust correlation between EBOB displacement and synergized housefly toxicity across several structurally diverse classes of convulsants.

In contrast to the uniformly convulsive actions of the cyclodienes, the avermectins cause a different poisoning syndrome and have different effects on the $GABA_a$ receptor (Fisher and Mrozik 1992; Bloomquist 1993). A lethal dose of abamectin or ivermectin in mice causes hyperexcitability, incoordination, and tremor that gives way to ataxia and coma-like sedation. In mammalian dorsal root ganglion neurons, avermectins can block the action of GABA at low concentrations and irreversibly activate the chloride channel at high concentrations (Robertson 1989). Abamectin displayed some ability to augment ^{36}Cl uptake in rat brain vesicles that was sensitive to block by bicuculline, but acted as a pure noncompetitive antagonist of mouse brain $GABA_a$ receptors. GABA antagonism is consistent with tremor in exposed animals, whereas ataxia and sedation suggest potentiation of GABA action or direct activation of the chloride channel. Binding of $[^3H]ivermectin$ displays 100-fold lower affinity for the rat brain receptor compared with the nematode *Caenorhabditis elegans*, consistent with the wide therapeutic index of this compound for controlling nematode parasites.

3.1
Polychlorocycloalkanes and Related Norbornanes

Recent studies have provided more detailed knowledge of the action of cyclodienes. A surprising effect (reviewed in Narahashi et al. 1998) is the observation that dieldrin enhances the peak amplitude of GABA-induced current in dorsal root ganglion neurons, before the onset of current suppression. The potency for inhibition (EC_{50} = 92 nM) is greater than that of the enhancing effect (EC_{50} = 754 nM). The significance of this effect for the neurotoxicity of cyclodienes is unclear; however, it is dependent on GABA receptor subunit composition. Dieldrin enhancement occurs in GABA receptors having the γ2s subunit, whereas blockage of the channel occurs in the presence or absence of this subunit. Subunit-specific effects on blocker pharmacology have also been observed with cloned insect subunits (Zhang et al. 1995), most significantly in terms of cyclodiene resistance (ffrench-Constant et al. 1995). Single channel studies of dieldrin and PTX in dorsal root ganglion neurons found that three open states were induced by GABA, and that the kinetics of these states were essentially unchanged by application of dieldrin or PTX (Ikeda et al. 1998). Four closed states were also observed in the presence of GABA, having time constants of 0.4, 1.6, 4.8, and 33.5 ms. The corresponding time constants in the presence of dieldrin were 0.5, 3.8, 32.8, and 200 ms, and with PTX were 0.6, 3.8, 46, and 141 ms. Thus, the overall effect of the toxicants was a lengthening of several closed channel states, with the shortest closed state being unaffected (Ikeda et al. 1998).

Cyclodiene compounds continue to be explored as possible new insecticides and receptor probes. Attempts were made to use [^{3}H]α-endosulfan as a radioligand (Fig. 1), but it was found that it had high levels of nonspecific binding in housefly head membrane preparations (Cole et al. 1994). There was some correlation between cyclodiene displacement of [^{3}H]α-endosulfan binding and potency for displacing [^{3}H]EBOB binding; however, the binding of [^{3}H]α-endosulfan was not displaced by EBOB, TBPS, phenylpyrazoles, or PTX, indicating a difference in the binding sites of cyclodienes and these other ligands. The experimental norbornane, 3,3-bistrifluoromethylbicyclo[2.2.1]heptane-2,2-dicarbonitrile (BIDN, Fig. 1), blocks GABA-induced currents in *Xenopus* oocytes expressing homomultimeric wild-type Rdl GABA receptors (Hosie et al. 1995b). In addition, the blockage showed mixed kinetics, both depressing the maximal effect of saturating concentrations of GABA and shifting its dose-response curve to the right. When labeled with tritium, BIDN shows specific binding in insects with K_d values in the low nanomolar range, but 200–300 nM in rat brain membranes (Rauh et al. 1997). Greatest affinity for [^{3}H]BIDN was as a radioligand in the southern corn rootworm, *Diabrotica undecimpunctata*

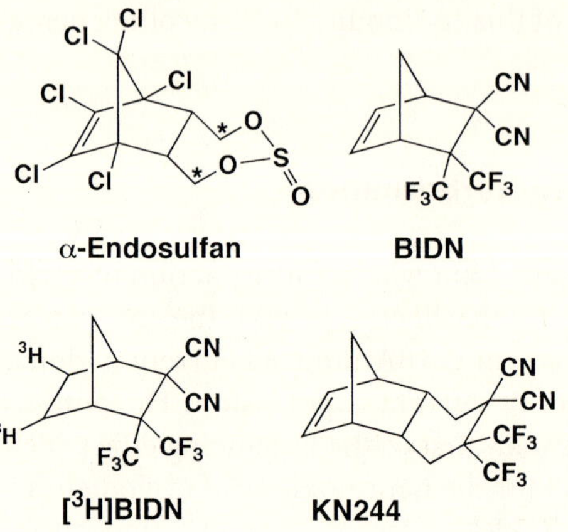

α-Endosulfan   BIDN

[³H]BIDN   KN244

Fig. 1. Structures of currently used poly-chlorocycloalkane insecticides and structurally related experimental compounds which act as GABA antagonists, including those used as experimental radioligands. In the case of α-endosulfan, introduction of tritium atoms is indicated by *asterisks*

howardi (K_d = 26 nM), and the binding was displaced competitively by dieldrin. Binding of this ligand was displaced <50% by 100 µM concentrations of EBOB, TBOB, TBPS, and PTX, again suggesting that there is a distinction between the binding domain for these compounds and the cyclodiene/BIDN binding domain (Rauh et al. 1997). Matsuda et al. (1999) reported that a tricyclic dinitrile closely related to BIDN (KN244, Fig. 1) suppressed GABA-stimulated currents with an EC_{50} of about 42 nM, but was nearly a competitive inhibitor, depressing the maximal effect of GABA only 9% when KN244 was present at 2.7 µM, well above its IC_{50} value. In contrast, BIDN suppressed the maximal effect of GABA by about 40% (Hosie et al. 1995b).

3.2
Picrodendrin and Silphinene Natural Products

As discussed above, PTX has effects identical to those of dieldrin on GABA receptor function, and recent studies have provided some interesting results with natural products having actions similar to those of PTX (Fig. 2). Two papers (Anthony et al. 1994; Shirai et al. 1995) discussed the activity of natural and synthetic picrotoxinin analogs, including those having 6-fluoro and 6-acetate substitutions for the hydroxyl group that normally resides in this position. The order of effectiveness for blocking responses to GABA was PTX > fluoro-PTX > acetate-PTX. Thus, the 6-hydroxyl group is not a limiting factor in the action of PTX. Newer PTX analogs are a series of terpenoids, including various tutins and picrodendrins isolated from the Euphorbiaceae plant, *Picrodendron baccatum*. The picrodendrins displayed a biphasic blockade of GABA-gated currents in Rdl homomultimers expressed in *Xenopus* oocytes, first accelerating the decay of the current, and at higher concentrations reducing peak amplitude (Hosie et al. 1996). The most active compound was

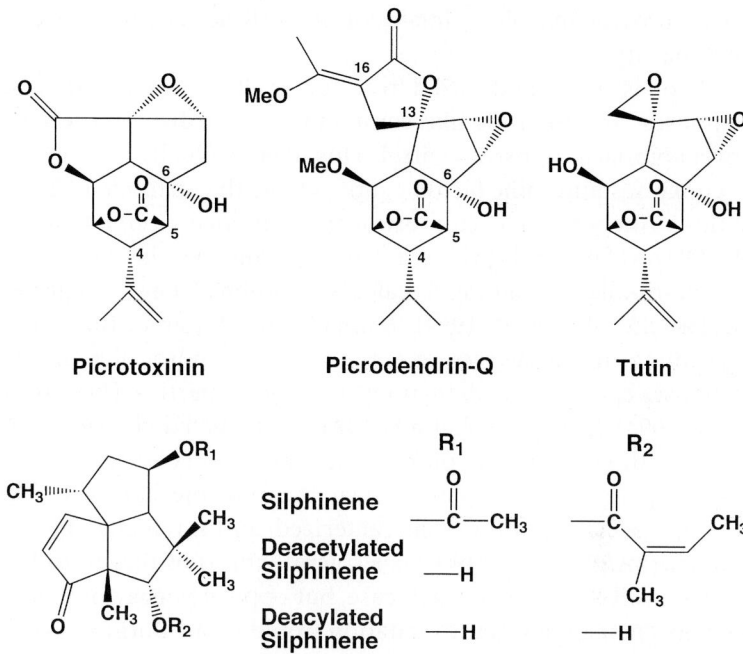

Fig. 2. Picrodendrin and silphinene natural products that interact with GABA receptors. Deacetylated silphinene is hydroxyl at R_1, with R_2 the same as the parent. Deacylated silphinene is hydroxyl at R_1 and R_2

picrodendrin-Q (IC_{50} = 17 nM), followed in rank order by picrodendrin-B (23 nM), picrodendrin-A (44 nM), picrodendrin-G (140 nM), picrodendrin-F (317 nM), and picrodendrin-O (1006 nM). Thus, the picrodendrins are quite potent GABA antagonists in this system. The tutins were uniformly less active, with tutin itself the most active compound in this series (IC_{50} = 880 nM). More detailed structure-activity studies (Ozoe et al. 1998a) found that picrodendrins were competitive inhibitors of [^{3}H]EBOB binding in rat and housefly membranes. Comparative molecular field analysis identified structural determinants that differentiated between rat and housefly receptors. The γ-butyrolactone at the 13 position and certain 4-position substituents play important roles in the rat binding site, while identical structural features are not strictly required by the housefly receptor (Ozoe et al. 1998a). The rank order of potency for picrodendrin displacement of EBOB binding in housefly membranes showed a better correlation with their toxicity to *Blattella germanica* (r = 0.918, lethality by injection, synergized with piperonyl butoxide), than it did for blocking GABA-gated currents in Rdl homomultimers (Ozoe et al. 1998a). This finding suggests that native receptors are heteromultimers, and that heterologous expression studies of homomultimeric receptors may not always provide reliable estimates of intrinsic activity on native receptors or in vivo toxicity.

There is an hypothesized role for GABA receptors in insect gustatory responses, and for PTX-like compounds to mediate antifeedant action as a novel approach to insect control. This idea originally arose from the isolation of epoxy-sesquiterpine lactone antifeedants that exhibited PTX-like neurotoxicities in the western corn rootworm, *Diabrotica virgifera virgifera* (Mullin et al. 1994). GABA and glycine are phagostimulants in the western corn rootworm, stimulating spiking of galeal chemosensilla and having a sweet taste in humans (Mullin et al. 1994). Similarly, the bitter tasting compounds PTX, strychnine, and β-hydrastine excite galeal chemosensilla and the intensity of responses correlated with their antifeedant properties. These results led Mullin et al. (1994) to suggest that a GABA/glycine-gated chloride channel receptor with binding sites for picrotoxinin and other blockers mediated these responses. As the authors make clear, the ionic mechanisms of taste transduction in insects are not well characterized, and it remains unclear how activation (GABA/glycine) or blockage (picrotoxinin) of the chloride channel leads to sense cell discharge in each case, but opposite behavioral outcomes. A possible mechanism for neuroexcitation would be a chloride channel that is open most of the time, suppressing activity, and where binding of agonist either closes or desensitizes the receptor, and with PTX having a channel-blocking action. A similar mechanism explains unusual responses to cyclodienes blocking homomultimeric GABA receptors from a cloned molluscan (*Lymnea stagnalis*) subunit expressed in *Xenopus* oocytes (Darlison 1992). More research is needed to clarify the cellular and ionic mechanisms involved in these chemoreceptor responses. A more recent study (Eichenseer et al. 1998) compared the feeding deterrence of PTX, strychnine, and β-hydrastine in susceptible and cyclodiene-resistant beetles (140-fold to aldrin). There was two- to four-fold less sensitivity in the threshold for antifeedant action of these compounds in the resistant field-collected beetles. However, there was no difference in the dose-response curves for PTX in the two beetle strains, which is surprising given the well-documented cross-resistance to PTX in cyclodiene-resistant strains (Bloomquist 1993).

The silphinenes (Fig. 2) are natural products isolated from the plant *Senecia palmensis* (Gonzalez-Coloma et al. 1997). These compounds have documented antifeedant and toxic effects on the Colorado potato beetle (*Leptinotarsa decimlineata*) and western corn rootworm (Gonzalez-Coloma et al. 1997; Mullin et al. 1997). In choice assays with Colorado potato beetle on treated and untreated food, the parent silphinene (S) was at least tenfold more active than the deacetylated (DS) or deacylated (DAS) analogs (Gonzalez-Coloma et al. 1997). In feeding choice assays with western corn rootworm, DS was the most active by at least 17-fold, indicating a difference between beetle species (Mullin et al. 1997). Molecular modeling suggested a structural similarity between S

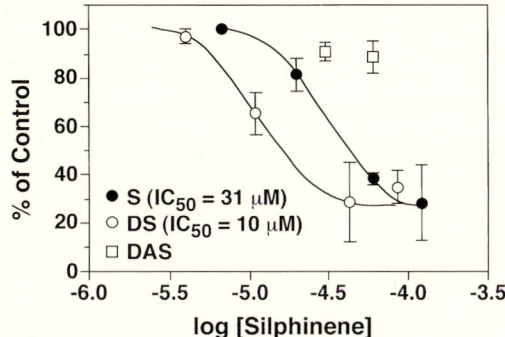

Fig. 3. Silphinenes as mammalian GABA$_a$ receptor antagonists, comparing the activity of the parent silphinene (*S*) with its deacetylated (*DS*) and deacylated (*DAS*) analogs. For calculating IC$_{50}$s, curves were fit to a four parameter logistic equation using Prism by GraphPad Software, San Diego, California

and PTX (Mullin et al. 1997), and my laboratory has recently confirmed that silphinenes are GABA antagonists at mammalian receptors (Fig. 3). In these studies, mouse brain vesicles were incubated with S, DS, or DAS, and the preparation challenged with 100 µM GABA in the presence of ^{36}Cl. DS was the most potent inhibitor, being threefold more active than the parent compound, while DAS was essentially inactive. These results parallel their effects on excitation of rootworm galeal chemoreceptors, where the EC$_{50}$s for excitation were 1.4, 3.0, and 40 µM for DS, S, and DAS, respectively (Mullin et al. 1997). The greater potency associated with the presence of an hydroxyl group in DS parallels findings with PTX, where fluorine or acetate replacement of the hydroxyl reduces activity (Anthony et al. 1994; Shirai et al. 1995). Greater activity with the hydroxyl group suggests the formation of an additional hydrogen bond between the ligand and the GABA receptor. Future studies will define silphinene effects in established insect GABA receptor preparations.

3.3
Fipronil and Fipronil Analogs

Fipronil (Fig. 4) is the newest commercial insecticide that can act on the GABA-gated chloride channel. Fipronil increased the frequency of discharges in insect CNS preparations and reversed the action of GABA (Gant et al. 1998). Fipronil is also an effective inhibitor of GABA-induced currents in Rdl homomultimers expressed in *Xenopus* oocytes, with an IC$_{50}$ value of around 10 µM in some studies (Buckingham et al. 1994; Hosie et al. 1995a), whereas Gant et al. (1998) reported 50% inhibition at around 100 nM. Similar experiments with cloned GABA receptor Rdl subunits from *D. melanogaster* and *H. virescens* found that

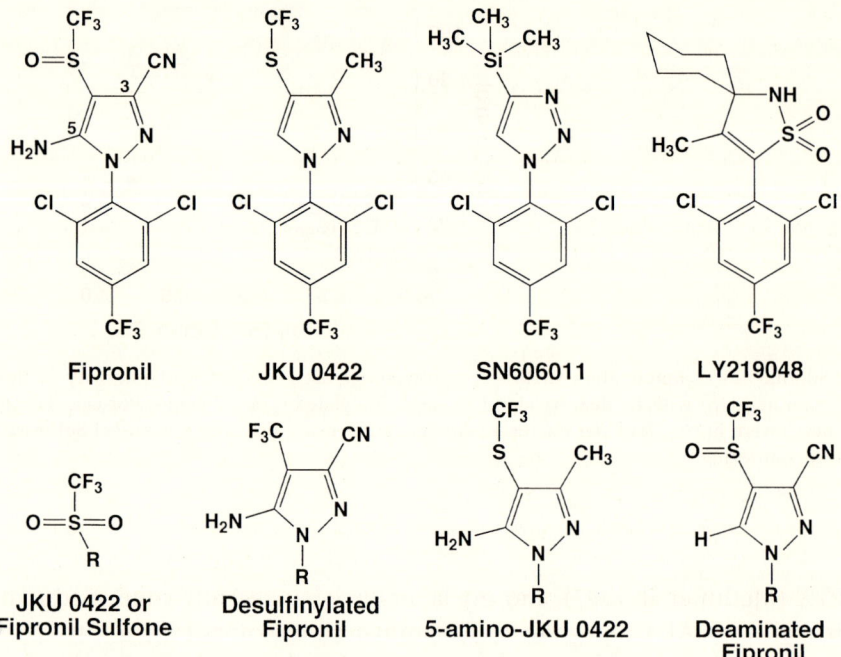

Fig. 4. Fipronil and related compounds discussed in the text

the IC_{50} values of fipronil for blocking these receptors were 250 and 110 nM, respectively (Wolff and Wingate 1998). Fipronil blocks ^{36}Cl uptake into mouse (Cole et al. 1993) and rat (Gant et al. 1998) brain vesicles with IC_{50} values >10 µM, consistent with its somewhat lower mammalian toxicity compared to the older cyclodienes (Bloomquist 1998). Fipronil also blocks [^{3}H]EBOB, [^{35}S]TBPS, and [^{3}H]BIDN binding with IC_{50}s in the 0.5–1.0 µM range in rat, and an IC_{50} for EBOB binding in housefly head membranes of 8 nM (Gant et al. 1998). Cole et al. (1993) reported that fipronil displaced [^{3}H]EBOB binding with IC_{50}s of 2.3 nM in housefly and 4.3 µM in mouse brain. Moreover, fipronil was a noncompetitive inhibitor of EBOB binding in the housefly, suggesting that its binding is poorly reversible or allosteric to the EBOB site. Fipronil is oxidized, in vivo, to the corresponding sulfone (Fig. 4), which is slightly more toxic and two- to sixfold more active on the receptor, especially in mammals (Gant et al. 1998; Hainzl et al. 1998). Thus, relative rates of conversion and intrinsic activity in different species impact overall toxicity and selectivity.

The commercial success of fipronil stimulated an investigation of related arylheterocycles with a 2,6-dichloro-4-trifluoromethyl moiety (Fig. 4). The compounds JKU 0422 (pyrazole), SN606011 (triazole), and LY219048 (spiro-sultam) variously increase spontaneous nerve activity, generate prolonged high frequency discharges following nerve stimulation, and reverse the inhibitory effect of GABA in electrophysiological preparations (Bloomquist 1993, 1994; Bloomquist et al. 1993). Similarly, LY219048 blocks GABA-dependent chloride uptake into bovine brain vesicles and is a potent inhibitor of [^{3}H]TBOB binding to bovine brain membranes ($IC_{50} = 42$ nM), without effect on muscimol or benzodiazepine binding (Bloomquist et al. 1993). Another important derivative of fipronil is its desulfinylated photoproduct (Fig. 4), which retains high levels of toxicity and blocking activity at the GABA receptor (Hainzl and Casida 1996; Hainzl et al. 1998).

3.4
Trioxabicyclooctanes and Related Compounds

Recent studies have provided more information on the interaction of EBOB with mammalian GABA receptors. [^{3}H]EBOB binding had high affinity for GABA receptors in cultured cerebellar granule cells ($K_d = 0.51$ nM), and was displaced by convulsants (Huang and Casida 1996). Subsequent studies explored binding displacement by GABA agonists, and found that compared to cerebellar granule cells, sensitivity to this displacement was about three-fold lower in cerebellum, and as much as 20-fold lower in cerebral cortex, midbrain, pons, and medulla (Huang and Casida 1997a). The authors speculated that the difference was the specific expression of the $\alpha 6$ subunit in granule neurons, a conclusion supported by the finding that forskolin, which down-regulates expression of this subunit, markedly decreased the sensitivity of EBOB binding to GABA.

Structural modification of trioxabicyclooctanes has continued in the search for new insecticides and additional probes of the convulsant binding site (Fig. 5). These studies resulted in various substituted thiazines (Pulman et al. 1996), arylpyrimidines (Pulman et al. 1996), oxathianes (Palmer and Casida 1995), new trioxabicyclooctanes and dithianes (Weston et al. 1995), modified cyclic phosphonates (Ozoe et al. 1995), derivatized thiophosphonic acids (Ozoe et al. 1998b), and 3,4-disubstituted phosphorothionates (Ju and Ozoe 1999). Particularly active compounds were produced by replacing phenyl with a cyclohexyl moiety in the trioxabicyclooctanes and dithianes, where some compounds exceeded by 10-fold or more the toxicity of permethrin (Weston et al. 1995). Generally, all of these compounds showed signs of intoxication similar to known GABA antagonists, displaced radioligands that label the convulsant site, or blocked GABA-stimulated ^{36}Cl uptake. They also required synergism with piperonyl butoxide for expression of insect toxicity and generally displayed poor selectivity. However, the recent studies of Ju and Ozoe (1999) have shown that bulky alkyl side chains in the 3-position (propyl and butyl isomers) of the trioxabicyclooctane moiety are well accepted by insects, but not mammals. This finding resulted in compounds with good insecticidal activity and showed selectivity ratios over 50 for [^{3}H]EBOB binding (rat IC_{50}/housefly

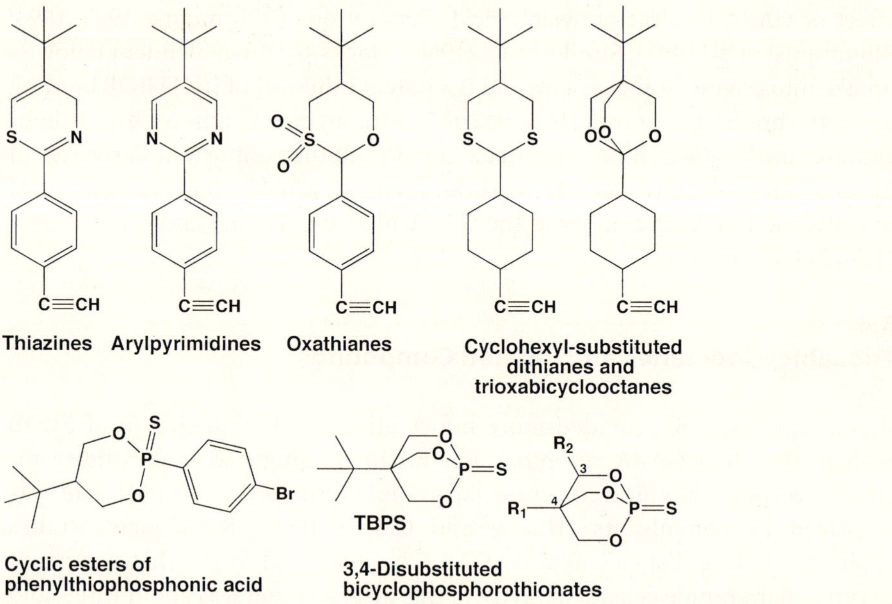

Fig. 5. Trioxabicyclooctanes and related compounds discussed in the text. Examples of high activity compounds for each series are shown, which typically required a 4-ethynylphenyl substituent. For analogs of TBPS (Ju and Ozoe 1999), effective combinations of alkyl group substituents were placed at the R_1 and R_2 positions

IC_{50}). However, toxicity to insects still required synergism with piperonyl butoxide.

3.5
New Avermectins and the Mammalian GABA Receptor

The recent emphasis on the glutamate-gated chloride channel as the primary target site of the avermectins, and the fact that this receptor is absent in mammals, has led to some new research on the interaction of avermectins with the mammalian GABA receptor. Binding studies in cultured cerebellar granule cell neurons with [³H]avermectin B_{1a} found high ($K_d = 5$ nM) and low (866 nM) affinity sites (Huang and Casida 1997b). These binding site affinities were correlated with the ability of avermectin B_{1a} to stimulate ³⁶Cl uptake at 3–100 nM and to block uptake at 1–3 µM. From these data the authors concluded that avermectins bind to two different sites on the GABA-gated chloride channel, activating the channel upon binding to the high affinity site and blocking it upon further binding of the low affinity site (Huang and Casida 1997b). This hypothesis is at odds with electrophysiological studies in both vertebrate and invertebrate preparations, where avermectins tend to block the action of GABA

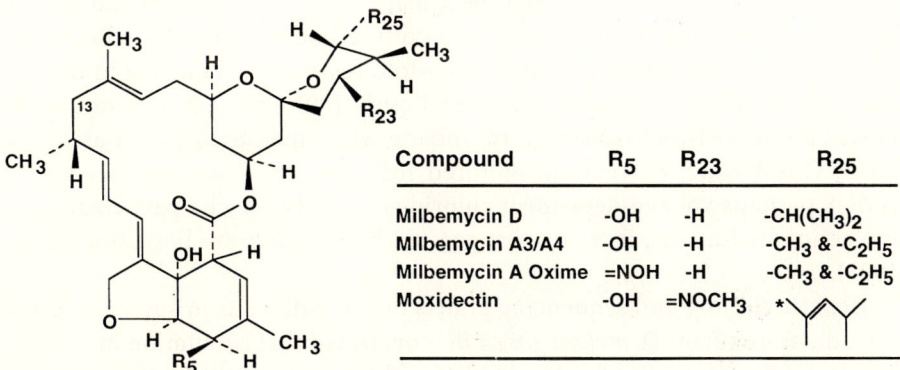

Fig. 6. Structures of avermectins and milbemycins discussed in the text. Avermectin B_{1a} is equivalent to abamectin, which in technical form is supplied as a mixture that is at least 80% avermectin B_{1a} (shown) and not more than 20% of the isopropyl analog at the 25-position, which is known as avermectin B_{1b}. Milbemycins and 13-difluoroavermectin are aglycones at the 13-position, and lack the oleandrose disaccharide

at low concentrations, and then activate the channel irreversibly at higher concentrations. Further research is required to clear up this discrepancy.

Certain new *gem*-difluoroavermectins have been synthesized and their anticonvulsant activity assessed in mammals (Meinke et al. 1998). These compounds are difluoro-substituted at the 13-position (difluoro-aglycone), 4''-position, 4'-position (difluoromonosaccharide), and 23-position (difluoroivermectin). For numbering of the avermectin molecule, see Fig. 6. Using the pentylenetetrazole-treated mouse as a seizure model involving the GABA receptor, 23-difluoroivermectin was three times as active as ivermectin, but about 30-fold less active than diazepam.

3.6
Altered GABA Receptors in Resistance

The following is a summary of knowledge on the mechanisms underlying reduced neuronal sensitivity as the primary mechanism of resistance to cyclo-dienes and other chloride channel blockers. The discussion is taken from reviews (Oppenoorth 1985; Bloomquist 1993, 1998; Deng 1995; ffrench-Constant et al. 1995, 1996), along with some recent papers. Resistance at the GABA receptor is expressed in a large number of insect species and provides varying levels of protection (4.8-fold for lindane to several 1000-fold for dield-rin) against a wide range of compounds, including all the cyclodienes, lindane, and the trioxabicyclooctanes, but not to the avermectins. Neurophysiological assays in several species documented that the resistance reduced the sensitiv-ity of the nervous system and that the sensitivity of GABA-evoked currents to blockage by convulsants was reduced in cultured neurons from the Rdl strain of *D. melanogaster*. Although there is 4.8-fold resistance to lindane in toxicity assays with Rdl (ffrench-Constant and Roush 1991), ffrench-Constant et al. (1995) found 973-fold resistance to lindane when measured as the ability to block GABA-gated currents in cultured Rdl neurons. Thus, another site of action, perhaps voltage-dependent chloride channels, may be participating in the action of lindane, in vivo, as suggested by previous studies (Bloomquist 1993).

Genetic cloning and sequencing efforts on the Rdl locus in susceptible and cyclodiene-resistant *D. melanogaster* demonstrated that a mutation in a GABA receptor subunit gene was responsible for the resistance. Reduction in sensi-tivity to PTX and dieldrin was associated with a single Ala to Ser mutation at position 302 in the M_2 membrane-spanning region of the resistant gene. Using polymerase chain reaction and nucleotide diagnostic techniques, the identical mutation was found in several orders of cyclodiene-resistant insects world-wide, with a number of additional variants. An Ala to Gly substitution was present in *Drosophila simulans*, and six variants were found in the German cockroach, only one of which, Ala to Ser in area B, was clearly associated with resistance (Kaku and Matsumura 1995). In the aphid *Myzus persicae*, Ala to Gly was present, as well as two different Ala to Ser nucleotide variants (Anthony et al. 1994). Detailed analysis of 141 strains of the red flour beetle (*Tribolium castaneum*) indicated that the resistance gene arose independently, and not as a single mutation that spread, which is thought to be the case for Rdl resistance in *Drosophila* (Andreev et al. 1999). Genetic analysis of a susceptible and endosulfan-resistant strain of *Helicoverpa armigera* found that both possessed a Gln at the Ala-equivalent position, suggesting that the resistance had a metabolic origin (ffrench-Constant et al. 1996). Recent screening studies of field strains of *B. germanica* collected in Virginia found several strains that possessed high levels of chlordane resistance (up to 33-fold), even though this compound has not been used for cockroach control for many years (Bloomquist and Robinson 1999). The mechanism of this resistance remains to be investigated.

The Rdl resistance mutation (Ala 302 to Ser) is hypothesized to have a dual effect (Zhang et al. 1994). The mutation apparently reduces convulsant binding affinity, and is associated with longer channel open times and shorter closed times. This latter mechanism reflects a net stabilization of the open channel by a factor of five and a destabilization of the insecticide-favored desensitized receptor conformation by a factor of 29. [^{3}H]EBOB binding affinity was reduced 4-fold in a cyclodiene-resistant strain of housefly that displayed over 60-fold resistance to EBOB in synergized toxicity bioassays, supporting a dual mechanism of resistance. Because the mutation reduces binding affinity and is within the M_2 membrane-spanning domain that makes up the ion pore, it suggests that channel blockers bind within the channel lumen.

3.7
Resistance to New and Experimental Insecticides

Bioassays and electrophysiological studies demonstrate that cyclodiene resistance extends to new synthetic insecticides and natural products. Resistant CNS preparations of *D. melanogaster* (Rdl) were almost completely insensitive to concentrations of LY219048 (Bloomquist et al. 1993) and JKU 0422 (Bloomquist 1994) that excited 100% of susceptible preparations. Expression of susceptible and resistant isoforms of the Rdl gene in *Xenopus* oocytes observed high levels of resistance expressed toward BIDN (Hosie et al. 1995b), and similar studies with KN244 found IC_{50}s for blocking GABA-gated currents of 42 nM and 4.5 µM for susceptible and resistant receptors, respectively (Matsuda et al. 1999). In the same experimental system, picrodendrin-O completely inhibited GABA-induced currents through the susceptible Rdl isoform, but had little effect on resistant receptors (Hosie et al. 1996). Based on these results, one would predict that Rdl cross resistance should also extend to the silphinenes.

Additional studies with cyclodiene-resistant insects and cloned receptors found variable cross-resistance to fipronil. Early toxicity bioassays with fipronil in a cyclodiene-resistant strain of the housefly (2900-fold resistance to dieldrin) found about 90-fold resistance to fipronil (Bloomquist 1993). In a recent study, surface contact bioassays with the Rdl strain of *D. melanogaster* showed 35-fold resistance to fipronil (Bloomquist and Robinson 1999). Resistance to fipronil has also been studied in insect homomultimeric GABA receptors expressed in *Xenopus* oocytes. Hosie et al. (1995a) observed high levels of resistance to fipronil and picrotoxinin, where 50% inhibition of GABA-evoked currents in resistant *Drosophila* Rdl receptors could not be attained, even at concentrations >100 μM. Wolff and Wingate (1998) showed 15-fold resistance to fipronil for blocking GABA-gated currents in a Ser-containing *H. virescens* receptor, compared to its more sensitive Ala-containing counterpart. In contrast, they observed no difference in sensitivity to fipronil blockage of currents through the homologous *D. melanogaster* GABA receptor pairs. This latter finding is puzzling in light of the results of Hosie et al. (1995a) and the resistance to fipronil in Rdl flies (Bloomquist and Robinson 1999). Much lower levels of resistance to fipronil are observed in cockroaches. For example, Valles et al. (1997) reported that a lindane- and dieldrin-tolerant strain of *B. germanica* also displayed tolerance to fipronil, but only 1.6-fold at the LD_{50} level compared with a susceptible strain. Resistance ratios <2.4 were found by Bloomquist and Robinson (1999) in unselected field strains of *B. germanica* showing up to 28-fold resistance to chlordane. Scott and Wen (1997) found that a highly selected dieldrin-resistant strain of *B. germanica* (LD_{50} resistance ratio >17,000) was only 7.7-fold resistant to fipronil.

Cross-resistance studies with the fipronil analog JKU 0422 have raised some interesting structure-activity questions. The 7.7-fold resistance to fipronil observed by Scott and Wen (1997) stands in contrast to the findings of Bloomquist (1994), where over 553-fold resistance to the fipronil analog JKU 0422 was observed in the same resistant strain. Similarly, 1253-fold resistance to JKU 0422 was observed in the Rdl strain of *D. melanogaster* (Bloomquist and Robinson 1999). Thus, relatively low resistance to fipronil and high resistance to JKU 0422 are observed in cyclodiene-resistant laboratory strains of *B. germanica* and *D. melanogaster*. The structural dependence of this large difference apparently resides in the chemical substituents at the 3- and 5-positions of the pyrazole ring (Fig. 4), since the sulfur atoms of both compounds are likely to be oxidized in vivo to the corresponding sulfones. Thus, the structure-activity dependence of the resistance could be addressed by measuring the toxicity and receptor-mediated effects of 5-amino-JKU 0422 and deaminated fipronil (Fig. 4). The methyl group in the 3-position of JKU 0422 is a tetrahedral moiety, unlike the narrow and electron-rich cyano group present in fipronil (Fig. 4). In addition, the presence of the 5-amino group in fipronil provides the possibility of hydrogen bonding at this position, which could account for the greater toxicity of fipronil in resistant cockroaches through a more avid interaction with the Ser isoform GABA receptor. The large difference in resistance between JKU 0422 and fipronil in light of their close structural similarity is reminiscent of target site resistance to pyrethroids, where super-*kdr* resistance provides several hundred fold resistance to α-cyanopyrethroids, but only about 30- to 50-fold resistance to structurally similar pyrethroids lacking an α-cyano group (Sawicki et al. 1986).

4
Glutamate-Gated Chloride Channels

4.1
Physiology, Pharmacology, and Molecular Structure

A comprehensive review of inhibitory glutamate receptor channels, with extensive citation of papers on glutamate-gated chloride channels, has appeared (Cleland 1996) and the following discussion is taken largely from this excellent work. The presence of glutamate-gated chloride channels is lacking in vertebrates, but they are found on skeletal muscle and in the CNS of insects and other invertebrates. A recent paper has also described their presence on caterpillar visceral muscle (Walker and Bloomquist 1999). These receptors are extrajunctional, that is, not clustered at synaptic sites, and were termed H-type receptors because receptor activation hyperpolarized the membrane potential. It has been proposed that they may play a role in regulating muscle excitability to circulating levels of plasma glutamate. Native glutamate receptors are activated by two molecules of agonist and typically display desensitizing responses. Ibotenic acid is a specific agonist and the receptors are blocked by PTX.

The glutamate receptor/chloride ionophore complex is structurally similar in many ways to other ligand-gated chloride channels. Recent gene structure and phylogenetic analyses indicate that glutamate-gated chloride channels are a distinct branch of the ligand-gated ion channel superfamily and may represent genes orthologous to vertebrate glycine receptors (Vassilatis et al. 1997). These glutamate receptors are ionotropic, with multiple cloned subunits identified, including α and β subtypes from *C. elegans* (Cully et al. 1994). When expressed in *Xenopus* oocytes, the β subunit carries the binding site for glutamate, while the α subunit alone confers sensitivity to avermectins. Moreover, glutamate sensitivity in β homomultimers is about 3-fold greater than heteromultimeric $\alpha+\beta$ receptors, whereas α homomultimers and $\alpha+\beta$ receptors are equally sensitive to avermectin. Glutamate responses under these conditions show little evidence of desensitization (Cully et al. 1994). An α-type subunit has been identified in *D. melanogaster* and differs from the *C. elegans* subunit in that it has dual gating with glutamate and avermectin, and a strongly desensitizing glutamate response that is not potentiated by avermectin (Cully et al. 1996). PTX blockade of expressed *C. elegans* receptors was localized to the β subunit, and was reduced >10,000-fold by an Ala to Thr substitution in the M_2 domain, suggesting PTX binding within the channel pore (Etter et al. 1999).

4.2
Effects of the Avermectins

Much of the older literature on avermectin action is summarized via the following reviews (Bloomquist 1993; Duce et al. 1995; Rohrer and Arena 1995; Cleland 1996). In insects and nematodes poisoned by these compounds, ataxia and paralysis predominate, and signs of hyperexcitation are largely absent. Muscular paralysis occurs through suppression of electrical activity in muscle and nerve through activation of an irreversible chloride permeability. The first suggestion that a GABA receptor might not be involved was the documentation of this effect on muscles that lacked GABA receptors, but were responsive to ibotenic acid. Abamectin was also found to paralyze caterpillar visceral muscle; the potency of this effect depended on diet, and may be due to fluctuating glutamate levels (Walker and Bloomquist 1999). Effects on glutamate-gated chloride channels have been investigated in detail by using *C. elegans* receptors expressed in *Xenopus* oocytes. These studies found that avermectin and its analogs activated the glutamate-gated chloride channel with a threshold concentration of 10 nM. When applied with glutamate, avermectins at >5 nM increased its potency and reduced its Hill coefficient to one, suggesting that only one molecule of glutamate was now required to open the channel. Nematicidal activity correlates with ability to activate the glutamate-gated chloride channel. High affinity binding of [^{3}H]ivermectin ($K_d = 0.26$ nM; B_{max} = 3.5 pmol/mg) was observed in membrane preparations from *C. elegans* and kinetic studies showed that the binding was initiated by a rapid, reversible complex formation followed by essentially irreversible binding. These results are consistent with electrophysiological studies showing reversible effects at low concentrations, followed by irreversible channel activation at higher concentrations. In structure-activity studies, the potencies of a series of eight avermectin analogs for inhibiting [^{3}H]ivermectin binding showed excellent correlation with their potency for producing immobility of *C. elegans*.

There are some unresolved issues concerning the relative contributions of glutamate-gated and GABA-gated chloride channels to intoxication by avermectins. Initial claims for an action on GABA receptors supported the observation that avermectins were inactive on cestodes and trematodes, worms that apparently lack the GABA system (Putter et al. 1981). Moreover, Holden-Dye and Walker (1990) found that functional responses of *Ascaris* muscle cells to GABA were blocked by avermectin analogs in a manner that paralleled their anthelmintic activity. The issue of which site is most important in poisoning is related to the relative affinities of these channels for avermectins, as well as their overall contribution for mediating paralysis. In locust extensor tibia muscle, Duce et al. (1995) reported that 100 pM ivermectin blocked GABA-induced conductance increase by 63%, while 1 nM ivermectin was required to have a similar effect on ibotenate responses mediated by glutamate receptors. However, there was a similar dose-response relationship for causing an irreversible increase in conductance on muscle fibers with or without GABAergic innervation. Thus, if paralysis in the animal is caused by this irreversible conductance increase, the presence of GABAergic innervation would seem to be moot. Moreover, glutamate-gated chloride channels may be more widely distributed in invertebrate muscle, and therefore more available as sites mediating avermectin-induced paralysis. One could address this problem by treating animals with toxicologically relevant doses of avermectins, analyzing electrical and mechanical measures of neuromuscular physiology in various muscles, and correlating effects with the presence or absence of GABA receptors. Structure-activity analysis of avermectin analogs on native glutamate-gated chloride channels should also be performed, to complement studies performed in *Xenopus* oocytes.

4.3
New Avermectins and Their Uses

The original commercial products ivermectin and abamectin have been joined by several derivatives (Fig. 6). The search for new avermectins and milbemycins (aglycone avermectins) has been well reviewed by Shoop et al. (1995). These new compounds include semi-synthetic derivatives which exhibit essentially the same mode of action, but with subtle shifts in the relative potency of actions against nematodes, acarines, and insects. The structural change which bestows the excellent lepidoptericide activity of emamectin compared with the parent abamectin is well known (Fisher and Mrozik 1992). Another derivative is eprinomectin, which is apparently optimized for use on lactating dairy cattle (Meinke et al. 1998). Doramectin is a cyclohexyl derivative of avermectin B_{1a} with good activity as an enterocide for cattle (Goudie et al. 1993), where the cyclohexyl group increases lipophilicity and tissue half life. The 4'-difluoro-, 4''-difluoro-, 13-difluoro-, and 23-difluoroavermectins displayed essentially identical activity against *Hemonchus contortus* in an anthelmintic assay (Meinke et al. 1998). In contrast, 13-difluoroivermectin was the most toxic compound to brine shrimp, with the others less active. Substituted milbemycins have included oxime derivatives, such as milbemycin A oxime, which has an oxime at the 5-position instead of hydroxyl. Replacement of this hydroxyl in the avermectins usually reduces potency or spectrum of activity, and does so with milbemycin A as well. However, there is an increase in safety, as milbemycin oxime is better tolerated by dogs for nematode control, even in collies that are especially sensitive to ivermectin (Shoop et al. 1995).

4.4
Target Site Resistance to the Avermectins

There are few reports of avermectin resistance via an altered target site (Clark et al. 1995). Scott (1995) has described a strain of housefly that displays >60,000-fold resistance to abamectin that was reduced to 35-fold after injection, suggesting a major penetration mechanism and a lesser target site component, since no difference in abamectin metabolism was detected. Binding studies with [^{3}H]avermectin in housefly thoraces lacking ganglia showed no change in K_d, but there was a 32% decrease in the density of avermectin binding sites in the resistant strain. In addition, the change in B_{max} is inherited in the same way as the resistance in backcrossing studies. Selection experiments coupled with genetic polymorphism analysis showed that ivermectin resistance in *Haemonchus contortus*, a common livestock parasite, was associated with increased frequency of an α-subunit allele, with no change in the frequency of the β-subunit gene (Blackhall et al. 1998). These findings suggest that target site resistance to the avermectins is specifically associated with altered α-subunits, which are known to bestow sensitivity to these compounds in assembled receptors (Cully et al. 1994).

5
Conclusions

The continued development of new GABA receptor-directed insecticides modeled on the cyclodienes possesses both advantages and disadvantages. The GABA receptor is an established insecticide target site with proven utility, as epitomized by the continued commercial use of the less persistent chlorinated compounds (e.g., endosulfan and lindane) and the introduction of fipronil. There are a number of known structural series to serve as starting points for chemical synthesis efforts. The recent work of Ju and Ozoe (1999) has begun to unravel some of the structural determinants of selectivity, addressing the high level of mammalian toxicity that has been a problem with many of the trioxabicyclooctanes. However, these compounds still need synergism with piperonyl butoxide to observe toxicity in insects. There is also the problem of high levels of cross resistance displayed by cyclodiene-resistant insects to the action of experimental compounds. The problem of cross-resistance is compounded by the wide distribution of the Rdl mutation in pest species and the preselection for resistance from use of the cyclodienes. The introduction of fipronil shows that the resistance situation does not preclude development of new compounds targeting the GABA receptor and, for reasons that remain unclear, cross resistance to this compound in cyclodiene-resistant insects is less than expected. Moreover, fipronil has reduced mammalian toxicity compared with the older cyclodienes, especially via the dermal route. These properties, coupled with high activity on insect pests, gives it greater selectivity than the cyclodienes and a somewhat reduced acute poisoning hazard to mammals. The improved properties of fipronil suggest that continued research and development should produce additional commercial compounds with lower levels of cross resistance, less environmental persistence, and better selectivity than that provided by the cyclodienes. None of these problems affects the avermectins, which are selective and not affected by Rdl resistance; however, they are complicated molecules and somewhat expensive to produce. Given the scope of chemical modifications and the small effects on biological activity, I concur with Shoop et al. (1995) that none of the new avermectins qualify as second-generation compounds. It will be difficult to improve on the already outstanding properties of this class of insecticide.

Acknowledgements. I would like to thank Ms. Becky Barlow for proofreading the manuscript and Dr. Azucena Gonzalez-Coloma for kindly providing the silphinenes used in some of these studies.

References

Albert J, Lingle C, Marder E, O'Neil M (1986) A GABA-activated chloride-conductance not blocked by picrotoxin on spiny lobster neuromuscular preparations. Br J Pharmacol 87: 771–779

Andreev D, Kreitman M, Phillips T, Beeman R, ffrench-Constant R (1999) Multiple origins of cyclodiene insecticide resistance in *Tribolium castaneum* (Coleoptera: Tenebrionidae). J Mol Evol 48:615–624

Anthony N, Holyoke C Jr, Sattelle D (1994) Blocking actions of picrotoxinin analogues on insect (*Periplaneta americana*) GABA receptors. Neurosci Lett 171:67–69

Bai D, Sattelle D (1995) A GABA$_B$ receptor on an identified insect motor neurone. J Exp Biol 198:889–894

Blackhall W, Pouliot J-F, Prichard R, Beech R (1998) *Haemonchus contortus*: selection at a glutamate-gated chloride channel gene in ivermectin- and moxidectin-selected strains. Exp Parasitol 90:42–48

Bloomquist J (1993) Toxicology, mode of action, and target site-mediated resistance to insecticides acting on chloride channels. Comp Biochem Physiol 106C:301–314

Bloomquist J (1994) Cyclodiene resistance at the insect GABA receptor/chloride channel complex confers broad cross resistance to convulsants and experimental phenylpyrazole insecticides. Arch Insect Biochem Physiol 26:69–79

Bloomquist J (1998) Chemistry and toxicology of the chlorinated cyclodienes and lindane. Rev Toxicol 2:333–355

Bloomquist J, Robinson W (1999) Prevalence and magnitude of resistance to cyclodiene and phenylpyrazole insecticides in *Blattella germanica* and *Drosophila melanogaster*. In: Robinson W, Rettich F, Rambo G (eds) Proceedings of the 3rd international conference on insect pests in the urban environment. Graficke zavody Hronov, Czech Republic, pp 27–34

Bloomquist J, Grubs R, Soderlund D, Knipple D (1991) Prolonged exposure to GABA activates GABA-gated chloride channels in the presence of channel-blocking convulsants. Comp Biochem Physiol 99C:397–402

Bloomquist J, Jackson J, Karr L, Ferguson H, Gajewski R (1993) Spirosultam LY219048: a new chemical class of neurotoxin acting upon the GABA receptor/chloride ionophore complex. Pestic Sci 39:185–192

Buckingham S, Hosie A, Roush R, Sattelle D (1994) Actions of agonists and convulsant antagonists on a *Drosophila melanogaster* GABA receptor (Rdl) homo-oligomer expressed in *Xenopus* oocytes. Neurosci Lett 181:137–140

Casida J (1993) Insecticide action at the GABA-gated chloride channel: recognition, progress, and prospects. Arch Insect Biochem Physiol 22:13–23

Casida J, Nicholson R, Palmer C (1988) Trioxabicyclooctanes as probes for the convulsant site of the GABA-gated chloride channel in mammals and arthropods. In: Lunt GG (ed) Neurotox'88: molecular basis of drug and pesticide action. Elsevier, Amsterdam, pp 125–144

Clark J, Scott J, Campos F, Bloomquist J (1995) Resistance to avermectins: extent, mechanisms, and management implications. Annu Rev Entomol 40:1–30

Cleland T (1996) Inhibitory glutamate receptor channels. Mol Neurobiol 13:97–136

Cole L, Nicholson R, Casida J (1993) Action of phenylpyrazole insecticides at the GABA-gated chloride channel. Pestic Biochem Physiol 46:47–54

Cole L, Saleh M, Casida J (1994) Housefly head GABA-gated chloride channel: [^{3}H]α-endosulfan binding in relation to polychlorocycloalkane insecticide action. Pestic Sci 42:59–63

Corronc H, Hue B (1999) A native picrotoxin-resistant GABA-gated chloride channel receptor subtype in cockroach neurons. Pestic Sci 55:1007–1011

Cully D, Vassilatis K, Liu K, Paress P, Van der Ploeg L, Schaeffer J, Arena J (1994) Cloning of an avermectin-sensitive glutamate-gated chloride channel from *Caenorhabditis elegans*. Nature 371:707–711

Cully D, Paress P, Liu K, Schaeffer J, Arena J (1996) Identification of a *Drosophila melanogaster* glutamate-gated chloride channel sensitive to the antiparasitic agent avermectin. J Biol Chem 271:20187–20191

Darlison M (1992) Invertebrate GABA and glutamate receptors: molecular biology reveals predictable structures but some unusual pharmacologies. Trends Neurosci 15:469–474

Deng Y (1995) Insecticide binding sites in the housefly head γ-aminobutyric acid gated chloride-channel complex. In: Clark JM (ed) Molecular action of insecticides on ion channels. ACS Symp Ser 591. American Chemical Society, Washington, DC, pp 230–250

Duce I, Bhandal N, Scott R, Norris T (1995) Effects of ivermectin on γ-aminobutyric acid and glutamate-gated chloride conductance in arthropod skeletal muscle. In: Clark JM (ed) Molecular action of insecticides on ion channels. ACS Symp Ser 591. American Chemical Society, Washington, DC, pp 250–263

Eichenseer H, Mullin C, Chyb S (1998) Antifeedant discrimination thresholds for two populations of western corn rootworm. Physiol Entomol 23:220–226

Etter A, Cully D, Liu K, Reiss B, Vassilatis D, Schaeffer J, Arena J (1999) Picrotoxin blockade of invertebrate glutamate-gated chloride channels: subunit dependence and evidence for binding within the pore. J Neurochem 72:318–326

ffrench-Constant R, Roush R (1991) Gene mapping and cross-resistance in cyclodiene insecticide-resistant *Drosophila melanogaster* (Mg.). Genet Res Camb 57:17–21

ffrench-Constant R, Zhang H-J, Jackson M (1995) Biophysical analysis of a single amino acid replacement in the resistance to dieldrin γ-aminobutyric acid receptor. In: Clark JM (ed) Molecular action of insecticides on ion channels. ACS Symp Ser 591. American Chemical Society, Washington, DC, pp 192–204

ffrench-Constant R, Anthony N, Andreev D, Aronstein K (1996) Single versus multiple origins of insecticide resistance: inferences from the cyclodiene resistance gene Rdl. In: Brown T (ed) Molecular genetics and evolution of pesticide resistance. ACS Symp Ser 645:106–116

Fischer M, Mrozik H (1992) The chemistry and pharmacology of avermectins. Annu Rev Pharmacol Toxicol 32:537–553

Gant D, Chalmers A, Wolff M, Hoffman H, Bushey D (1998) Fipronil: action at the GABA receptor. Rev Toxicol 2:147–156

Ghiasuddin S, Matsumura F (1982) Inhibition of gamma-aminobutyric acid (GABA)-induced chloride uptake by gamma-BHC and heptachlor epoxide. Comp Biochem Physiol 73C:141–144

Gonzalez-Coloma A, Gutierrez C, Cabrera R, Reina M (1997) Silphinene derivatives: their effects and modes of action on Colorado potato beetle. J Agric Food Chem 45:946–950

Goudie A, Evans N, Gration K, Bishop B, Gibson S, Holdom K, Kaye B, Wicks S, Lewis D, Weatherly A, Bruce C, Herbert A, Seymour D (1993) Doramectin – a potent novel endectocide. Vet Parasitol 49:5–15

Hainzl D, Casida J (1996) Fipronil insecticide: novel photochemical desulfinylation with retention of neurotoxicity. Proc Natl Acad Sci USA 93:12764–12767

Hainzl D, Cole L, Casida J (1998) Mechanisms for selective toxicity of fipronil insecticide and its sulfone metabolite and desulfinyl photoproduct. Chem Res Toxicol 11:1529–1535

Holden-Dye L, Walker R (1990) Avermectin and avermectin derivatives are antagonists at the 4-aminobutyric acid (GABA) receptor on the somatic muscle cells of *Ascaris*; is this the site of anthelmintic action? Parasitology 101:265–271

Hosie A, Baylis H, Duckingham S, Sattelle D (1995a) Actions of the insecticide fipronil on dieldrin-sensitive and -resistant GABA receptors of *Drosophila melanogaster*. Br J Pharmacol 115:909–912

Hosie A, Shirai Y, Buckingham S, Rauh J, Roush R, Baylis H, Sattelle D (1995b) Blocking actions of BIDN, a bicyclic dinitrile convulsant compound, on wild-type and dieldrin-resistant GABA receptor homo-oligomers of *Drosophila melanogaster* expressed in *Xenopus* oocytes. Brain Res 693:257–260

Hosie A, Ozoe Y, Koike K, Ohmoto T, Nikaido T, Sattelle D (1996) Actions of picrodendrin antagonists on dieldrin-sensitive and -resistant *Drosophila* GABA receptors. Br J Pharmacol 119:1569–1576

Hosie A, Aronstein K, Sattelle D, ffrench-Constant R (1997) Molecular biology of insect neuronal GABA receptors. Trends Neurosci 20:578–583

Huang J, Casida J (1996) Characterization of [^{3}H]ethynylbicycloorthobenzoate ([^{3}H]EBOB) binding and the action of insecticides on the γ-aminobutyric acid-gated chloride channel in cultured cerebellar granule neurons. J Pharmacol Exp Ther 279:1191–1196

Huang J, Casida J (1997a) Role of cerebellar granule cell-specific GABA$_A$ receptor subtype in the differential sensitivity of [^{3}H]ethynylbicycloorthobenzoate binding to GABA mimetics. Neurosci Lett 225:85–88

Huang J, Casida J (1997b) Avermectin B1a binds to high- and low-affinity sites with dual effects on the γ-aminobutyric acid-gated chloride channel of cultured cerebellar granule neurons. J Pharmacol Exp Ther 281:261–266

Hue B (1991) Functional assay for GABA receptor subtypes of a cockroach giant interneuron. Arch Insect Biochem Physiol 18:147–157

Ikeda T, Nagata K, Shono T, Narahashi T (1998) Dieldrin and picrotoxinin modulation of GABA$_A$ receptor single channels. NeuroReport 9:3189–3195

Ju X, Ozoe Y (1999) Bicyclophosphorothionate antagonists exhibiting selectivity for housefly GABA receptors. Pestic Sci 55:971–982

Kaku K, Matsumura F (1995) Chloride-channel gene probes from cyclodiene-resistant and -susceptible strains of *Blattella germanica*. In: Clark JM (ed) Molecular action of insecticides on ion channels. ACS Symp Ser 591. American Chemical Society, Washington, DC, pp 216–229

Knipple D, Henderson J, Soderlund D (1995) Structural and functional characterization of insect genes encoding ligand-gated chloride-channel subunits. In: Clark JM (ed) Molecular action of insecticides on ion channels. ACS Symp Ser 591. American Chemical Society, Washington, DC, pp 205–215

Lummis S (1990) GABA receptors in insects. Comp Biochem Physiol 95C:1–8

Matsuda K, Hosie A, Holyoke C Jr, Rauh J, Sattelle D (1999) Cross-resistance with dieldrin of a novel tricyclic dinitrile GABA receptor antagonist. Br J Pharmacol 127:1305–1307

Meinke P, Shoop W, Michael B, Blizzard T, Dawson G, Fisher M, Mrozik H (1998) Synthesis of gem-difluoroavermectin derivatives: potent anthelmintic and anticonvulsant agents. Bioorg Med Chem Lett 8:3643–3646

Mullin C, Chyb S, Eichenseer H, Hollister B, Frazier J (1994) Neuroreceptor mechanisms in insect gustation: a pharmacological approach. J Insect Physiol 40:913–931

Mullin C, Gonzalez-Coloma A, Gutierrez C, Reina M, Eichenseer H, Hollister B, Chyb S (1997) Antifeedant effects of some novel terpenoids on Chrysomelidae beetles: comparisons with alkaloids on an alkaloid-adapted and nonadapted species. J Chem Ecol 23:1851–1866

Narahashi T, Ginsburg K, Nagata K, Song J, Tatebayashi H (1998) Ion channels as targets for insecticides. NeuroToxicol 19:581–590

Newland C, Cull-Candy S (1992) On the mechanism of action of picrotoxin on GABA receptor channels in dissociated sympathetic neurones of the rat. J Physiol (Lond) 447:191–231

Oppenoorth FJ (1985) Biochemistry and genetics of insecticide resistance. In: Kerkut GA, Gilbert LI (eds) Comprehensive insect biochemistry, physiology, and pharmacology, vol 12. Pergamon Press, Oxford, pp 731–773

Ozoe Y, Matsumoto K, Mochida K, Nakamura T (1995) Six-membered cyclic phosphonate GABA antagonists: 2,5-disubstituted 1,3,2-dioxaphosphorinanes. Biosci Biotech Biochem 59: 2314–2316

Ozoe Y, Arkamatsu M, Higata T, Ikeda I, Mochida K, Koike K, Ohmoto T, Nikaido T (1998a) Picrodendrin and related terpenoid antagonists reveal structural differences between ionotropic GABA receptors of mammals and insects. Bioorg Med Chem 6:481–492

Ozoe Y, Niina K, Matsumoto K, Ikeda I, Mochida K, Ogawa C, Matsuno A, Miki M, Yanagi K (1998b) Actions of cyclic esters, S-esters, and amides of phenyl- and phenylthiophosphonic acids on mammalian and insect GABA-gated chloride channels. Bioorg Med Chem 6:73–83

Palmer C, Casida J (1995) Insecticidal 1,3-oxathianes and their oxides. J Agric Food Chem 43:498–502

Pulman D, Smith I, Larkin J, Casida J (1996) Heterocyclic insecticides acting at the GABA-gated chloride channel: 5-alkyl-2-arylpyrimidines and -1,3-thiazines. Pestic Sci 46:237–245

Putter I, Mac Connell J, Preiser F, Haidiri A, Ristich S, Dybas R (1981) Avermectins: novel insecticides, acaricides, and nematicides from a soil microorganism. Experientia 37:963–964

Rauh J, Lummis S, Sattelle D (1990) Pharmacological and biochemical properties of insect GABA receptors. Trends Pharmacol Sci 11:325–329

Rauh J, Benner E, Schnee M, Cordova D, Holyoke C, Howard M, Bai D, Buckingham S, Hutton M, Hamon A, Roush R, Sattelle D (1997) Effects of [^{3}H]BIDN, a novel bicyclic dinitrile radioligand for GABA-gated chloride channels of insects and vertebrates. Br J Pharmacol 121:1496–1505

Robertson B (1989) Actions of anaesthetics and avermectin on GABA$_A$ chloride channels in mammalian dorsal root ganglion neurones. Br J Pharmacol 98:167–176

Rohrer S, Arena J (1995) Structural and functional characterization of insect genes encoding ligand-gated chloride-channel subunits. In: Clark JM (ed) Molecular action of insecticides on ion channels. ACS Symp Ser 591. American Chemical Society, Washington, DC, pp 264–283

Sawicki R, Farnham A, Denholm I, Church V (1986) Potentiation of super-kdr resistance to deltamethrin and other pyrethroids by an intensifier (Factor 161) on autosome 2 in the housefly (Musca domestica L.). Pestic Sci 17:483–488

Scott J (1995) Resistance to avermectins in the housefly, Musca domestica. In: Clark JM (ed) Molecular action of insecticides on ion channels. ACS Symp Ser 591. American Chemical Society, Washington, DC, pp 284–292

Scott J, Wen Z (1997) Toxicity of fipronil to susceptible and resistant strains of German cockroaches (Dictyoptera: Blattellidae) and houseflies (Diptera: Muscidae). J Econ Entomol 90:1152–1156

Shirai Y, Hosie A, Buckingham S, Holyoke C Jr, Baylis H, Sattelle D (1995) Actions of picrotoxinin analogues on an expressed, homo-oligomeric GABA receptor of Drosophila melanogaster. Neurosci Lett 189:1–4

Shoop W, Mrozik H, Fisher M (1995) Structure and activity of avermectins and milbemycins in animal health. Vet Parasitol 59:139–156

Valles S, Koehler P, Brenner R (1997) Antagonism of fipronil toxicity by piperonyl butoxide and S,S,S-tributyl phosphorotrithioate in the German cockroach (Dictyoptera: Blattellidae). J Econ Entomol 90:1254–1258

Vassilatis D, Elliston K, Paress P, Hamelin M, Arena J, Schaeffer J, Van der Ploeg L, Cully D (1997) Evolutionary relationship of the ligand-gated ion channels and the avermectin-sensitive, glutamate-gated chloride channels. J Mol Evol 44:501–508

Walker L, Bloomquist J (1999) Pharmacology of contractile responses in the alimentary system of caterpillars: implications for insecticide development and mode of action. Annu Ent Soc Am 92:902–908

Weston J, Larkin J, Pulman D, Holden I, Casida J (1995) Insecticidal isomers of 4-tert-buty1-1-(4-ethynylcyclohexyl)-2,6,7-trioxabicyclo[2.2.2]octane   and   5-tert-butyl-2-(4-ethynylcyclohexyl)-1,3-dithiane. Pestic Sci 44:69–74

Wolff M, Wingate V (1998) Characterization and comparative pharmacological studies of a functional γ-aminobutyric acid (GABA) receptor cloned from the tobacco budworm, *Heliothis virescens* (Noctuidae: Lepidoptera). Invertebr Neurosci 3:305–315

Zhang H-G, ffrench-Constant R, Jackson M (1994) A unique amino acid of the *Drosophila* GABA receptor with influence on drug sensitivity by two mechanisms. J Physiol (Lond) 479:65–75

Zhang H-G, Lee H-J, Rocheleau T, ffrench-Constant R, Jackson M (1995) Subunit composition determines picrotoxin and bicuculline sensitivity of *Drosophila* γ-aminobutyric acid receptors. Mol Pharmacol 48:835–840

Insecticides Affecting Voltage-Gated Ion Channels

Eli Zlotkin[1]

1
Insecticides and Ion Channels

1.1
Scope and Aim

In this chapter the term insecticides is used in its broadest sense and refers to any category of chemical compounds (from low molecular weight organics to globular polypeptides) of either synthetic or natural origin which is processed/designed by either chemical or genetic manipulations to provide a device for large-scale insect control.

The nervous system provides a critical target for chemical attack since it comprises a small fraction of the body mass but fulfills essential regulatory roles that dictate the physiology and behavior of the entire organism. In other words, a fatal effect can be induced by a relatively small amount of chemistry directed to the nervous system. This essential aspect is exemplified by the abundance and multiplicity and potency of naturally occurring neurotoxins employed in either defensive or offensive allomonal natural systems (Zlotkin 1987; see below).

Thus it is not surprising that the chemical control of insect pests has been primarily aimed at affecting the function of the insect nervous system. The latter is targeted by four major groups of synthetic insecticides. These comprise the vast majority of insecticide production (96% in 1980; Lund 1985), namely, the organophosphates, carbamates, chlorinated hydrocarbons and the expanding group of pyrethroids.

The neurotoxicity induced by the above chemicals covers a wide range of the central and peripheral nervous functions, including an effect on receptor voltage-gated ion channels (see below) as well as transmitter receptors, release sites and its metabolic processing (Lund 1985).

The present chapter focuses on insecticidal chemicals which specifically affect ion conductance mediated by voltage-gated ion channels. Disturbance in the action of the above channels results in the malfunction of axons, dendrites

---

[1] Department of Cell and Animal Biology, Institute of Life Sciences, The Hebrew University, 91904 Jerusalem, Israel

Isaac Ishaaya (Ed.): Biochemical Sites of
Insecticide Action and Resistance
© Springer-Verlag Berlin, Heidelberg 2001

and synaptic membranes. The various substances can be roughly subdivided into two categories: (1) synthetic organic compounds which form the existing commercial insecticides and (2) the emerging and expanding neurotoxic polypeptides derived from arthropod venoms.

1.2
Voltage-Gated Ion Channels

This section provides a condensed summary concerning the structure and function of the cation channels responsible for the generation of action potentials in electrically excitable cells (Hille 1992) and some, less specified, information concerning the voltage-gated chloride channels.

Ions cross membranes through specialized holes (pores, channels) located in certain membrane proteins (channel peptides) in cell plasma membranes. According to an accepted definition (Hille 1992; Gordon 1997), an ion channel is a transmembrane protein complex that forms a water-filled pore across the lipid bilayer through which specific inorganic ions can diffuse down their electrochemical gradient (Fig. 1A). This ion movement underlies a variety of cellular functions such as cell volume regulation, muscle contraction and the production of electrical signals. The ionic channels in excitable membranes are commonly divided into two classes, depending on the mechanism by which they open and close (or "gate"). One class of channels gate in response to changes in the membrane potential and are named voltage-dependent/sensitive/gated ion channels. The second class of channels are those which are allosterically coupled to chemical receptors and gate in response to the binding

▶

Fig. 1. A The voltage gated ion channel. Cartoon from a model of a generalized voltage-gated cation channel. The channel is presented as a transmembranous macromolecule which is crossed by a hole in its center. The externally directed surface is glycosylated. The functional elements, namely, selectivity filter, gate and voltage sensor are shown (From Hille 1992). **B** Schematic presentation of the transmembrane organization of the main subunits of Na, Ca and K channels. Sodium and calcium channels share the basic structure of four homologous repeat domains (*I–IV*) that presumably function as pseudo subunits in the membrane and form a central ion pore. The K channel is composed of four peptides, each homologous to one domain of the Na and Ca channels. Four peptides form a functional K channel. The short segments suggested to possess a β hairpin arrangement in the membrane (SS_1 and SS_2 – indicated by the *dark triangles*) from each repeat form the ion-conducting pore. The most highly conserved transmembrane and helix segment, S_4, is present in each domain and contains a unique motif of positively charged amino acid residues followed by two nonpolar residues that repeat four to eight times in the helix. This structure is present in a similar location in each of the Na, Ca, and K voltage-dependent channels and is suggested to function as the voltage sensor of the channel. S_4 segments are marked in black with a (+) sign. Domain *IV* is suggested to include two functional segments on the intracellular side of the channel: the sequence connecting repeats *III* and *IV* are thought to play an important role in the Na$^+$ channel inactivation mechanisms ("III–IV linker"); the short segment connecting the transmembrane segments S_4 and S_5 in each domain ("S_4-S_5 linker") is thought to play a role in the activation and inactivation mechanism (Durell and Guy 1992) (From Gordon 1997)

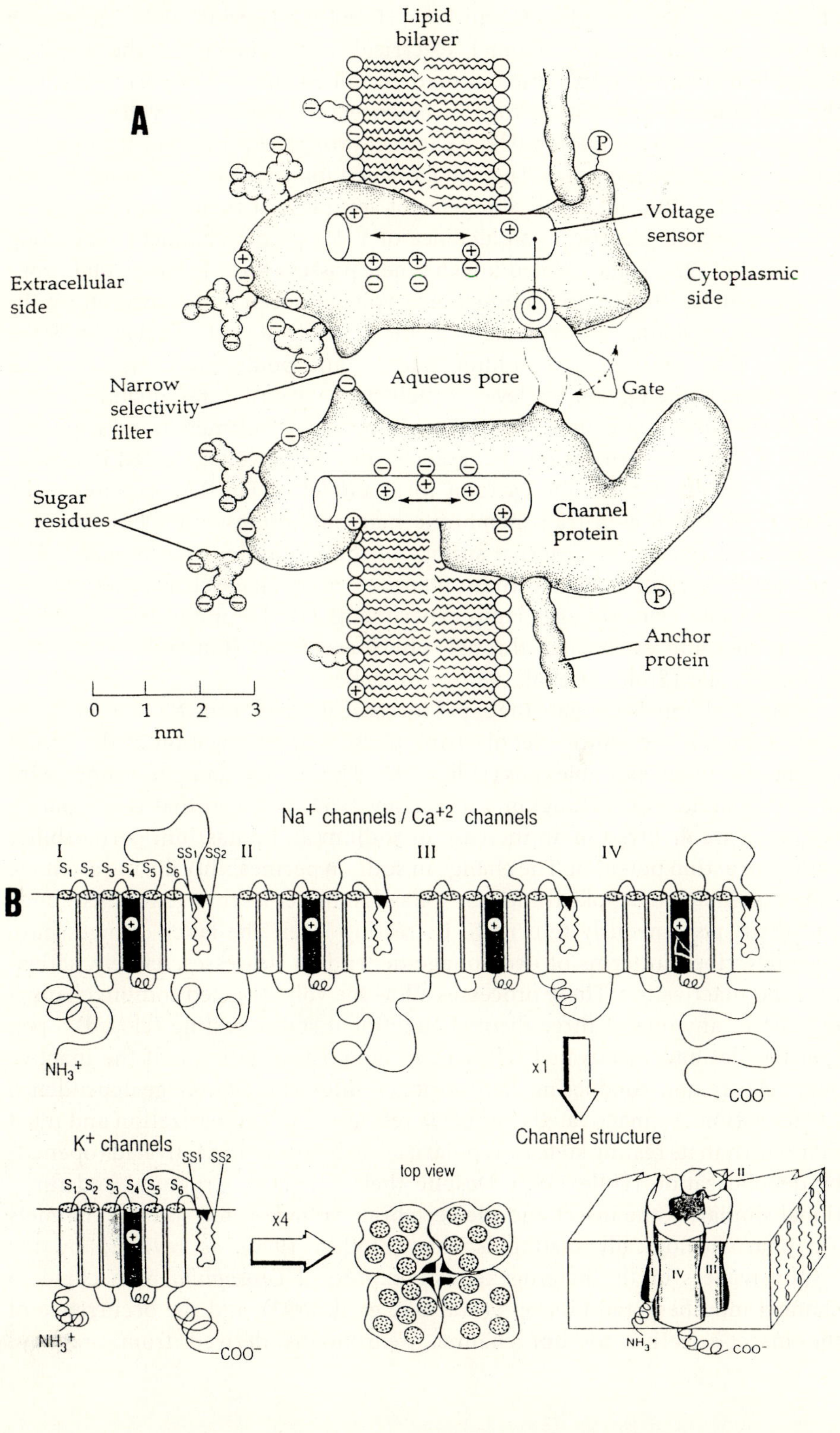

A

Lipid
bilayer

Extracellular
side

Cytoplasmic
side

Voltage
sensor

Narrow
selectivity
filter

Aqueous pore

Gate

Sugar
residues

Channel
protein

Anchor
protein

0   1   2   3
nm

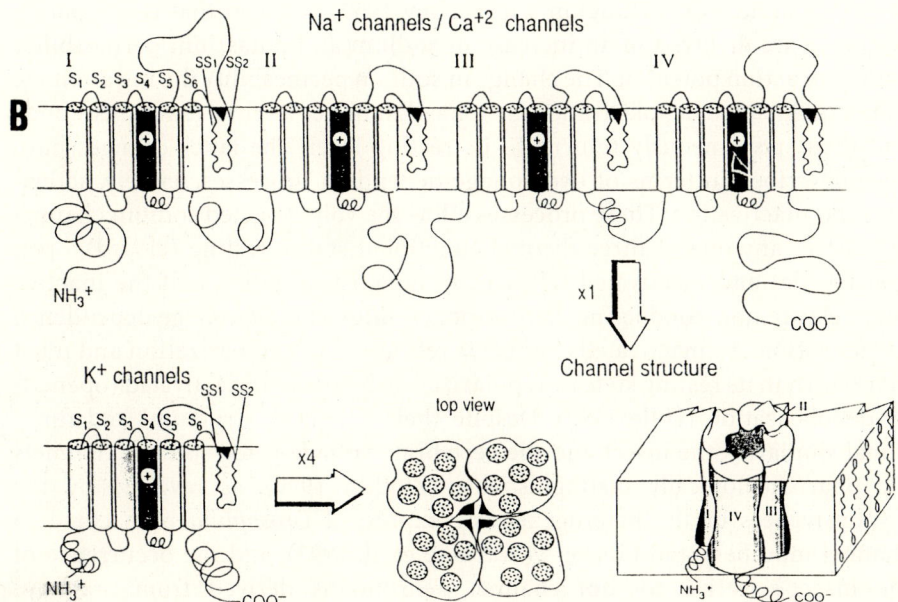

B

Na$^+$ channels / Ca^{+2} channels

I
S_1 S_2 S_3 S_4 S_5 S_6 SS$_1$/SS$_2$ II   III   IV

NH$_3^+$

COO$^-$

x1

Channel structure

K$^+$ channels   SS$_1$ SS$_2$

S_1 S_2 S_3 S_4 S_5 S_6

x4

top view

NH$_3^+$

COO$^-$

NH$_3^+$ COO$^-$

of a ligand to these receptors (Lund 1985; Hille 1992; Gordon 1997). The gating process controlling the channel electrical conductance in the voltage-dependent channels is exhibited in predetermined changes in permeability of the individual pores. The gating in essence refers to small voltage-driven changes in the conformation of the channel protein expressed in the opening and closing of the ion pores. The voltage-gated ion channels commonly reveal densities of 5–20/μm^2, a pore diameter of several Å at their narrowest parts, and a modifiable electrical conductance of 1–150 pS per channel (Armstrong 1992; Hille 1992). Voltage-gated ion channels possess three fundamental properties, namely, selective ion conductance, voltage-dependent activation resulting in pore opening, and voltage-dependent inactivation that terminates the ion flux. The voltage-gated sodium, calcium and potassium channels reveal an obvious homology on the level of their primary structures, including their subdivision into domains composed of six putative transmembrane segments (Fig. 1B). The above-mentioned characteristics of the voltage-gated ion channels are exemplified in the cartoon form model in Fig. 1A. As shown, the channel is drawn as a transmembrane macromolecule with a hole or pore crossing its center. The external surface of the molecule is glycosylated. The functional parts, namely, selectivity, filter, gate and voltage sensor, were essentially deduced from voltage clamp experiments (Hodgkin and Huxley 1952), but verified and established through pharmacological (Catterall 1980, 1992) and molecular (Stuhmer et al. 1989) approaches.

Sodium channels mediate the rapid increase in membrane Na^+ conductance responsible for the rapidly depolarizing phase and propagation of the action potential in many excitable cells (Hille 1992). The voltage clamp technique with the giant squid axon (Hodgkin and Huxley 1952) revealed that two separate pathways are involved in an increase in sodium and potassium permeability within an action potential. The change in sodium permeability during a voltage clamp maintained depolarization is biphasic. It increases for a few milliseconds and then spontaneously returns to its resting level. The latter changes have been described in terms of two voltage-dependent processes, namely, activation and inactivation. These processes allow the voltage-gated sodium channel to exist in any one of three distinct functional states: resting (closed), open (permeable) and inactivated (closed). Although the resting and the inactive channels are non-conducting, they strongly differ in their voltage dependence for activation. An inactivated channel is refractory to depolarization and must first return to its resting state by repolarization before being activated (opened) by depolarization (Hille 1992). Despite their chemical, structural and functional similarity, the insect and the vertebrate voltage-gated sodium channels are pharmacologically distinguishable (Zlotkin 1999), as revealed by the responsiveness of the heterogeneously expressed *Drosophila* para clone to channel modifiers and blockers (Warmke et al. 1997), and the occurrence of the insect-selective sodium channel neurotoxins derived from arachnid venoms, presently used for the design of recombinant baculovirus – mediated selective biosecticides (Zlotkin 1999 and below).

Voltage-gated Ca^{2+} channels are activated by depolarization. The resulting Ca^{2+} flux triggers excitation-secretion coupling, neurotransmitter/hormone release, excitation-contraction coupling, muscle contraction and various intracellular functions (Catterall 1988, 1991). At least three distinct classes of voltage-activated Ca channels exist: the low threshold, dihydropyridine-sensitive L-type, present in most cell types, and the N- and P-type present only in neurons (Beam 1989; Miller 1992). The L-type Ca^{2+} channels in the traversed tubule membrane of the skeletal muscle reveal a multi-subunit complex struc-ture – based on a central transmembrane $\alpha 1$ (175 kDa) subunit, associated with four other polypeptides: The disulfide-linked $\alpha 2$ (143 kDa), the δ (27 kDa) subunit and the γ (30 kDa) subunit, which is non-covalently linked with the others. The hydrophilic β-subunit (54 kDa) (which, unlike the others, is not a glycoprotein) is located on the intracellular site. As shown in Fig. 1B, the sequence of the α-subunit reveals a very high similarity in its deduced primary structure and organization to those of the α-subunits of the voltage-gated Na^+ and K^+ channels. Similarly to sodium channels (see above), a pharmacological distinction between the vertebrate and the insect voltage-gated calcium channels is revealed by the occurrence of the insect-selective calcium blockers (omega atracotoxins) derived from an Australian funnel-web spider venom (see below).

Voltage-gated K^+ channels function in setting the resting membrane poten-tial, duration of action potentials, and generation of firing patterns (Hille 1992). The K channels reveal a broad diversity in their conductance and gating mechanisms in spite of their sequence similarity. The voltage-dependent K^+ channels can be functionally subdivided into types such as the delayed recti-fier (which repolarizes the action potential) K^+ selective channel which is prac-tically non-inactivating and is found in the majority of nerve and muscle cells. The other type is the rapidly inactivating transient A-channels which shape the action potential. Other K channels reveal intermediate properties and consist of several subtypes (Pongs 1992).

Similarly to the cation-selective channels the chloride channels can be sub-divided into ligand gated (those involved in inhibitory synapses; Bloomquist, this volume) and voltage gated. The latter were specifically studied in verte-brate twitch muscles that revealed a high resting chloride permeability (Hodgkin and Horowicz 1959, 1960). In spite of the fact that, due to their minor and slow voltage dependence, these channels do not play a critical role in excitability, they are selective to chloride currents as demonstrated by specific blockers (Palade and Barchi 1977).

The study of the structure of voltage-gated chloride channels did not reach the level of the cation channels (Hille 1992; Foskett 1998). The types of chlo-ride channels which were sequenced and expressed by cloning are a phosphorylation controlled secretory Cl⁻ channel (the cystic fibrosis transmembrane conductance regulator [CFTR]; Schwiebert et al. 1999; Kartner et al. 1991), a chloride channel of Torpedo electric organ (Jentsch et al. 1990) and renal chloride channels (Steinmeyer and Jentsch 1998). The latter, renal channel, a 77-kDa monoglycosylated protein (Maulet et al. 1999), does not reveal much similarity to any other known family of channels except for the multiplicity of putative transmembrane hydrophobic segments. The secretory CFTCR channel is composed of two transmembrane domains (TMD 1 and 2), two cytoplasmic nucleotide-binding domains (NBD 1 and 2) and the regulatory (R) domain – which are devoid of any sequence homology with the cation channels (Devidas and Guggino 1997).

2
Industrial Insecticides Targeting Ion Channels

The topic of ion channel insecticides has recently been covered in a series of review articles and books, such as Hassall (1990), Narahashi (1992, 1996), Clark (1995), Bloomquist (1996) and Zlotkin (1999), allowing a condensed presentation of the essential information in a tabular form leaving room for a more specified treatment of certain specific aspects (see below).

Industrial insecticides are presented in Tables 1 and 2. The structural formulas of representative compounds are given in Fig. 2. Receptor-gated chloride channels are affected by two groups of industrial insecticides: Blockers of the cyclodiene and BHC groups and activators belonging to the anthelmintic avermectins. These substances are beyond the scope of the present review (see Bloomquist, this Vol.).

2.1
Insecticides of the Voltage-Gated Sodium Channels

The voltage-gated sodium channel as the key structure of excitability in biological systems is targeted by a long series of neurotoxins derived from plant and animal allomonal systems (Caterall 1980), as well as four major categories of industrial insecticides, namely, the pyrethroids, DDT analogs, dihydropyrasoles and alkylamides (Tables 1 and 4). Their specific action on sodium conductance was revealed by the aid of "classical" electrophysiology coupled to neurochemistry and molecular techniques (Warmke et al. 1997). This information was achieved through the employment of various arthropod- as well as vertebrate-derived experimental systems (Table 1).

Table 1. Insecticides that act on voltage-gated sodium channels[a,b]

Insecticide group	Properties and uses	Chemistry	Effects on ion conductance
Pyrethroids	More than 30% of the world market; high potency and speed of paralysis; low toxicity to mammals and birds; limited soil persistence; chemical diversity and modifiability. Arthropods are susceptible exhibiting hyperexcitabilty, discoordination, tremor, convulsions-paralysis, death. Uses (see toxicity data [1]): dipterous disease vectors [2,3]; domestic pests [4,5]; veterinary pests [6,7]; stored products wool fabrics [8–10]. The most common application: agricultural pests [9,11,12] especially lepidopterous larvae [13–15]	Cyclic aromatic hydrocarbons which are either synthesized analogs of photostabilized pyrethins derived from pyrethrum flowers (Fig. 2a,b) or entirely synthetic products [16,17]. In the latter, incorporation of the α-cyano group (Fig. 2c,d) on the α-C-atom has enhanced toxicity (LD_{50} 10–30 μg/kg for deltamethrin, Fig. 2c [1]). Pyrethroids which are devoid of α-cyano groups are classified as type I, these elicit tremor syndrome followed by flaccidity (prostration) [24–26]. Type II pyrethroids (Fig. 2c,d) elicit writhing syndrome in rodents and salivation[18,23]	Voltage clamp, single channel patch clamp, functional expression of channel cRNA analyses and binding assays (with crosslinking photosensitive radioactive derivatives [19–22]) reveal: (1) Inhibition of channel inactivation and shift of activation to hyperpolarized membrane potentials. The resulting prolonged sodium "tail" currents – reach threshold depolarization to initiate repetitive discharges and synaptic disturbances leading to hyperexcitability (Fig. 3). (2) The latter can be induced by modification of a small fraction (~1%) of the sodium channel population [20,22]. (3) The sodium channel α-subunit reveals high affinity specific binding to pyrethroids [21]
DDT and analogs	(1) Used in the past substantially for control of disease vectors. (2) Their usage in crop protection is limited due to environmental hazards (persistence in nature, killing beneficial insects, resistance) thus encouraging their replacement by organophosphorus, carbamate and pyrethroid insecticides [23]	Chlorinated aromatic hydrocarbons. DDT is termed as 2,2-bis(p-chlorophenyl)-1,1,1-trichloroethane (Fig. 2e) [27]. Solubility in water 1–2 ppb, very soluble in organic solvents, natural oils and body lipids (accumulation!). Methoxychlor (Fig. 2f) is favorable to DDT environmentally (reduced fat accumulation and persistence increased water solubility)	Studies based on sodium fluxes, voltage clamp and radioligand binding assays indicate that DDT analogs and pyrethroids: (1) Occupy an identical binding site (Table 4: site 7 [32–37]) on the voltage-gated sodium channel. (2) Exhibit identical electrophysiological manifestations [31,33–37], (Fig. 3). (3) Reveal a very similar form of allosteric coupling with other classes of neurotoxins ([32–34], Table 4). (4) Possess a closely similar toxicity to the KDR and sKDR point mutagenesis in the insect

Table 1. *Continued*

Insecticide group	Properties and uses	Chemistry	Effects on ion conductance
		Dicofol (Fig. 2g; a metabolic product of DDT) and chlorobenzilate (Fig. 2h) both have low toxicity to insects and were marketed as acaricides [23]	voltage-gated sodium channel (Fig. 5) [28–30,38]
N-Alkylamides	The crude, petroleum-ether plant extract containing affinin (Fig. 2i) was toxic by topical application to adult and larval stages of various dipterous, lepidopterous hemipterous and coleopterous insects [38,39] revealing (a) symptoms of neurotoxicity, and (b) depolarization, repetitive firing and conduction block of motor neurons [40]	Designed according to certain unsaturated aliphatic amidated hydrocarbons isolated from roots, stems and fruits of the plant families Compositae, Piperaceae and Rutaceae [39,41]. As exemplified in Fig. 2i,k, are divided into two major forms: aliphatic and aromatic.	Effect on sodium conductance was revealed: (1) TTX suppressed the excitatory effect of IPT-alkylamide [42]. (2) Whole cell patch recordings of locust CNS neuronal cell bodies reveal pyrethroid-like slowly decaying "tail" currents and a hyperpolarizing shift in the sodium activation curve [42]. (3) The indication that alkylamides act on Na$^+$ channel site 2 and allosterically interact with site 3 (Table 4) was provided by: (a) antagonizing the veratridine-stimulated choline release; (b) stimulation of ^{22}Na uptake in scorpion venom dependent fashion; (c) inhibition of BTX-induced sodium uptake in a scorpion venom dependent fashion; (d) competitive inhibition of [^{3}H]BTX binding (IC50~2 pM) by BTG 502 (Fig. 2) [43]

Dihydropyrazoles	Recently synthesized substances (Rhom and Haas Co.) cause a strong prostrate-flaccid paralysis of insects and possess a mammalian toxicity [32]	Dihydropyrazoles are derivatives of 1-(phenylcarbamoyl)-2-pyrazolines (Fig. 2l,m) [32]	Effects on sodium conductance revealed (see also Table 4):

Effects on sodium conductance revealed (see also Table 4):

(1) Blockage of induced action potentials in a crayfish neuron without affecting resting potential and input resistance [44].

(2) In insect and mammalian synaptosomes blockage of sodium influx, sodium channel dependent neurotransmitter release and sodium uptake [45].

(3) Noncompetitive inhibitors of BTX binding to mammalian synaptosomes [46].

(4) Voltage-dependent block of sodium conductance in crayfish giant axons [47].

(5) A close correlation between the ability to inhibit sodium uptake, batrachotoxin binding and insecticidal potency of six dihydropyrazoles [48]

a Complementary information is presented in Table 4

b References: [1] Ruight (1985); [2] Nassif et al. (1980); [3] Smies et al. (1980); [4] Schreck et al. (1978). [5] Chadwick (1979); [6] Blackman and Hodson (1977); [7] Knapp and Herald (1981); [8] Mayfield and O'Loughlin (1980); [9] Elliot et al. (1978); [10] Berry (1977); [11] Webb et al. (1974); [12] Griffiths (1997); [13] Crowder et al. (1979); [14] Moore (1980); [15] Plapp (1981); [16] Elliot et al. (1974a); [17] Elliott et al. (1974b); [18] Miyamoto et al. (1995); [19] Narahashi (1992); [20] Narahashi (1996); [21] Trainer et al. (1997); [22] Narahashi et al. (1998); [23] Hassall (1990); [24] Gammon et al. (1981); [25] Adams and Miller (1979); [26] Adams and Miller (1980); [27] Osborne (1985); [28] Lee et al. (1999); [29] Ingles et al. (1996); [30] Pittendrigh et al. (1997); [31] Hassall (1990); [32] Bloomquist (1996); [33] Soderlund and Knipple (1995); 34 Viyerberg and Van den Bercken (1990); [35] Zlotkin (1999); [36] Pauron et al. (1989); [37] Lazdunski et al. (1988); [38] Zlotkin et al. (1999); [39] Elliott et al. (1987); [40] Blade et al. (1985); [41] Su (1985); [42] Lees and Burt (1988); [43] Ottea et al. (1990); [44] Salgado (1990); [45] Nicholson and Zhang (1995); [46] Deecher and Soderlund (1991); [47] Salgedo (1992c); [48] Payne-Gregory et al. (1998)

Table 2. Insecticides that affect K^+ and Ca^{+2} voltage-gated channels[a]

Channel	Insecticides	Toxicity mode
Potassium	Diacylhydrazines. 1,2-Dibenzoyl-1-*tert*-butylhydrazine (RH-5849, Fig. 2l)	(1) Initially developed as a nonsteroidal ecdysone agonist and showed a biphasic (excitatory-flaccid) paralysis attributed to an affect on CNS [1,2]. (2) Its blocking effect on K^+ conductance was revealed by: (a) Prolongation of the repolarization phase of action potentials [2]. (b) Its effect was minimized by 4-aminopyridine, a K^+ blocker [3,4]. (c) Shown to block the steady state (sustained) component of potassium current in a voltage clamp assay [2,5]
	Azadirachtin	When assayed on cultured rat dorsal root ganglion neurons by the whole cell patch technique it was shown to modify in a dose-dependent manner potassium conductance at micromolar range of concentrations [6]
Calcium	Ryanodine. Rayonoids are derived from the tropical shrub Ryania [7,8] (Fig. 2m).	The effect on calcium conductance: (1) Induction of muscle contracture without depolarization [9] attributed to activation of the sacroplasmic reticulum (SR) calcium channels [12]. (2) Increase in the peak Ca^{2+} current in voltage clamp of crab muscle membrane [9]. (3) Increase in Ca^{2+} release in SR demonstrated by patch clamp [10], ion flux [11] and radioligand binding assay (Kda in the nanomolar range) [8,13]. (4) Toxicity was correlated to binding affinity (muscle preparations and assays on competitive displacement) [13] as well as calcium release using 22 various ryanodine modificants [14]

[a] References: [1] Bloomquist (1996); [2] Salgado (1992a); [3] Nishimura et al. (1996); [4] Salgado (1992b); [5] Salgado (1998); [6] Scott et al. (1999); [7] Ware (1982); [8] Lehmberg and Casida (1994); [9] Ghiasudkin and Soderlund (1985); [10] Fill and Coronado (1988); [11] Meissner (1986); [12] Usherwood (1962); [13] Pessah et al. (1985); [14] Jeffries et al. (1993)

2.2
Insecticides of the Potassium and Calcium Channels

The data presented in Table 2 reveal that, in addition to Na^+, commercial insecticides interfere with the function of other voltage-gated ion channels. The table presents three groups of compounds shown to affect specifically potassium and calcium conductance.

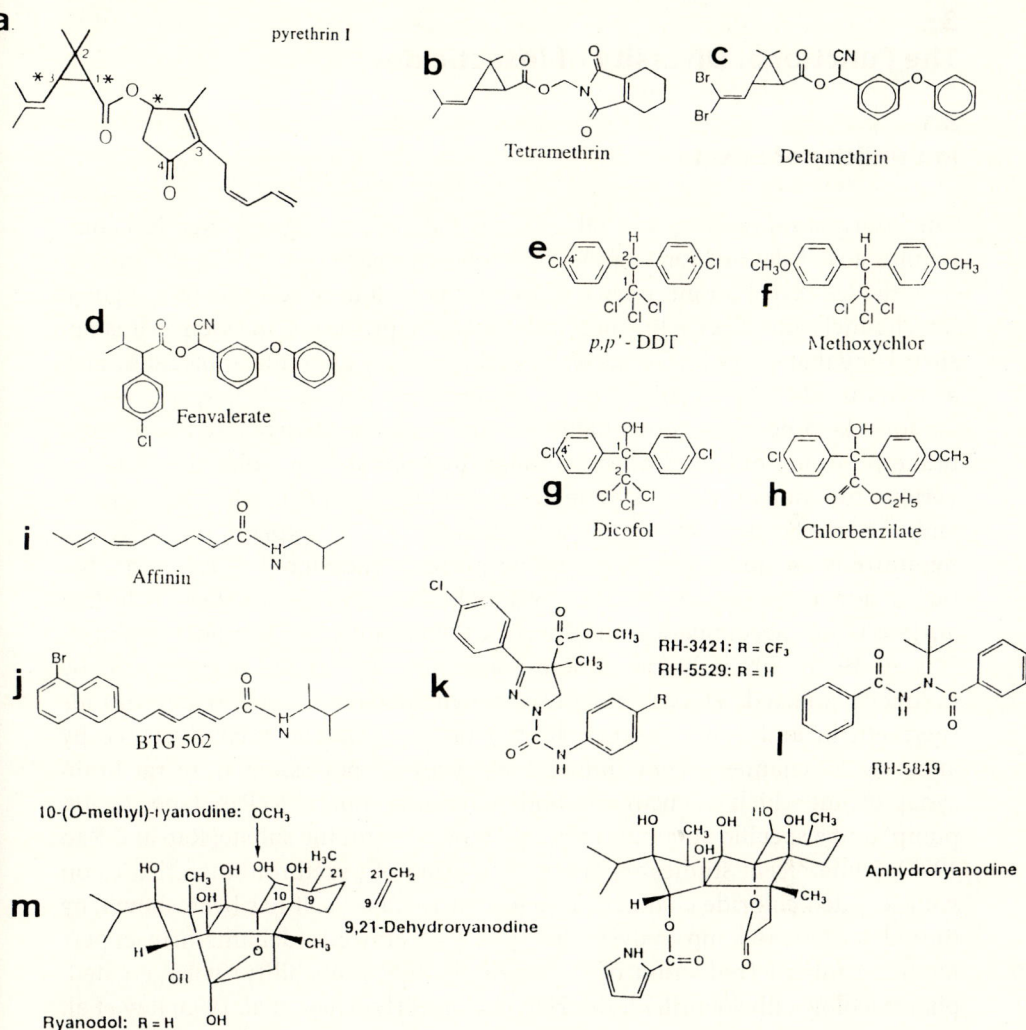

Fig. 2. Structural formulas of insecticides that affect voltage-gated ion channels. Sodium channel: pyrethroids: **a, b** type I, **c, d** type II. **e–h** DDT and analogs; **i, j** alkylamides; **k** dihydropyrazoles. Potassium channel: **l,** diacylhydrazine. Calcium channel: **m,** ryanodines

3
The Functional Diversity of Insecticides

3.1
Multiplicity of Effects

The information presented in Tables 1 and 2 attributes to given chemical compounds a well-defined ion channel selective pharmacology. In fact, the various insecticides reveal a multiplicity of functions including several voltage-gated ion channels and diverse biochemical metabolic processes. In essence, it is not surprising that stable low molecular weight strongly lipophilic molecules, such as pyrethroids (Fig. 2; Ruigt 1985) – which carry polar and charged sites (in the form keto, epoxy cyano or halide groups) – can be incorporated into membrane phospholipids and affect, through hydrophobic or polar interactions, various functions mediated by membrane-bound proteins such as enzymes and ion channels. Such simple considerations may explain the effect of pyrethroids in modifying GTP-binding proteins (Rossignol 1995), perturbation of adenosine triphosphatases, protein kinases and phosphatases (Clark et al. 1995), and accumulation in biolipids ("mechanical or physical toxicity", Hassall 1990). Similarly, membrane-bound ion-conducting proteins can be indirectly affected. Thus it has been shown that type I and II pyrethroids (permethrin and cypermethrin, 40 μM) modify potassium conductance by affecting K^+ channels. They induce the release of potassium from rat brain synaptosome which is sensitive to both membrane-bound ATPase (specific ion pump) and a specific potassium channel blocker (quinine sulfate; Rao and Rao 1997). Some recent studies report on the action of pyrethroid insecticides on voltage-gated chloride channels in neuroblastoma cells. It has been shown, by the aid of a patch-clamp analysis, that type II pyrethroids (deltamethrin, cypermethrin) inhibit (reduction of the open channel probability) voltage-gated, pharmacologically identified, chloride channels (Forshaw et al. 1993; Ray et al. 1997).

As demonstrated by the data presented in Table 1, there is no doubt that the voltage-sensitive sodium channels serve as the principal target of pyrethroid–DDT action as a consequence of a direct interaction with the channel protein. The latter provides the receptor binding site of the insecticide. This conclusion is strongly supported by the fact that point mutations on the *para* type voltage-gated insect sodium channel (Pittendrigh et al. 1997; Zlotkin 1999; see below) confer the well-known KDR type of insect resistance to pyrethroids and DDT analogs. With this background, it is noteworthy that pyrethroid–DDT insecticides were shown to affect, in a receptor-mediated fashion, the voltage-gated calcium channels as well. Such a possibility was hinted at by the complexity of the symptoms induced by insect intoxication (hyperexcitation, convulsions, seizures, paralysis), the myocardial positive intropism revealed by intoxicated mammals (Berlin et al. 1984), and the following information:

1. Indirect clues that DDT (Fig. 2) affects calcium conductance were provided by the induction of repetitive firing in insect neurosecretory neurons (Osborne 1985); inhibition of a calcium uptake by SR (Huddart 1977, 1978) and stimulation of calcium efflux from muscles (Huddart 1977, 1978).
2. Tetramethrin (type I; Fig. 2b) was shown to suppress a certain class of voltage-dependent calcium channels in neuroblastoma cells (Narahashi 1996).
3. Deltamethrin (type II; Fig. 2c) induced rat brain synaptosomal neurotransmitter (norepinephrin) release was blocked by a phenethylamine type of calcium channel antagonist and only partially affected by TTX (Clark and Brooks 1989).
4. DDT and deltamethrin were shown to act as potent agonists of a protozoan voltage-sensitive calcium channel (Clark et al. 1995).
5. A DDT analog (p,p-DDD: 50, 100 μM) was shown to greatly increase intracellular calcium in mammalian smooth muscle cells. The above increase was dependent on extracellular calcium and sensitive to blockers of voltage-gated calcium channels (nifedipine, cadmium chloride) (Juberg et al. 1995).
6. Occurrence of mutual competitive displacability between verapamil (specific Ca^{2+} ligand) and deltamethrin in rat brain synaptic membranes (Kadous et al. 1994) indicating a possibility that, similarly to the sodium channel (Trainer et al. 1997), the calcium channel provides the receptor.

It is noteworthy that the duality of Na^+-Ca^{2+} shown by the DDT–pyrethroid insecticides is similarly revealed by the dihydropyrazoles. The dihydropyrazoles RH3421 and RH5529 (Fig. 2k) when applied to mouse brain synaptosomes were shown to inhibit neurotransmitter release, calcium release and uptake induced by either Na^+ or Ca^{2+} channel activation. The distinction between the two categories of channel activation was based on the employment of specific blockers (such as TTX and cobalt). The two categories of effects, the sodium as well as the calcium mediated, were affected by equal concentrations (several μM) of the insecticides. It has been concluded that the dihydropyrazoles inhibit both the sodium as well as the calcium channels (Nicholson and Zhang 1995).

Presently, we do not have a satisfactory molecular explanation of the pharmacological similarity between the sodium and calcium channels as revealed by their closely resembling interaction with the pyrethroids–DDT insecticides and dihydropyrazoles. The simplest explanation is provided by Fig. 2, which reveals a close chemical topological similarity between the sodium and calcium channel α-subunits.

3.2
Distinction Between Mammals and Insects

Chemical, structural, physiological and immunological similarities among insect and vertebrate ion channels, as exemplified by the Na^+ channels (Zlotkin 1999), can explain the similarity in the effects of insecticides on insect and vertebrate organisms and the fact that mammalian systems are often employed to study insecticidal modes of action (Bloomquist 1996; Narahashi 1992, 1996). However, certain distinctions between the insect and vertebrates were noticed in the action of sodium channel insecticides.

Pyrethroids and DDT analogs were shown to exhibit, by their toxicity, preferability to insects which is attributed to:

1. Temperature dependence in the action. It has been shown (Yamasaki and Ishi 1952) that the toxic potency to nerves increases greatly with lowering of temperature (Q_{10} = 0.2). Thus, even with the same degree of intrinsic nerve sensitivity to pyrethroids and DDT in mammals and insects, the difference in their body temperature (~10 °C) will make the insect nerve much more sensitive.
2. Metabolic rate. Because of the temperature difference insects metabolize the insecticides more slowly than mammals. Furthermore, due to their small size, insects have less chance to metabolize the insecticides before they reach the target sites (Narahashi 1996).
3. Increased binding affinity of pyrethroids to invertebrate sodium channels when compared with mammalian sodium channels (Song et al. 1996). It is noteworthy that the 100-fold increased potency of permethrin to the functionally heterologously expressed insect voltage-gated *para* sodium channel when compared with the rat brain IIA sodium channel was attributed to an increased affinity (Warmke et al. 1997).

A curious distinction between insects and vertebrates was recently reported by Nicholson and Zhang (1995) and Zhang et al. (1996). They suggest that dihydropyrazoles may affect mammals by interfering with the operation of presynaptic calcium channels. Such conclusions were based on the employment of two dihydropyrazoles (the highly insecticidal RH3421 and the weakly insecticidal RH 5529, Fig. 2k) coupled with mouse brain synaptosomes and the monitoring of the sodium as well as calcium channel mediated transmitter release and synapsin phosphorylation. The two dihydropyrazoles, in spite of their differences in insecticidal potency, reveal a similar capacity to affect calcium channels and this action, expressed in the synaptic effects, was not observed in insect systems. The above example may indicate that the toxicity of an insecticide to mammals and insects is not, necessarily, mediated by an identical mechanism. Such differences in action may reflect pharmacological differences between insect and vertebrates ionic channels (see below).

4
Neurotoxic Polypeptides

A strict and convincing distinction between insect and vertebrate voltagegated ion channels was revealed by the presence of the insect-selective neurotoxic polypeptides derived from arachnid venoms (Zlotkin 1999; see below).

4.1
Animal Group Specificity

Venom is defined as a mixture of substances produced in specialized glandular tissues in the body of venomous animals and injected by the aid of a stinging-piercing apparatus into the body of its prey in order to paralyze it. The majority of venomous animals (such as snakes, spiders, scorpions, venomous snails, cnidarians) are slow and even static predators which feed on freshly killed prey of, relatively, vigorous animals. The locomotor inferiority of the venomous predator is largely compensated by the neurotoxic components of its venom, the neurotoxins, which are able to induce a rapid paralysis of the prey at a low range of concentrations (10^{-9}–10^{-12} M) (Zlotkin 1987).

When classified according to effect and site of action in the nervous system, venom neurotoxins are commonly divided into ion channel toxins which modify ion conductance, presynaptic toxins which affect neurotransmitter release, and postsynaptic toxins which block neurotransmitter receptors (Dolly 1988). The ion channel toxins are subdivided into sodium, potassium, calcium and chloride channel toxins. Some of these neurotoxins are able to distinguish between subtypes of ion channels such as the omega (ω) and μ conotoxins which distinguish between the voltage-gated axonal and muscle calcium as well as sodium channels (Gray 1988).

Where the vital prey-predator relationships serve as a target of preference for evolutionary selective pressure, additional specificity occurs in the form of the animal group (phyletic) selective toxins. In this context, animal group specificity refers to an evolutionary specialization where animal venom and/or its derived toxins reveal specific toxicity to a given group of organisms which serve as the prey of the venomous predator. In cases where a venomous animal feeds on a given group of organisms, the animal group selectivity is manifested already on the level of the whole venom. For example, the venomous marine Conidae snails are subdivided according to their feeding habits into fish, mollusc and worm eaters. A series of early bioassays (Endean and Rudkin 1963) revealed that the above feeding preference is associated with the specific toxicity of their venom aimed at the respective group of prey animals. In cases where venomous predators, such as cnidarians, nemertines or scorpions, feed on a wide range of organisms, the phyletic specificity is revealed on the level of the isolated purified neurotoxins.

4.2
Insect-Selective Neurotoxins Affecting the Voltage-Gated Sodium Channels

As the key structure of excitability in biological systems (see above), the sodium channel is targeted by a long series of neurotoxins derived from plant and animal allomonal systems (Catterall 1980; Zlotkin 1999; Table 4). Several kinds of animal group selective sodium channel polypeptide neurotoxins, such as the mollusc-selective δ-conotoxins, which affect sodium current inactivation (Fainzilber et al. 1994, 1995), have been detected and characterized. In this context the insect-selective toxins derived from arachmid venoms are the most relevant.

4.2.1
Scorpion Venom Toxins

The venoms of Afro-Asian ("Old World") buthid scorpions (classified as Buthinae) possess a series of long-chain (60–70 amino acid; 4 disulfide bridges) toxins, which affect the voltage-gated sodium channels. As revealed by early studies (Zlotkin et al. 1978), in addition to their neuropharmacological medical/public health importance, the above toxins were shown to possess an additional pharmacological peculiarity in the form of the animal group specificity. When classified according to their toxicity to insects, the above long-chain sodium channel toxins can be classified into two categories: the insect selective and the insect nonselective. The insect-selective neurotoxins can be further subdivided into the excitatory and depressant toxins. The chemistry and mode of action of these toxins have recently been reviewed (Martin-Euclaire and Couraud 1995; Zlotkin et al. 1995; Fig. 4A) allowing their brief presentation:

1. The above toxins are single chained polypeptides of about 70 amino acids cross-linked by four disulfide bridges (Fig. 4A).
2. Their exclusive action on insects was revealed through toxicity, electro-physiology and binding assays.
3. They are nontoxic to mice up to concentrations of several milligrams per animal but toxic to insects as the range of nanogram amounts per animal (Fishman et al. 1997).
4. They have been shown to affect exclusively insect voltage-gated sodium channels. However, despite their mutual competitive displaceability of the excitatory and depressant toxins, their receptor binding sites were shown to be distinct (Gordon et al. 1992).

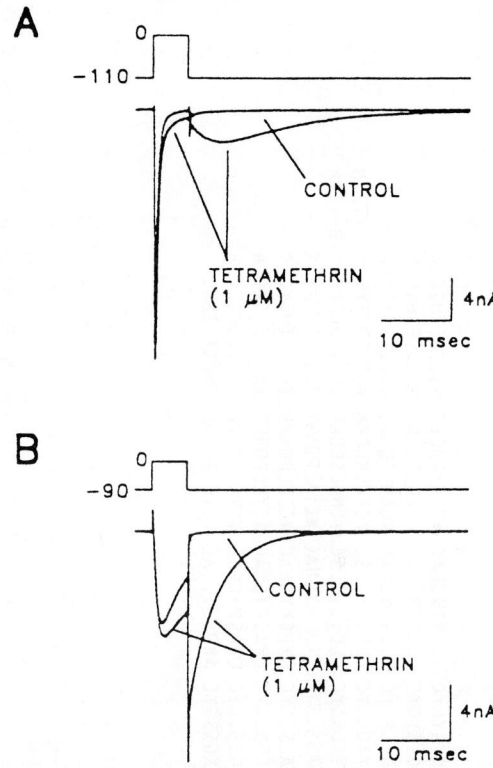

Fig. 3A,B. Effects of tetramethrin on tetrodotoxin (TTX)-sensitive sodium current (**A**) and TTX-resistant sodium current (**B**) in rat dorsal root ganglion neurons. A step depolarization to 0 mV was applied from a holding potential of –110 mV (**A**) or –90 mV (**B**) in control and in the presence of 1 μM tetramethrin. (Tatebayashi and Narahashi 1994)

5. The binding of the toxins to the sodium channel occurs through multi-point attachment sites which include segments of external loops in domains I, III and IV (Fig. 1; Gordon et al. 1992).
6. The toxins serve as unique and exclusive probes of the insect voltage-gated sodium channels and they distinguish between sodium channels of different groups of insects (Moskowits et al. 1994).
7. They are presently employed as pharmacological tools for the study of insect excitability and the design of selective bioinsecticides (see below) based on recombinant baculoviruses (Maeda et al. 1991; McCutchen et al. 1991; Stewart et al. 1991; Gershburg et al. 1998).
8. The insect nonselective, "long-chain", buthid toxins substantially comprise of the extensively studied alpha-toxins (Gordon et al. 1998), which are responsible for human envenomation (Zlotkin et al. 1978), but reveal an obvious toxicity (Zlotkin et al. 1978) or even preferability (see Fig. 4 LghαIT; Eitan et al. 1990) for insects.

A

```
             1          2           3           4           5           6           7
             °          °           °           °           °           °           °
LqhIT2  ...DGYIKRR  DCCKVACLIG  NEG.CDKECK  AYGG.SYGC   ...WTWGLAC  WCEGLPDDET  WK..SETNTC  G
LqqIT2  ...DGYIRKR  DGCKLSCLFG  NEG.CNKECK  SYGG.SYGC   ...WTWGLAC  WCEGLPDEKT  WK..SETNTC  G
BjIT2   ...DGYIRKK  DGCKVSCIIG  NEG CRKECV  AHGG.SFGYC  ...WTWGLAC  WCENLPDAVT  WK..SSTNTC  G
AaIT    .KKNGYAVDS  SGKAPECLLS  N..YCNNQCT  KVHYADKGYC  CLL.....SC  YCFGLNDDKK  VLEISDTRKS  YCUTIIN
LqqIT1  .KKNGYAVDS  SGKAPECLLS  N..YCYNECT  KVIIYADKGYC CLL.....SC  YCNGLSDDKK  VLEISDARKK  YCDFVTIN
LqhaIT  .VRDAYIAKN  YNCVYEC.FR  DA.YCNELCT  KNGASS.GYC  QWAGKYGNAC  WCYALPDNVP  IR...VPGKC  R
Lqq4    GVRDAYIADD  KNCVYTC.GS  NS.YCNTECT  KNGAES.GYC  QWLGKYGNAC  WCIKLPDKVP  IR...IPGKC  R
Lqq5    .LKDGIVDD   KNCTFFC.GR  NA.YCNDECK  KKGGES.GYC  QWASPYGNAC  WCYKLPDRVS  IK...EKGRC  N
AaH2    .VKDGYIVDD  VNCTYFC.GR  NA.YCNEECT  KLKGES.GYC  QWASPYGNAC  YCYKLPDHVR  TK...GPGRC  H
Ts7     ..KEGYLMDH  EGCKLSCFIR  PSGYCGRECG  .IKKGSSYC   AWP.....AC  YCYGLPNMVK  VWDRA.TNKC
```

B

```
                10             20             30
a  S  P  T  C  I  P  S  G  Q  P  C  P  Y  N  E  N  C  C  S  Q  S  C  T  F  K  E  N  E  N  G  N  T  V  K  R  C  D
b  S  S  T  C  I  P  S  G  Q  P  C  P  Y  Y  N  E  N  C  C  S  Q  S  C  T  Y  K  E  N  E  N  G  N  T  V  V  K  R  C  D
c  S  S  S  C  I  P  S  G  Q  Q  C  P  Y  N  E  N  C  C  S  Q  K  C  T  F  K  E  N  E  N  G  N  T  V  V  K  R  C  D
d  S  P  T  C  I  P  S  G  Q  Q  C  P  Y  N  E  N  C  C  S  K  S  C  T  Y  K  E  N  E  N  G  N  T  V  V  Q  R  C  D
e  S  S  P  T  C  I  P  S  G  Q  P  C  P  Y  S  K  Y  C  C  S  G  S  C  T  Y  K  T  N  E  N  G  N  S  V  Q  R  C  D
f  S  A  V  C  I  P  S  G  Q  P  C  P  Y  S  K  Y  C  C  S  G  S  C  T  Y  K  T  N  E  N  G  N  S  V  Q  R  C  D
```

Fig. 4A,B. Primary structures of insect-selective neurotoxic polypeptides. **A** Amino acid sequences of various scorpion neurotoxins that affect sodium conductance. The toxins that specifically affect insects are the depressant insect-selective toxins (LghIT2, BjIT2 and LqqIT2; Zlotkin et al. 1991), the excitatory insect-selective toxins (AaIT, Darbon et al. 1982; LqqIT1, Kopeyan et al. 1990) and LghαIT (Eitan et al. 1990), which possesses a limited toxicity to mammals and resembles in its primary structure the alpha scorpion toxins which are responsible for human envenomation by the Afro-Asian scorpions (Buthinae). The latter, typical α-toxins are represented by Lgg4 (Kopeyan et al. 1985) and Lgg5 (Kopeyan et al. 1985) and AaH$_2$ (Rochat et al. 1972). Ts7 (Possani et al. 1985) is a representative of the β-toxins that occur in the venoms of American scorpions and are responsible for human envenomation. The various scorpions are represented by the initials of their Latin names: *Aa* the North African scorpion (*A. australis*); *Bj* the Israeli black scorpion (*B judaicus*); *Lqh* the Israeli yellow scorpion (*L.q. hebraeus*); *Lqq* the African yellow scorpion (*L.q. quinquestriatus*); *Ts* the South American scorpion (*T. serrulatus*). **B** Primary structures of omega-atracotoxin isoforms isolated from the Australian funnel-web spider *H. versuta* (ω-ACTX-Hv1a-f). (After Atkinson et al. 1993; Wang et al. 1999)

4.2.2
Spider Venom Toxins

An additional group of insect-selective polypeptide neurotoxins that affect sodium conductance was revealed in the venom of the primitive weaving spider *Diguetia canities* (Krapcho et al. 1994). These are polypeptides of 6–7 kDa in molecular mass toxic to insects but not vertebrates. One of those toxins (DTX9.2) was shown (Bloomquist 1996) to induce a fast paralysis of insects, elevated spike discharges in insect sensory nerves and at the neuromuscular junction and dipolarization of axonal membrane. DTX9.2 was unable to competitively displace the insect-selective excitatory neurotoxin [^{125}I]-AaIT from insect neuronal membranes but its effects were reversed by TTX. These findings suggest that the toxin acts upon voltage-gated sodium channels of insect nerve membranes via receptors that differ from those of the AaIT toxin (Bloomquist 1996). Three related potent insecticidal peptide toxins were isolated from the *Diguetia* spider with masses ranging from 6371 to 7080 daltons. In tobacco budworm larvae, the toxins cause a progressive spastic paralysis, with 50% paralytic doses in the range of 0.3–2.0 µg/100 mg of body weight. These toxins appear to be quite selective for insects since they had no effects on mice by intraperitoneal and intracerebroventricular injection.

4.3
Insect-Selective Neurotoxins Affecting the Voltage-Gated Calcium Channel

The phenomenon of the insect-selective sodium channel neurotoxins raised an expectation that a similar phenomenon should occur with other ion channels – especially the voltage-gated calcium channels which reveal a considerable amount of variability in the vertebrate system (Gordon 1997). Such expectation was fulfilled by the recent finding of the insect-selective calcium channel blockers (omega-atracotoxins) derived from the venom of the Australian funnel-web spider (Atkinson et al. 1993; Fletcher et al. 1997; Wang et al. 1999).

The funnel-web spiders of the genera *Atrax* and *Hadronyche*, when compared with other Australian spiders, were shown to be the most potent against larvae of the cotton bollworm (*Helicoverpa*), a species which is becoming increasingly important as an agricultural pest (Atkinson et al. 1996).

The omega-atracotoxins are a family of 36 to 37 residue peptide neurotoxins derived from the venom of the Blue Mountains funnel-web spider (*H. versuta*) that block insect but not mammalian voltage-gated calcium channels (Fig. 4B). In addition to the omega-atracotoxin Hv1a (ω-ACTX-Hv1a) (Atkinson et al. 1993), five additional homologous isoforms (ω-ACTX-Hv1a to f) were isolated and sequenced (Wang et al. 1999). As shown in Fig. 4B, five of the six omega-atracotoxins (Hv1a-e) differ from one another by only one to three residues and possess similar insecticidal potencies. In contrast, omega-ACTX-Hv1f differs from the other toxins by up to ten residues (Fig. 4B) and it has a markedly reduced insecticidal potency, thus providing information on key functional residues.

The pharmacological uniqueness of the insect nervous system is not limited to an ion-conducting mechanism (as demonstrated by the above sodium as well as calcium channel insect-selective neurotoxins), but includes also mechanisms of synaptic transmission as exemplified by the insect-specific presynaptic neurotoxin derived from the venom of the black-widow spider, delta-latroinsectotoxin (Magazanik et al. 1992). In other words, the neurotoxic polypeptides derived from venomous insect predators may, in fact, mimic all the neurotoxic mechanisms of the industrial insecticides. The polypeptide neurotoxins, when compared with the lipophilic/amphipathic, stable low molecular weight insecticides, are devoid of the accessibility provided by topical delivery. On the other hand, the polypeptide neurotoxins possess the pharmacological selectivity derived from their variable primary structures and the resulting spatial molecular arrangements. The simplest, most common way to provide accessibility to the insect-selective polypeptides is to incorporate their genes into an insect-specific microorganism.

5
Recombinant Baculovirus Bioinsecticides

The main complication that follows from the use of classical industrial pesticides is the growing public awareness of the *environmental damage* caused by agrochemicals and the evolution of *insect resistance*. This has raised the necessity to develop environmentally friendly measures to combat the highly resistant insect species (see below). One approach, which has been recently introduced, is the production of transgenic crops that express toxins such as engineered potato, corn and cotton crops that express insecticidal proteins derived from the soil bacterium *Bacillus thuringensis*.

An alternative strategy, which has been successfully field-tested (Cory et al. 1994), is the release of baculoviruses that have been genetically engineered to express insecticidal neurotoxins. Baculoviruses are insect specific and do not infect vertebrates or plants. The infectivity of baculoviruses is limited to a few closely related species within a single insect family. Thus, an important factor in the environmental safety of baculoviral insecticides derived from their species selectivity, is the fact that they complement natural predators instead of replacing them, as in the case of many chemical insecticides.

Naturally occurring viruses are evolutionarily adapted for self-preservation and propagation. Thus, a part of the baculovirus replication strategy is to keep its host alive as long as possible to maximize progeny production (Bonning and Hammock 1996). Genetic engineering is used to improve the efficacy of the viruses as pesticides by reducing the time it takes the virus to stop the feeding of their insect hosts. Recently, the insect pathogenic virus *Autographa californica* nuclear polyhedrosis virus (AcNPV) has been genetically modified to increase speed of kill by the introduction of genes encoding insect-selective toxins into the viral genome. These genes are *tox34* (or the *tox21a* homologue), which encodes the paralytic TxPi toxin found in the venom of the strew itch mite *Pyemotes triciti* (Tomalski et al. 1989), and the AaIT gene (Fig. 4), which encodes the excitatory insect-selective neurotoxin AaIT. Expression of these genes with a very late viral promoter, the p10 promoter or an enhanced poly-hedrin promoter, results in a 25–40% reduction in the time (ET_{50}) it takes for 50% of infected insects to be paralyzed or die after infection. The interest sur-rounding these genetically engineered biopesticides has triggered some recent developments in design of insecticidal recombinant baculoviruses. Synthetic genes of site 3 sodium toxins (Zlotkin 1999), namely, those of the sea anemone, *Anemonia sulcata* (AtxII), and *Strichodactyla helianthus* (ShI), when inserted into a nonessential site in the genome of the Ac-NPV under the control of a polyhedrin promoter, have resulted in recombinant viruses that produce toxins in infected cells and induce an enhanced early lethality in infected lepidopter-ous larvae (Prikhodko et al. 1996). A similar approach, with the μ-Aga IV aga-toxin derived from the spider *Agelenopsis aperta*, which has been shown to shift sodium channel activation, has similarly yielded a recombinant AcNPV virus with about 37–38% enhancement of insecticidal activity when compared with that of wild-type virus (Prikhodko et al. 1996). A recent study (Gerschburg et al. 1998) claims that the Ac-NPV baculovirus engineered by the scorpion venom derived, depressant, insect-selective toxin (LghIT2) is superior in its insecticidal properties to the one that expresses the excitatory AaIT toxin. LghIT2 baculovirus is already being used in industry (Dupont-H 66638 1996). In addition to the above scorpion venom derived toxins, a recombinant virus, expressing the insect-selective toxin NPS-901 from the venom of the weaving spider *Diguetia canities* (see above), has been constructed. The cDNA for NPS-901 was combined with a lepidopterous secretory signal sequence, and this hybrid gene was cloned into the AcNPV under the control of the polyhedrin promoter. In preliminary assays, the resulting recombinant virus (vAc 901) exhibited significantly enhanced insecticidal activity compared with that of the parental wild-type virus (Krapcho et al. 1994).

To summarize, the insect voltage-gated ion channel may be targeted by two categories of insecticidal substances: industrial synthetic organic compounds and baculovirus-mediated neurotoxic polypeptides. As shown below they may be used in a complementary fashion.

6
Allosteric Coupling and Allosteric Antagonism

Resistance is officially defined by the World Health Organization as the "development of an ability in a strain of an organism to tolerate doses of toxicant which would prove lethal to the majority of individuals in a normal (susceptible) population of the species" (Mullin and Jeffrey 1992). Insecticide resistance, documented in almost all arthropod species in which it has been studied, is a major obstacle to the control of agriculturally and medically important pests (Hassall 1990). Resistance can be acquired through the relatively rare behavioral or the abundant physiological action (see final seven chapters, this Vol.). The physiological resistance may be, largely, subdivided into two major categories: the *metabolic* (modified penetrability, excretion, detoxification-degradation) and the *altered target site* resistance. The knockdown resistance (KDR) phenomenon, which represents the most common form of resistance against DDT and pyrethroids, can be induced by various metabolic factors (Mullin and Jeffrey 1992; Scott and Wheelock 1992) but is substantially attributed to the altered target site, namely, the point mutations in the insect *para* type (Zlotkin 1999) voltage-gated sodium channel (Taylor et al. 1993; Williamson et al. 1996; Pittendrigh et al. 1997; Dong 1997; Fig. 5, Table 3).

A similar pattern of resistance, namely, increased detoxification and modified target site, was revealed in the preexisting tolerance of lepidopterous larvae and certain tenebrionid beetles to the insect-selective AaIT polypeptide (Herrmann et al. 1990; Fishman et al. 1997). The detoxification was exhibited by increased degradation, excretion, and binding to pharmacologically irrelevant tissues. A change in the target site was examined by binding assays. Dissociation kinetics revealed a destabilization of the toxin receptor complex. The disassociation of toxin receptor complexes is fast ($t_{0.5}$ of seconds) in tolerant insects compared with the slow disassociation ($t_{0.5}$ of minutes) revealed by susceptible insects (Fishman et al. 1997).

The insect voltage-gated sodium channel closely resembles its vertebrate counterpoint in two major aspects: Firstly by its structure and function (primary structure, electrophysiology, ion conductance allocation of all functional domains; Fig. 1) and, secondly, by its pharmacological diversity and flexibility. The latter is exhibited by the occurrence of at least nine different allosterically coupled receptor-binding sites of various natural neurotoxins and insecticides (Table 4). Despite their similarity, the insect and the vertebrate channels are distinguishable as exemplified by the occurrence of insect-selective neurotoxins (see above) and the derived recombinant baculoviral biopesticides (see above).

The diversified and flexible pharmacology of the voltage-gated insect sodium channel (Table 4), when viewed from an applied agrotechnical per-

Table 3. The paralytic effect (PD_{50}) of AaIT toxin to wild-type (WT) and pyrethroid-knockdown resistant (KDR) houseflies[a]

Housefly strain[b]	PD_{50} (ng/insect $\pm$ SD)[c]	Susceptibility ratio[d]
WT-"Cooper"	21.1 $\pm$ 1.35 (3)[e]	1.0
KDR[f]	2.3 $\pm$ 0.39 (3)	9.2
Super-KDR[g]	1.5 $\pm$ 0.25 (3)	14.0

[a] Taken from Zlotkin et al. (1999). Represents the "knockdown" assay: determined on mobile (alert) flies 5 min following injection
[b] Bred in IACR, Rothamsted, UK
[c] PD_{50} – paralytic dose 50%
[d] Susceptibility ratio: PD_{50} of WT/PD_{50} of strain
[e] Numbers in parentheses indicate the number of independent assays performed
[f] KDR-stock 579 homozygous for a point mutation (L1014F) in the hydrophobic S6 transmembrane segment of homology domain II of the para sodium channel which confers resistance of about one order of magnitude to pyrethroid insecticides (Williamson et al. 1996; Fig. 5)
[g] Super-KDR-stock 530, possessing two point mutations in the *para* gene: the first is the above KDR (L1014F) mutation. The second point mutation (M918T) is located in the intracellular loop, connecting segments 4 and 5 in homology domain II (Pittendrigh et al 1997). The super-KDR strain is about 500-fold resistant to deltamethrin (Williamson et al. 1996; Fig. 5)

spective, strongly suggests a pharmacological approach to pest control and resistance management. The occurrence of positive cooperativity (Table 4) between two insecticidal substances that occupy separate but allosterically coupled sites can be useful. In such cases, field delivery of a synergistic mixture may allow a significant reduction in the required dosages and concentrations of the coupled insecticides, which lowers the cost and may, at least, delay the onset of resistance. A recent examination of the insecticidal potency of various scorpion venom neurotoxins (Hermann et al. 1995) exemplifies this principle. It has been shown that a combination of two toxins increased the insecticidal activity five- to tenfold with no apparent increase in mammalian toxicity. Synergistic combinations were obtained between toxins (such as the AaIT and the LghαIT) that do not reveal competitive displaceability in binding assays and do not belong to the same site category. With this background one would expect an insecticidal synergism between N-alkylamides and dihydropyrazoles, which were shown to interact on the neurophysiological level (Table 4). In other words, the increased affinity of dihydropyrazoles to a partially depolarized and thus inactivated conformation (Salgado 1992c) may yield a positive cooperativity with sodium channel activators such as N-alkylamides which, similarly to veratridine (Table 4), shift activation and depolarize the axonal membrane.

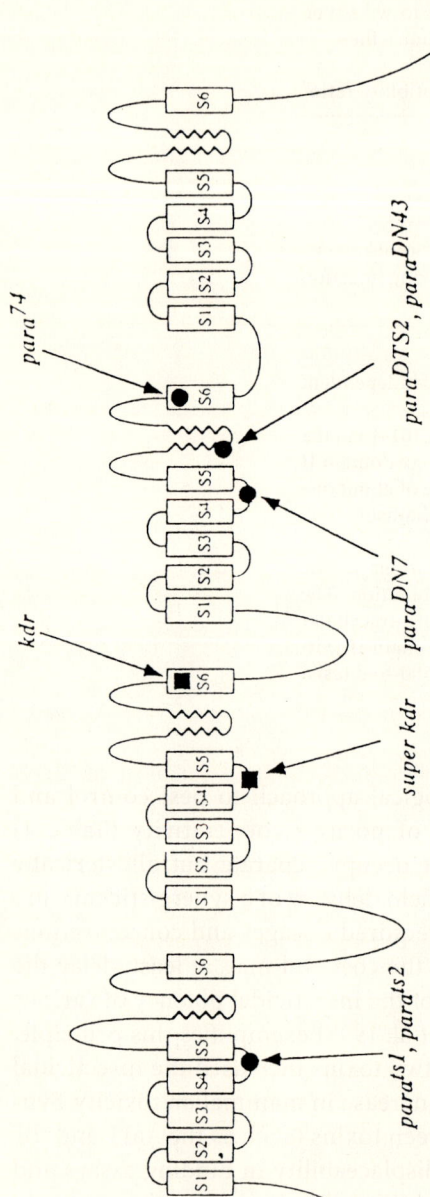

Fig. 5. A two-dimensional model of the topology of the *para* voltage gated sodium channel, indicating the locations of the four amino acid replacements in the DDT-pyrethroid resistant *para* mutants alongside the relative location of the kdr and super-kdr replacements documented in homologues from the housefly and the German cockroach. (Pittendrigh et al. 1997)

Table 4. Neurotoxin and insecticide binding sites on the voltage-gated sodium channel[a]

Site[b]	Toxicant[c]	Physiological effect	Allosteric coupling[d]
1	Tetrodotoxin (TTX) Saxitoxin (STX) μ-Conotoxin	Inhibit transport	+3, 5, −2
2	Batrachotoxin (BTX) Veratridine Aconitine Grayanotoxin *N*-Alkylamides	Cause persistent activation	+3, −6
3	α-Scorpion toxins Sea anemone II toxin (ATXII)	Inhibit inactivation Enhances persistent activation	+2
4	β-Scorpion toxins	Shift voltage dependence of activation	
5	Brevetoxins Ciguatoxins	Shift voltage dependence of activation	+2, 4, −3
6	δ-Conotoxins (δ-TxVIA)	Inhibit inactivation	
7	DDT and analogs Pyrethroids	Inhibit inactivation[e] Shift voltage dependence and rate of activation	+2, 3, 5
8	*Goniopora* coral toxin *Conus striatus* toxin	Inhibit inactivation	
9	Local anesthetics Anticonvulsant Dihydropyrazoles	Inhibit transport	+2

[a] Taken from Zlotkin (1999)
[b] Sites 1–5 according to Catterall (1992); site 6 according to Fainzilber et al. (1994); site 7 according to Pauron et al. (1989) and Soderlund and Knipple (1995). Sites 8 and 9 were numbered arbitrarily in order to distinguish them from the others. Site 9 according to Suderland and Knipple (1995) and Payne-Gregory et al. (1998). Substances are considered to belong to a given site if they compete with each other in binding assay and induce similar electrophysiological effects
[c] Insecticides are shown in italics
[d] Allosteric coupling refers to modulations induced by occupancy of a receptor site by the corresponding neurotoxin on the binding of neurotoxins at the indicated receptor site(s). Positive modulation (+) indicates enhancement of binding of the toxin at its indicated receptor site and/or stimulation of Na⁺ influx. Negative modulation (−) refers to decrease in toxin binding at the indicated receptor site (Gordon 1997)
[e] See Soderlund and Knipple (1995)

A second possibility for insecticidal cooperativity is based on the assumption that, if a certain population (or strain) of insects has developed through insecticidal-selective pressure, a modified voltage-gated ion channel, such as that involved in KDR (see above), the presumable conformational change induced by the mutation can modify channel pharmacology.

Such an assumption can be exemplified both by three earlier and more recent findings that suggest the occurrence of an *allosteric antagonism* between the KDR phenomenon and increased susceptibility to the insect sodium channel neurotoxicants. Firstly, a phenomenon of a negatively correlated resistance was previously revealed by KDR houseflies and *N*-alkylamides (such as BTG 501, 502). The latter were shown to be up to four times more effective against the super KOR strain (530) of houseflies (Elliot et al. 1986). Secondly, it has been shown (McCutchen et al. 1997) that neonate pyrethroid-resistant KDR tobacco budworms (*Heliothis virescens*) exhibited a significantly reduced (by 33%) median lethal time (LT_{50}) when infected with a recombinant baculovirus expressing the AaIT toxin (AcAaIT) and compared with the wild-type budworms. Thirdly, as shown in Table 3, KDR and super-KDR houseflies injected with the AaIT toxin revealed a 9- and 14-fold increase in their susceptibility to the AaIT toxin, respectively (Zlotkin et al. 1999). It is noteworthy that similar allosteric antagonism was revealed by the acquisition of hypersensitivity to pyrethroids by TTX-resistant dorsal root ganglion sodium channels (Fig. 3).

As a polar polypeptide of molecular mass of approx. 8 kDa, AaIT is expected to interact with the external loops of the sodium channel (Gordon 1997) at receptor binding sites different from those of the low molecular weight and lipophilic DDT and pyrethroids (Soderlund and Knipple 1995). The latter presumably act at the intramembranous region of the channel similarly to the lipid-soluble alkaloid gating modifiers, such as veratridine or batrachotoxin (Hille 1992). With this background of binding site diversity, the data presented in Table 3 demonstrate the pharmacological flexibility of sodium channels, which has implications for the management of insecticide resistance.

Briefly, the suitability of the insect sodium channel to serve as a target for future insecticides depends on its:

1. Critical role in excitability (Hille 1992),
2. Pharmacological diversity exhibited in the multiplicity of binding sites for various neurotoxicants including four groups of industrial insecticides (Soderlund and Knipple 1995; Gordon 1997; Table 4),
3. Being pharmacologically distinguishable from vertebrate sodium channels as revealed by heterologous expression of the *Drosophila para* sodium channel (Warmke et al. 1997) and the insect selectivity of neurotoxins derived from arachnid venoms (Zlotkin et al. 1995; Bloomquist 1996 and see above), and

4. Pharmacological flexibility, as exhibited by "allosteric coupling" (Table 4).

"Allosteric coupling" indicates that binding of one ligand to the sodium channel protein induces a conformational change that affects its functionality and responsiveness to other toxicants acting at a different site on the channel protein (Soderlund and Knipple 1995; Gordon 1997). A similar conformational change in the channel protein could be induced by point mutations of the kind that confer insecticide resistance. Such a change may affect the responsiveness of the channel to various toxicants either by antagonizing or enhancing their effects. The decreased as well as increased susceptibilities to AaIT by the different sodium channel mutated fly strains (Zlotkin et al. 1999) provide evidence for such effects.

The obvious structural resemblance between the various ion channel proteins (Fig. 1) suggests that other ion channels share the pharmacological flexibility revealed by the sodium channel as exemplified by the above omega atracotoxins.

Finally, the above phenomenon of the allosteric antagonism (or negatively correlated resistance) should be viewed against the background of the increasing agroeconomical significance of pyrethroid resistance (Elliot et al. 1986) and the appearance of bioinsecticides based on recombinant baculovirus-expressing neurotoxins (Cory et al. 1994). These developments emphasize the need to clarify the molecular basis underlying the allosteric coupling between toxicant binding sites on the ion channel in order to assess its implications for insecticide resistance management.

Acknowledgements. Some of the studies reported were supported by grants from the Israel Science Foundation (ISF 329/96-1) and the United States/Israel Binational Science Foundation (BSF-9300294).

References

Adams ME, Miller TA (1979) Site of action of pyrethroids; repetitive backfiring in flight motor units of house fly. Pestic Biochem Physiol 11:218–231

Adams ME, Miller TA (1980) Neural and behavioural correlates of pyrethroid and DDT-type poisoning in the house fly *Musca domestica.* Pestic Biochem Physiol 13:137–147

Armstrong CM (1992) Voltage dependent ion channels and their gating. Physiol Rev 72:5–13

Atkinson RK, Howden MEH, Tyler MI, Vonarx EJ (1993) Insecticidal toxins derived from funnel web spiders (*Atrax* or *Hadronyche*). International patent application PCT/AU93/00039 (W093/15108)

Atkinson RK, Vonarx EJ, Howden MEH (1996) Effects of whole venom and venom fractions from several Australian spiders including *Atrax (Hadrohnyche)* species, when injected into insects. Comp Biochem Physiol 114c:113–117

Beam BP (1989) Classes of calcium channels in vertebrate cells. Annu Rev Physiol 51:367–384

Berlin JR, Akera T, Brody TM, Matsumura F (1984) The inotropic affects of a synthetic pyrethroid decamethrin on isolated guinea pig atrial muscle. Eur J Pharmacol 98:313–322

Berry RW (1977) The evaluation of permethrin for wood preservation. Pestic Sci 8:284–290

Blade RJ, Burt PE, Hart RJ, Mos MDV (1985) The action of insecticidal isobutylamide compounds on the insect nervous system. Pestic Sci 16:554–564

Blackman GG, Hodson MJ (1977) Further evaluation of permethrin for biting fly control. Pestic Sci 8:270–273

Bloomquist JR (1996) Ion channels as targets for insecticides. Annu Rev Entomol 41:163–190

Bonning BC, Hammock BD (1996) Development of recombinant baculoviruses for insect control. Annu Rev Entomol 41:191–210

Catterall WA (1980) Neurotoxins that act on voltage sensitive sodium channels in excitable membranes. Annu Rev Pharmacol Toxicol 20:15–43

Catterall WA (1988) Structure and function of voltage sensitive ionic channels. Science 242:50–61

Catterall WA (1991) Structure and function of voltage gated sodium and calcium channels. Curr Opin Neurobiol 1:5–13

Catterall WA (1992) Cellular and molecular biology of voltage gated sodium channels. Am J Physiol 72:15–48

Chadwick PR (1979) The activity of some pyrethroids against *Periplaneta americana* and *Blatella germanica*. Pestic Sci 10:32–38

Clark JM (ed) (1995) Molecular action of insecticides on ion channels. ACS Symp 591. American Chemical Society, Washington DC

Clark JM, Brooks MW (1989) Role of ion channels and intraterminal calcium homeostasis in the action of deltamethrin on presynaptic nerve terminals. Biochem Pharmacol 38:2233–2245

Clark JM, Edman SJ, Nagy SR, Canhoto A, Hecht F, Van Houten (1995) Action of DDT and pyrethroids on calcium channels in *Paramecium tetraurelia*. In: Clark JM (ed) Moleular action of insecticides on ion channels. ACS Symp 591. American Chemical Society, Washington, DC, pp 173–191

Cory JS, Hirst ML, Williams T, Halls RS, Gouson D, Green BM, Curty TM, Possee RD, Cayley PJ, Bishop DHL (1994) Field trial of a genetically improved baculovirus insecticide. Nature 370:138–140

Crowder LA, Tollefson MS, Watson TF (1979) Dosage mortality studies of synthetic pyrethroids and methyl parathion on the tobacco budworm in central Arizona. J Econ Entomol 72:1–3

Darbon H, Zlotkin E, Kopeyan C, Van Rietschoten J, Rochat H (1982) Covalent structure of the insect toxin of the North African scorpion *Androctonus australis* Hector. Int J Pept Protein Res 20:320–330

Deecher DC, Soderlund DM (1991) RH 3421, an insecticidal dihydropyrazole, inhibits sodium channel-dependent sodium uptake in mouse brain preparations. Pestic Biochem Physiol 39:130–137

Devidas S, Guggino WB (1997) CFTR: domains, structure and function. J Bioenerg Biomembr 29:443–451

Dolly JE (ed) (1988) Neurotoxins in neurochemistry. Ellis Horwood, Chichester

Dong K (1997) A single amino-acid change in the para sodium channel protein is associated with knock down resistance (KDR) to pyrethroid insecticides in German cockroach. Insect Biochem Mol Biol 27:93–100

Durell H, Guy HR (1992) Atomic scale structure and functional models of voltage gated potassium channels. Biophys J 62:238–250

Eitan M, Fowler E, Herrmann R, Duval A, Pelhate M, Zlotkin E (1990) A scorpion venom paralytic to insects which affects sodium current inactivation: purification, primary structure and mode of action. Biochemistry 29:5941–5947

Elliot M, Farnham AW, Janes NF, Needham PH, Pulman DA (1974a) Synthetic insecticides with a new order of activity. Nature 248:710–711

Elliot M, Farnham AW, Janes NF, Needham PH, Pulman DA (1974b) Insecticidally active conformation of pyrethroids. In: Kohn GK (ed) Mechanisms of pesticide action. Am Chem Soc Symp Ser 2, American Chemical Society, Washington, DC, pp 80–91

Elliot M, Janes NF, Porter C (1978) The future of pyrethroids in insect control. Annu Rev Entomol 23:443–469

Elliot M, Farnham AW, Janes NF, Johnson DM, Pulman DA, Sawicki RM (1986) Insecticidal amides with selective potency against a resistant (Super-KDR) strain of houseflies. Agric Biol Chem 50:1347–1349

Elliot M, Farnham AW, Janes NF, Johnson DM, Pulman DA(1987) Synthesis and insecticidal activity of lipophilic amides. I. Introductory survey and discovery of an active synthetic compound. Pestic Sci 18:191–201

Endean R, Rudkin C (1963) Studies of the venom of some Conidae. Toxicon 1:49–64

Fainzilber M, Kofman O, Zlotkin E, Gordon D (1994) A new neurotoxin receptor site on sodium channels is identified by a conotoxin that affects sodium channel inactivation in molluscs and acts as an antagonist in rat brain. J Biol Chem 269:2574–2580

Fainzilber M, Lodder H, Kits KS, Kofman O, Vinnitzky I, Van Rietschoten J, Zlotkin E, Gordon D (1995) A new conotoxin affecting sodium current inactivation interacts with the δ-conotoxin receptor site. J Biol Chem 270:1123–1129

Fill M, Coronado R (1988) Ryanodine receptor channel of sacroplasmic reticulum. Trends Neurosci 11:453–457

Fishman L, Herrmann R, Gordon D, Zlotkin E (1997) Insect tolerance to a neurotoxic polypeptide: pharmacokinetic and pharmacodynamic aspects. J Exp Biol 2000:1115–1123

Fletcher JI, Smith R, O'Donoghue SI, Nilges M, Connor M, Howden MEH, Christie MJ, King GF (1997) The structure of a novel insecticidal neurotoxin, ω-atracotoxin-HV1 from the venom of an Australian funnel web spider. Nature Struct Biol 4:559–566

Forshaw PJ, Lister T, Ray DE (1993) Inhibition of neuronal voltage dependent chloride channel by the type II pyrethroid, deltamethrin. Neuropharmacology 32:105–111

Foskett JK (1998) CIC and CFTR chloride channel gating. Annu Rev Physiol 60:689–717

Gammon DW, Brown MA, Casida JE (1981) Two classes of pyrethroid action in the cockroach. Pestic Biochem Physiol 15:181–191

Gerschburg E, Stockholm D, Froy O, Rashi S, Gurevitz M, Chejanovsky N (1998) Baculovirus mediated expression of a scorpion depressant toxin improves the insecticidal efficacy achieved with excitatory toxins. FEBS Lett 422:132–136

Ghiasudkin SM, Soderlund DM (1985) Pyrethroid insecticides: potent, stereospecific enhancers of mouse brain sodium channel activation. Pestic Biochem Physiol 24:200–206

Gordon D (1997) Sodium channels as targets for neurotoxins: mode of action and interaction of neurotoxins with receptor on sodium channels. In: Gutman Y, Lazorowici P (eds) Toxins and signal transduction. Harwood, Amsterdam, pp 119–149

Gordon D, Moskowitz H, Eitan M, Werner C, Catterall WA, Zlotkin E (1992) Localization of receptor sites for insect selective toxins on sodium channels by site directed antibodies. Biochemistry 31:7622–7628

Gordon D, Savarin P, Gurevitz M, Zinn-Justin S (1998) Functional anatomy of scorpion toxins affecting sodium channels. J Toxicol Toxin Rev 17:131–159

Griffiths DC (1977) The effectiveness of pyrethroid seed treatments against soil pests of cereals. Pestic Sci 8:258–263

Gray WR (1988) Conotoxins as probes of channel subtypes. In: Dolly JO (ed) Neurotoxins and neurochemistry. Ellis Horwood, Chichester, pp 151–161

Hassall K (1990) The biochemistry and uses of pesticides. VCH, Weinheim

Herrmann R, Fishman L, Zlotkin E (1990) The tolerance of lepidopterous larvae to an insect selective neurotoxin. Insect Biochem 20:625–637

Herrmann R, Moskovitz H, Zlotkin E, Hammock BD (1995) Positive cooperativity among insecticidal scorpion neurotoxins. Toxicon 33:1099–1102

Hille B (1992) Ionic channels of excitable membranes, 2nd edn. Sinauer, Sunderland

Hodgkin AL, Huxley AF (1952) A quantitative description of membrane current and its application to conduction and excitation in nerve. J Physiol (Lond) 117:500–544

Hodgkin AL, Horowicz P (1959) The influence of potassium and chloride ions on the membrane potential of single muscle fibres. J Physiol (Lond) 148:127–160

Hodgkin AL, Horowicz P (1960) The effect of sudden changes in ionic concentrations on the membrane potential of single muscle fibres. J Physiol (Lond) 153:370–385

Huddart H (1977) The effect of some organophosphorous and organo-chloride insecticides on contractability, membrane potential and calcium regulation in insect skeletal muscle. Comp Biochem Physiol 586:91–95

Huddart H (1978) Parathion and DDT-induced effect on tension and calcium transport in molluscan visceral muscle. Comp Biochem Physiol 61C:1–6

Ingles PJ, Adams PM, Knipple DC, Soderlund DM (1996) Characterization of voltage sensitive sodium channel gene coding sequences from insecticide susceptible and knockdown resistant housefly strains. Insect Biochem Mol Biol 26:319–326

Jeffries PR, Lehmberg E, Lam WW, Casida JE (1993) Bioactive ryanoids from nucleophilic additions to 4,12-reco-4,12-dioxoryanodine. J Med Chem 36:1128–1135

Jentsch TJ, Steinmeyer K, Schwartz G (1990) Primary structure of *Torpedo marmorata* chloride channel isolated by expression cloning of *Xenopus* oocytes. Nature 348: 510–514

Juberg DR, Stuenkel EL, Loch-Caruso R (1995) Dde chlorinated insecticide 1,1-dichloro-2,2-bis(4-chlorophenyl)ethane (p,p'-DDD) increases intracellular calcism in rat myometrial smooth muscle cells. Toxicol Appl Pharmacol 135:147–155

Kadous A, Matsumura F, Eran E (1994) High affinity binding of 3H-verapamil to rat brain synaptic membrane is antagonized by pyrethroid insecticides. J Environ Sci Health 29:855–871

Kartner N, Hanrahan JW, Jensen TJ, Naismith AL, Sun S, Ackerley CA, Reyes EF, Tsui LC, Rommens JM, Bear CE, Riordan JR (1991) Expression of the cystic fibrosis gene in non-epithelial invertebrate cells produces a regulated anion conductance. Cell 64:681–691

Knapp FW, Herald F (1981) Face fly and horn fly reduction on cattle with fenvalerate ear tags. J Econ Entomol 74:295–296

Kopeyan C, Martinez G, Rochat H (1978) Amino acid sequence of neurotoxin V from the scorpion *Leiurus quinquestriatus quinquestriatus*. FEBS Lett 89:54–59

Kopeyan C, Martinez G, Rochat H (1985) Primary structure of toxin IV of *Leiurus quinquestriatus quinquestriatus* characterization of a new group of scorpion toxins. FEBS Lett 181:211–217

Kopeyan C, Mansuelle P, Sampieri F, Brando T, Bahraoui ELM, Rochat H, Granier C (1990) Primary structure of scorpion anti insect toxins isolated from the venom of *Leiurus quinquestriatus quinquestriatus*. FEBS Lett 261:423–428

Krapcho KJ, Kral RM, Van Wagenen BC, Eppler KJ, Morgan TK (1995) Characterization and cloning of insecticidal peptides from the primitive weaving spider *Diguetia canites*, Insect biochem. Mol Biol 25:991–1000

Lazdunski M, Lombet A, Maurre C (1988) Specific binding sites for pyrethroids on the voltage-dependent sodium channel. In: Lunt GG (ed) Neurotox 88-Molecular basis of drug and pesticide action. Excerptia Medica, Amsterdam, pp 289–300

Lee SH, Smith TJ, Knipple DC, Soderlund DM (1999) Mutations in the housefly VSSC1 sodium channel gene associated with super-KDR resistance abolish the pyrethroid sensitivity of VSSC1/tip E sodium channels in *Xenopus* oocytes. Insect Biochem Mol Biol 29:185–194

Lees G, Burt PE (1988) Neurotoxic action in a lipid amide on the cockroach nerve cord and on locust somata maintained in short term culture: a novel preparation for the study of Na^+ channel pharmacology. Pestic Sci 24:189–191

Lehmberg E, Casida JE (1994) Similarity of insect and mammalian ryanodine binding sites. Pestic Biochem Physiol 48:145–152

Lund AE (1985) Insecticides: effects on the nervous system. In: Kerkut GA, Gilbert LI (eds) Comprehensive insect physiology, biochemistry and pharmacology, vol 12. Pergamon Press, Oxford, pp 9–56

Maeda S, Volrath SL, Hanzlik TN, Harper SA, Maddox DW, Hammock BD, Fowler E (1991) Insecticidal effects of an insect-specific neurotoxin expressed by a recombinant baculovirus. Virology 184:777–780

Magazanik LG, Federova IM, Kovalevskaya GI, Pashkov VN, Bulgakov DV, Grishin EV (1992) Selective presynaptic insectotoxin (alpha-latroinsectotoxin) isolated from black widow spider venom. Neuroscience 46:181–188

Martin-Eauclaire MF, Couraud F (1995) Scorpion neurotoxins: effects and mechanisms. In Chang LW, Dyer RS (eds) Handbook of neurotoxicology. Marcel-Dekker, New York, pp 683–716

Maulet Y, Lambert RC, Mykita S, Mouton J, Partisani M, Bailly Y, Bombarde G, Feltz A (1999) Expression and targeting to the plasma membrane of ClC-K, a chloride channel specifically expressed in distinct tubule segments of *Xenopus laevis* kidney. Biochem J 340:737–743

Mayfield RJ, O'Laughlin GJ (1980) Evaluation of decamethrin as an insect proofing agent for wool. J Agric Food Chem 28:886–887

McCutchen BF, Choudry PV, Crenshaw R, Maddox D, Kamita SG, Hammock BD (1991) Development of a recombinant baculovirus expressing an insect-selective neurotoxin: potential for pest control. Biotechnology 9:848–852

McCutchen BF, Hoover K, Preisler HK, Batana HD, Herrmann R, Robertson JL, Hammock BD (1997) Interaction of recombinant and wild-type baculoviruses with classical insecticides and pyrethroid resistant tobacco budworms (Lepidoptera, Noctuidae). J Econ Entomol 90:1170–1189

Meissner G (1986) Ryanodine activation and inhibition on the Ca^{++} release channel of sarcoplasmic reticulum. J Biol Chem 14:6300–6306

Miller RJ (1992) Voltage sensitive Ca^{+2} channels. J Biol Chem 267:1403–1406

Miyamoto J, Kaneko H, Tsuji R, Okuno Y (1995) Pyrethroids, nerve poisons: how their risks to human health should be assessed. Toxicol Lett 82/83:933–940

Moore RF (1980) Behavioral and biological effects of NRDC-161 as factors in control of the boll weevil. J Econ Entomol 73:265–267

Moskowitz H, Herrmann R, Zlotkin E, Gordon DC (1994) Variability among insect sodium channels revealed by binding of selective neurotoxins. Insect Biochem Mol Biol 224:13–19

Mullin CA, Jeffrey GS (eds) (1992) Molecular mechanisms of insecticide resistance. American Chemical Society, Washington, DC

Narahashi T (1988) Molecular and cellular approaches to neurotoxicology: past, present and future. In: Lunt GG (ed) Neurotox '88 molecular basis of drug and pesticide action. Excerpta Medica, Amsterdam, pp 269–288

Narahashi T (1992) Nerve membrane Na^+ channels as targets of insecticides. Trends Pharmacol Sci 13:236–241

Narahashi T (1996) Neuronal ion channels as the target sites of insecticides. Pharmacol Toxicol 78:1–14

Narahashi T, Ginsburg KS, Nagata K, Song JH, Tatebayashi H (1998) Ion channels as targets for insecticides. Neurotoxicology 19:581–590

Nassif M, Brooke JP, Hutchinson DBA, Kamel OM, Savage EA (1980) Studies with permethrin against body lice in Egypt. Pestic Sci 11:679–684

Nicholson RA, Zhang A (1995) Presynaptic actions of dihydropyrazoles. In: Clark JM (ed) Molecular actions of insecticides on ion channels. American Chemical Society, Washington, DC, pp 44–55

Nishimura K, Tada T, Nakagawa Y (1996) Effect of insect growth regulators, N-tert-butyl N,N-dibenzoylhydrazines, on neural activity of the American cockroach. Comp Biochem Physiol Pharmacol Toxicol Endocrinol 114:141–144

Osborne MP (1985) DDT, γ-HCH and the cyclodienes. In: Kerkut GA, Gilbert LI (eds) Comprehensive insect physiology, biochemistry and pharmacology, vol 12. Pergamon Press, Oxford, pp 131–182

Ottea JA, Payne GT, Soderlund DM (1990) Action of insecticidal N-alkylamides at site 2 of the voltage sensitive sodium channel. J Agric Food Chem 38:1724–1728

Palade PT, Barchi RL (1977) On the inhibition of muscle membrane chloride conductance by aromatic carboxylic acids. J Gen Physiol 69:879–965

Pauron D, Barhanin J, Amichot M, Pralavorio M, Berge JB, Lazdunski M (1989) Pyrethroid receptor in the insect Na^+ channel: alteration of its properties in pyrethroid-resistant flies. Biochemistry 28:1673–1678

Payne-Gregory T, Deecher-Darlene C, Soderlund DM (1998) Structure activity relationship for the action of dihydropyrazole insecticides on mouse brain sodium channels. Pestic Biochem Physiol 60:177–180

Pessah IN, Waterhouse AL, Casida JE (1985) The calcium-ryanodine receptor complex of skeletal and cardiac muscle. Biochem Biophys Res Commun 128:449–456

Plapp FW Jr (1981) Toxicity of synthetic pyrethroids to laboratory and field populations of the tobacco budworm in central Texas. J Econ Entomol 74:207–209

Pittendrigh B, Reenan R, ffrench-Constant RH, Ganetzky B (1997) Point mutations in the *Drosophila* sodium channel gene *para* associated with resistance to DDT and pyrethroid insecticides. Mol Gen Genet 256:602–610

Pongs O (1992) Molecular biology of voltage dependent potassium channels. Physiol Rev 72:69–88

Possani LD, Martin BM, Svendsen IB, Rude GS, Erickson BW (1985) Scorpion toxin from *Centruroides noxius* and *Tityus serrulatus*: primary structures and sequence comparison by metric analysis. Biochem J 229:739–750

Prikhodko GG, Robson M, Warmke JW, Cohen CJ, Smith MM (1996) Properties of three baculovirus-expressing genes that encode insect-selective toxins: μ-Aga-IV, As II and Sh. I. Biol Control 7:236–244

Rao GV, Rao KS (1997) Modulation of K^+ transport across synaptosomes of rat brain by synthetic pyrethroids. J Neurol Sci 147:127–133

Ray DE, Sutharsan S, Forshaw PJ (1997) Action of pyrethroid insecticides on voltage gated chloride channels in neuroblastoma cells. Neurotoxicol Little Rock 18:755–760

Rochat H, Rochat C, Sampieri F, Miranda F, Lissitzky S (1972) The amino acid sequence of neurotoxin II of *Androtonus australis* Hector. Eur J Biochem 28:381–388

Rossignol DP (1995) Possible role of guanosine S-triophosphate binding proteins in pyrethroid activity. In: Clark JM (ed) Molecular action of insecticides on ion channels. ACS Symp 591. American Chemical Society, Washington, DC, pp 149–161

Ruigt GSF (1985) Pyrethroids. In: Kerkut GA, Gilbert LI (eds) Comprehensive insect physiology, biochemistry and pharmacology, vol 12. Pergamon Press, Oxford, pp 183–262

Salgado VL (1990) Mode of action of insecticidal generation in dihydropyrazoles: selective block of impulse generation in sensory nerves. Pestic Sci 28:389–411

Salgado VL (1992a) The neurotoxic insecticidal mechanism of the nonsteroidal ecdysone agonist RH-5849: K^+ channel block in nerve and muscle. Pestic Biochem Physiol 43:1–13

Salgado VL (1992b) Block of voltage dependent K^+ channels in insect muscle by the diacylhydrazile insecticide RH-5849 4-aminopyridine and guanidine. Arch Insect Biochem Physiol 21:239–252

Salgado VL (1992c) Slow voltage-dependent block of sodium channels in crayfish nerve by dihydropyrazole insecticides. Mol Pharmacol 41:120–126

Salgado VL (1998) Block of neuronal voltage dependent K^+ channels by diacylhydrazine insecticides. Neurotoxicology 19:245–252

Schreck CE, Smith N, Weidhaas D, Posey K, Smith D (1978) Repellent vs. toxicants as clothing treatment for protection from mosquitoes and other biting flies. J Econ Entomol 71:919–922

Schwiebert EM, Benos DJ, Egan ME, Stutts MJ, Guggino WB (1999) CFTR is a conductance regulator as well as a chloride channel. Physiol Rev 79:5145–5166

Scott J, Wheelock GD (1992) Characterization of a cytochrome p450 responsible for pyrethroid resistance in the housefly. In: Mullin CA, Jeffrey CS (eds) Molecular mechanisms of insecticide resistance. American Chemical Society, Washington, DC, pp 16–30

Scott RH, O'Brien K, Roberts L, Mordue W, Mordue-Luntz J (1999) Extracellular and intracellular actions of azadirachtin on the electrophysiological properties of cultured rat DRG neurones. Comp Biochem Physiol Pharmacol Toxicol Endocrinol 123:85–93

Smies M, Everts RHJ, Rejnenburg FHM, Koeman A (1980) Environmental aspects of field trials with pyrethroids to eradicate tsetse fly in Nigeria. Exotoxicol Environ Safety 4:114–128

Soderlund DM, Knipple DC (1995) Actions of insecticides on sodium channels: multiple target sites and site specific resistance. In: Clark JM (ed) Molecular actions of insecticides on ion channels. American Chemical Society, Washington, DC, pp 97–108

Song JH, Nagata K, Tatebayshi H, Narahashi T (1996) Interactions of tetramethrin, fenvalerate and DDT at the sodium channel in rat dorsal root ganglion neurons. Brain Res 708:29–37

Steinmeyer K, Jentsch JJ (1998) Molecular physiology of renal chloride channels. Curr Opin Nephrol Hypertens 7:497–502

Stewart LMD, Horst M, Ferber ML, Merryweather AT, Cayley BJ et al (1991) Construction of an improved baculovirus insecticide containing an insect-specific toxin gene. Nature 352:85–88

Stuhmer W, Conti F, Suzuki H, Wang X, Woda M, Yahagi N, Kubott, Numa S (1989) Structural parts involved in activation and inactivation of the sodium channels. Nature 339:597–603

Su HCF (1985) N-Isobutylamides. In: Kerkut GA, Gilbert LI (eds) Comprehensive insect physiology, biochemistry and pharmacology, vol 12. Pergamon Press, Oxford, pp 237–289

Tatebayashi H, Narahashi T (1994) Differential mechanism of action of the pyrethroid tetramethrin on tetrodotoxin-sensitive and tetrodotoxin-resistant sodium channels. J Pharmacol Exp Ther 270:595–603

Taylor MFJ, Heckel DG, Brown TM, Kreitman ME, Black B (1993) Linkage of pyrethroid insecticide resistance to a sodium channel locus in the tobacco budworm. Insect Biochem Mol Biol 23:763–775

Tomalski MD, Kutney R, Bruce WA, Brown MR, Blum MS, Travis J (1989) Purification and characterization of insect toxins derived from the mite *Pyemotes triciti*. Toxicon 27:1151–1167

Trainer VL, McPhee JC, Boutelet-Bochan H, Baker C, Schener T, Catterall WA (1997) High affinity binding of pyrethroids to the alpha subunit of brain sodium channels. Mol Pharmacol 51:651–657

Usherwood PNR (1962) The action of the alkaloid ryanodine on insect skeletal muscle. Comp Biochem Physiol 6:181–195

Vijverberg HPM, Van den Bercken J (1990) Neurotoxicological effects and the mode of action of pyrethroid insecticides. Crit Rev Toxicol 21:105–126

Wang X, Smith R, Fletcher JI, Wilson H, Wood CI, Howden ME, King GF (1999) Structure function studies of omega atracotoxin, a potent antagonist of insect voltage gated calcium channels. Eur J Biochem 264:488–494

Ware GW (1982) Fundamentals of pesticides – a self instruction guide. Thompson, Fresno, pp 78–79

Warmke JW, Reenan RAG, Wang P, Qian S, Arena JP, Wang J, Wunderler D, Liu K, Kachorowski GJ, Van der Ploeg LHT, Genetzky B, Cohen CJ (1997) Functional expression of *Drosphilia* para sodium channels: modulation by the membrane protein Tip E and toxin pharmacology. J Gen Physiol 110:119–133

Webb RE, Smith FF, Sullivan WN, Schechter MS, Boswell AL, Ewing A (1974) Resmethrin evaluation against some common greenhouse pests. J Econ Entomol 67:295

Williamson MS, Martinez-Torres D, Hick CA, Devonshire AL (1996) Identification of mutations in the housefly *para* type sodium channel gene associated with knockdown resistance (kdr) to pyrethroid insecticides. Mol Gen Genet 252:51–60

Yamasaki T, Ishi T (1952) Studies on the mechanism of action of insecticides. IV. The effects of insecticides on the nerve conduction of insects. Oyo-Kontyu 7:157–164.

Zhang A, Towner P, Nicholson RA (1996) Dihydropyrazole insecticides: interference with depolarization-dependent phosphorylation of synapsin I and evoked release of L-glutamate in nerve terminal preparations from mammalian brain. Pestic Biochem Physiol 54:24–30

Zlotkin E (1987) Pharmacology of survival: insect selective neurotoxins derived from scorpion venom. Endeavour 11:168–174

Zlotkin E (1999) The insect voltage gated sodium channel as target of insecticides. Annu Rev Entomol 44:429–455

Zlotkin E, Miranda T, Rochat H (1978) Chemistry and pharmacology of Buthinae scorpion venoms. In: Bettini S (ed) Arthropod venoms. Handbook of experimental pharmacology, vol 48. Springer, Berlin Heidelberg New York, pp 317–369

Zlotkin E, Eitan M, Bindokas VP, Adams, Mayer M, Burkhart, Fowler E (1991) Functional duality and structural uniqueness of depressant insect selective neurotoxins. Biochemistry 30:4814–4821

Zlotkin E, Moskowitz H, Herrmann R, Pelhate M, Gordon D (1995) Insect sodium channel as the target for insect-selective neurotoxins from scorpion venom. In: Clark JM (ed) Molecular actions of insecticides on ion channels. American Chemical Society, Washington, DC, pp 56–85

Zlotkin E, Devonshire AL, Warke JW (1999) The pharmacological flexibility of the insect voltage gated sodium channel: toxicity of AaIT to knockdown resistant (kdr) flies. Insect Biochem Mol Biol 29:849–853

Acetylcholine Receptors as Sites for Developing Neonicotinoid Insecticides

R. NAUEN, U. EBBINGHAUS-KINTSCHER, A. ELBERT, P. JESCHKE, and K.TIETJEN[1]

1
Introduction

The world market for insecticides is still dominated, albeit declining, by compounds that irreversibly inhibit acetylcholinesterase (AChE), one of the most essential enzymes in the central nervous system of insects, responsible for the cleavage of the all-important neurotransmitter acetylcholine (Pittman 1971). However, as shown in Table 1, the market share of these AChE inhibitors, i.e. organophosphates and carbamates, decreased from 71% in 1987 to some 56% in 1997. Combining the AChE inhibitors and those insecticides that act on the voltage-gated sodium channel, in particular the pyrethroids, accounts for more than 75% of the world market only by these two modes of action (Table 1). One of the insecticide molecular target sites of growing importance is the nicotinic acetylcholine receptor (nAChR; Table 1); increasing considerably in value over the last decade. Only ten years ago insecticides that acted on the nAChR were of minor economic importance (<2% of the total insecticide market until 1991), and registered compounds were cartap (1964), bensultap (1968) and thiocyclam (1977). These compounds were metabolised in the insect's body to nereistoxin, a naturally occurring toxin described in the marine worm *Lumbriconereis heteropoda*.

One of the oldest known insecticides able to act on the nAChR is nicotine (Table 2). This compound is still used to control some homopteran pests such as aphids in greenhouses, e.g. in Japan. In the early 1970s, Shell invented a new class of compounds capable of acting on the nAChR. The most active member of that class was nithiazine; however, due to several drawbacks, this compound was never commercialized for agricultural use (Soloway et al. 1979; Kollmeyer et al. 1999). Today, these early compounds can be considered as the first generation of the so-called neonicotinoid insecticides (Table 2). By structural modification, i.e. the introduction of a heterocyclic aryl substituent, the properties of such compounds were considerably improved. This has led to a class of nAChR effectors with excellent biological efficacy against key homopteran and coleopteran pests, the so-called chloronicotinyls or second-generation

[1] Bayer AG, Agrochemicals Division, Research Insecticides, 51368 Leverkusen, Germany

Isaac Ishaaya (Ed.): Biochemical Sites of
Insecticide Action and Resistance
© Springer-Verlag Berlin, Heidelberg 2001

Table 1. Mode of action of the top 100 insecticides/acaricides and their world market share (excluding fumigants, endotoxins and those with unknown mode of action)

Mode of action	1987 (%)	1997 (%)	Change (%)
Acetylcholinesterase	71.2	56.1	−15.1
Voltage-gated Na channel	16.5	21.0	+4.5
Acetylcholine receptor	**1.5**	**8.6**	**+7.1**
GABA-gated Cl channel	5.0	7.7	+2.7
Chitin biosynthesis	2.1	2.7	+0.6
NADH dehydrogenase	0	1.4	+1.4
ATPase	3.1	1.3	−1.8
Uncouplers	0	0.4	+0.4
Octopamine receptor	0.5	0.4	−0.1
Ecdysone receptor	0	0.3	+0.3

neonicotinoids, a new class of chemical insecticides introduced onto the market in the early 1990s.

The first commercialized example – launched by Bayer – was imidacloprid (Table 2), and this product is currently the largest selling insecticide worldwide (Elbert et al. 1991; Elbert and Nauen 1998). One of the major problems in modern agriculture is the growing tendency of insect pests to become increasingly resistant to a wide variety of insecticidal classes. Therefore, imidacloprid was rapidly introduced into several market segments covered by conventional insecticides, but suffered from serious resistance problems.

Imidacloprid is registered in more than 75 countries and fulfills the ambition of an agrochemical company in developing fast-acting compounds with low use rates and a new or unconventional mode of action that can combat highly resistant insect pests (Elbert et al. 1996). The discovery of this new class of ligands of the nAChR can be considered a milestone in insecticide research and facilitates greatly the understanding of the functional properties of insect nAChRs (Tomizawa et al. 1999). In 1994, spinosyn, a compound structurally distinct from the neonicotinoids but which acts on the nAChR, was presented at the Annual Meeting of the Entomological Society in Dallas, USA (for an overview, see Thompson and Hutchins 1999). The toxicity of spinosyn to homopterans such as aphids is low, but it is particularly active against noctuid larvae and therefore complements the activity of the neonicotinoids in several cropping systems. Physiological observations of treated insects by Salgado (1998) suggested that spinosyn seems to act differently to imidacloprid. Electrophysiological data revealed that spinosyn as well as the neonicotinoids act as agonists of the nAChR (Salgado et al. 1997), but that spinosyn does not interact directly with the acetylcholine binding site of the nAChR (Dunbar et al. 1998).

The present chapter describes the current knowledge on the structure and function of insect nAChRs characterized by receptor binding studies, phylo-

Table 2. Evolution of chemical classes of insecticides that act on insect nAChRs

Insecticidal structure	Compound (chemical class)	Year of introduction; main target pests
	Nicotine (nicotinoids)	ca. 1890; Homoptera
	Cartap (nereistoxin analog)	1967; Homoptera/ Lepidoptera
	Nithiazine (first generation neonicotinoid)	1978[a]; Lepidoptera
	Imidacloprid (second generation neonicotinoid)	1991; Homoptera/ Coleoptera
	Spinosyn (macrocyclic lactone)	1997; especially Lepidoptera

[a] Never commercialized for agricultural use

genetic considerations regarding receptor homologies between orthologs from different animal species, electrophysiological investigations, as well as a description of the vast range of the structurally diverse array of receptor ligands, with particular emphasis on neonicotinoid insecticides.

2
Insect Nicotinic Acetylcholine Receptors

2.1
Structure

Most of the current information regarding nAChRs originated from research with vertebrate receptors. However, comparison of the known amino acid sequences and properties allows the safe prediction that insect receptors have a similar construction to vertebrate receptors. The best pictures of these are the electron-cryomicroscopic images of Unwin, now with a 4.6 Å resolution (Miyazawa et al. 1999).

nAChRs are composed of five single subunits, typically two identical alpha (α) subunits and three identical beta (β) subunits creating a pentameric transmembrane protein with a central cation-permeable ion channel (Fig. 1). However, other subunit compositions are also possible, e.g. in vertebrate muscles. The side view appears as a long channel structure embedded in the cell membrane. The real folding pattern of the protein is not known, i.e. Fig. 1 is only an estimate of the structure. Acetylcholine and insecticides like imidacloprid probably bind in a certain region where some amino acids cluster, which influences binding of these ligands. However, there are other physically distinct ligand binding sites which complicate the picture (see Fig. 1). Most important here, we do not know where the insecticidal spinosyns (Table 2) bind. Possibly there is a relation to the binding site for ivermectin, which is not only present in the distantly related insect glutamate gated chloride channels, but also in vertebrate α7-AChRs (Krause et al. 1998).

2.2
Diversity

About ten different nAChR genes have been discovered in insects. The complete genome of the nematode *Caenorhabditis elegans* identified an unexpectedly high number of nAChR genes (up to 40 subunits) (Mongan et al. 1998). The high number of genes discovered in *C. elegans* raises the question, are there additional subunits still to be discovered in insects? The complete *Drosophila* genome has recently been sequenced and will be published by the company Celera (http://www.celera.com) at the beginning of the year 2000 (Pennisi 1999). Currently, 13 subunits have been identified in vertebrates. We have constructed a phylogenetic tree of the known subunits (Fig. 2). The ten known insect subunits are based on six α- and two β-subunits found in the peach-potato aphid, *Myzus persicae* (Huang et al. 1999). Additionally, we have identified two to three α7-like subunits in *Heliothis virescens* (gb|AF143846, gb|AF143847). Subunits from other insects are orthologues of those already

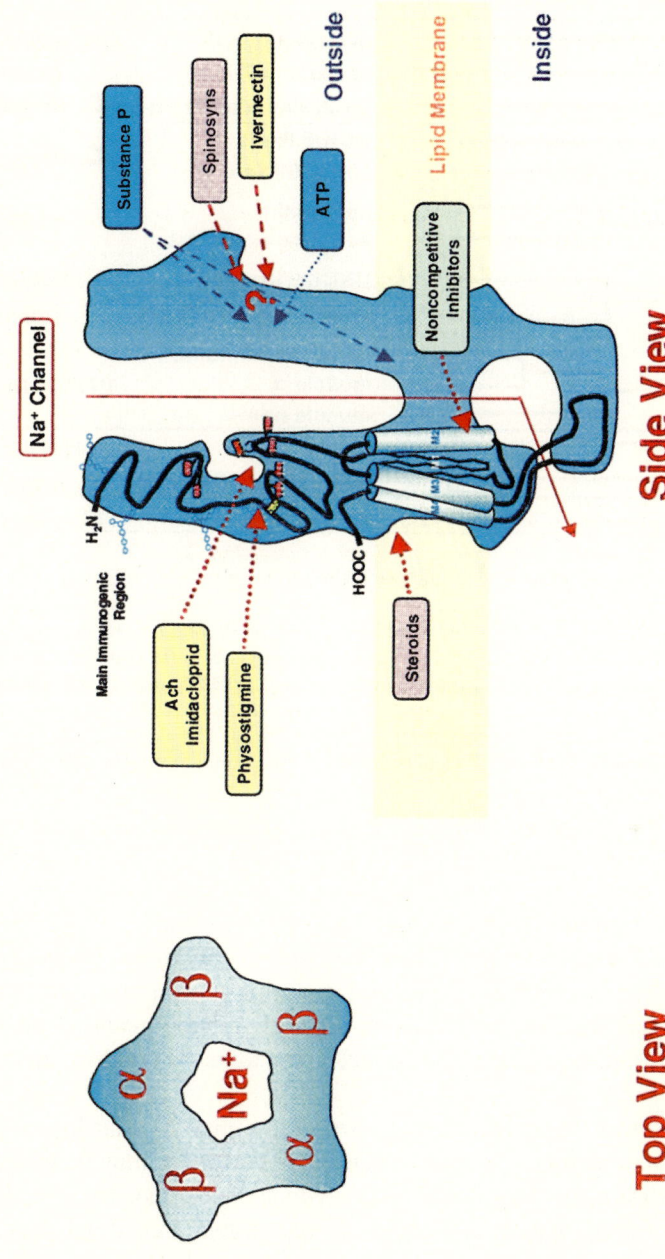

Fig. 1 General structure of nicotinic acetylcholine receptors (dimensions according to Miyazawa et al. 1999). *Left* Top view showing the pentameric arrangement of α- and β-subunits. *Right* Side view giving the tentative folding pattern and location of ligand binding sites. Ligands shown in *yellow boxes* are proposed to bind to α-subunits, ligands shown in *blue boxes* are proposed to bind to β-subunits. Amino acids involved in acetylcholine binding are labelled in *red*. Different ligands are specific for different individual subunits and glycosylation sites differ between individual subunits

no. α/β

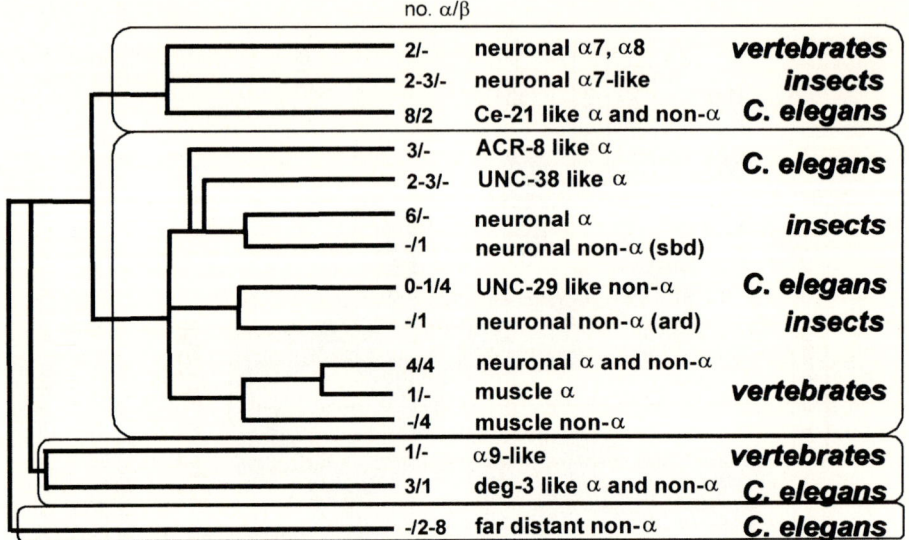

2/-	neuronal α7, α8	**vertebrates**
2-3/-	neuronal α7-like	**insects**
8/2	Ce-21 like α and non-α	**C. elegans**
3/-	ACR-8 like α	**C. elegans**
2-3/-	UNC-38 like α	
6/-	neuronal α	**insects**
-/1	neuronal non-α (sbd)	
0-1/4	UNC-29 like non-α	**C. elegans**
-/1	neuronal non-α (ard)	**insects**
4/4	neuronal α and non-α	
1/-	muscle α	**vertebrates**
-/4	muscle non-α	
1/-	α9-like	**vertebrates**
3/1	deg-3 like α and non-α	**C. elegans**
-/2-8	far distant non-α	**C. elegans**

Fig. 2. Phylogenetic tree of known nicotinic acetylcholine receptor subunits. Included are all known insect receptors from *Myzus*, *Drosophila* and *Heliothis*, but orthologues in the different species are scored only once. Also included are all identified *Caenorhabditis elegans* genes using the nomenclature from Mongan et al. (1998). The *C. elegans* genes were completed by further genes, identified by our own BLAST searches (Accession nos.: z93778, u88177, z81071, u88175, u39885, q22079, q22080, q22081, q22083, q22084, q2205, aac05108, aac05109). The vertebrate subunits are from humans, except α8 from chicken and α9 from rat. The exact location of the α9-like and *far distant subunits* in the phylogenetic tree is not certain and depends on the settings of the algorithms used

mentioned. The dendrogram shows excellent correlations between insects, *C. elegans* and vertebrates, except for deg-3 and α9-like subunits, which have so far not been described in insects. The correct phylogenetic position of deg-3 and α9-like subunits in relation to each other and to α7- and α8-like subunits is not unequivocally calculable. The more distantly related genes identified in *C. elegans* have not been sufficiently characterized to be certain that they really are nAChR subunits.

For non-α-subunits the distinction between subunits of other ligand-gated ion channels is increasingly difficult. We propose that there may be more subunits in insects waiting to be discovered. The suspicion that there exist still unknown subunits in insects is supported by the fact that it has not been possible to express functional insect nAChRs, except for acetylcholine and nicotine-responsive homo-oligomeric locust α1 and *Myzus* Mpa1/Mpa2 receptors expressed in *Xenopus* oocytes (Amar et al. 1995; Sgard et al. 1998; see also Sect. 4).

The number of insect nAChR genes implies the existence of a comparable or perhaps even greater number of nAChR proteins. Although some work has been done to study the in situ expression profiles of different genes (Schuster et al. 1993), information regarding the localization and role of single gene products is very limited. In contrast to vertebrates, no muscle nAChRs exist in insects. In vertebrates, nAChRs have been shown to occur not only postsynaptically, but also presynaptically and somatically. Some subtypes are more permissive for calcium ions than for sodium ions. Insect nAChRs are in general less well characterized and much work still remains in terms of functional subtype description. In the central nervous system, nAChRs facilitate the transduction of sensations, learning and memory.

In this context it is not surprising that the receptor subtypes and functions affected by nAChR insecticides are not well characterized. Affinity chromatography of *Drosophila* head homogenates using an insecticide (imidacloprid)-coupled Sepharose resin (Tomizawa et al. 1996) revealed three different nAChR subunits. Using the same protocol, we identified only two subunits (unpubl.). Since there is no sequence information on the purified subunits, the result only confirms that the insecticide does not bind unselectively to many receptors. The three co-purified subunits also bound to α-bungarotoxin, i.e., at least two different α-subunits are present in the mixture (see also Sect. 3.1). From the results available it is not possible to decide whether there are three (or two) subunits from a single heteropentameric receptor or possibly more.

3
Compounds Acting on the Nicotinic Acetylcholine Receptor

3.1
Radioligand Binding Studies

Radioligand binding studies contribute essentially to the characterization of insect nAChRs. There is no need for pure receptors since highly subtype-specific radioligands allow the investigation of different receptors. On the other hand, the methodology becomes less useful if the membrane preparations contain several similar receptors with overlapping pharmacology.

When the chemical exploration of neonicotinoid insecticides (see also Sect. 3.2) started in the early 1980s, only tritiated α-bungarotoxin was available as a tool to investigate the insecticide binding characteristics of receptors. However, neonicotinoid insecticides exhibit only a medium affinity to nAChRs if measured using α-bungarotoxin (Fig. 3a). Today, several more ligands are available (Table 3), tritiated imidacloprid (Liu et al. 1993b) being particularly important. The superior affinities obtained for neonicotinoids by using [³H]imidacloprid as a radioligand explains the high biological activity of these compounds (Fig. 3a). This also means that the α-bungarotoxin and imidacloprid sites are distinct sites on the same receptor, or exist on different receptors. Possibly there is even more than one imidacloprid site (Lind et al. 1998).

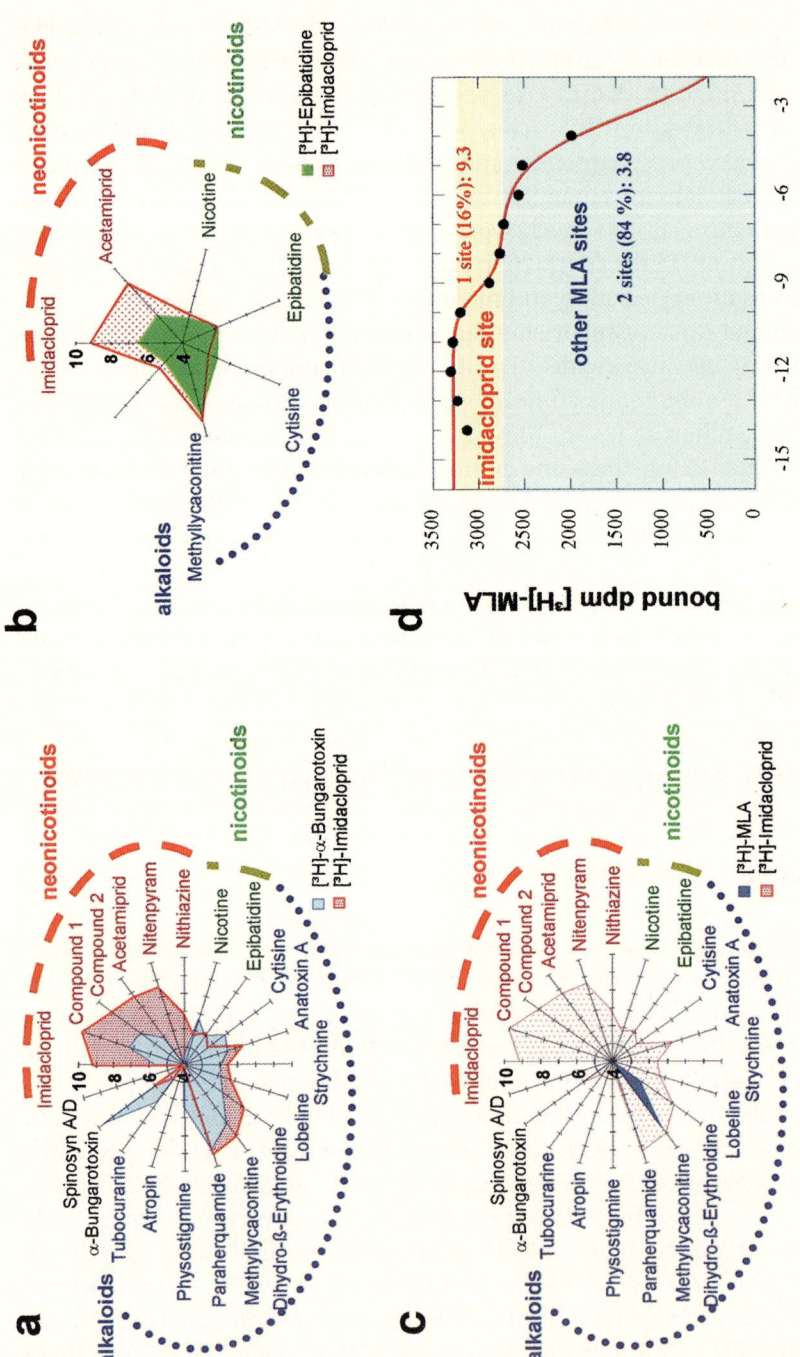

Fig. 3a–d. Binding affinities (measured by competition using different radioligands) of various compounds to housefly head (**a, b**) and cockroach (**c**) nicotinic acetylcholine receptors with different radioligands. Spider web graphs. **a:** [³H]α-bungarotoxin (*blue*) and [³H]imidacloprid (*red*). **b:** [³H]epibatidine (*green*) and [³H]imidacloprid (*red*); data taken from Orr et al. 1997). **c:** [³H]methyllycaconitine (*blue*) and [³H]imidacloprid (*red*). **d:** Competition curve for [³H]methyllycaconitine vs. unlabelled imidacloprid

Table 3. Characteristics of different radioligands used for the study of insect nicotinic acetylcholine receptors

Radiolabelled ligand	Characteristics (for insect receptors)		
	Affinity (range)	Subtype specificity	Usefulness in ligand binding assays
Acetylcholine	μM	Probably low	Not (low affinity)
Nicotine	μM	Limited	Limited
α-Bungarotoxin	nM	Probably high	High
Imidacloprid	nM	Probably high	High; very low non-specific binding
Epibatidine	nM	Probably limited	Probably limited
Methyllycaconitine	nM	Low	High–limited

Using epibatidine or methyllycaconitine (MLA) as radiolabelled ligands again gives unrelated results (Fig. 3b,c). This allows the conclusion that all these sites are distinct sites. The spinosyns, which are known by electrophysiology to act on nAChRs, do not compete with any of these tritiated ligands and therefore occupy still another site (Salgado et al. 1997; Salgado 1998).

Using tritiated MLA all ligands with the exception of dihydro-β-erythroidine – and of MLA itself – show only very weak competition. Considering the higher number of MLA sites, a detailed analysis of the competition between tritiated MLA and imidacloprid reveals a 16% fraction of MLA sites, which are sensitive to imidacloprid in the nM range (Fig. 3d).

In conclusion, radioligand binding experiments are useful for distinguishing several different AChRs and/or binding sites in insects. However, the type of receptor (subunit composition) affected/modulated is due to the fact that in vitro expression of insect nAChRs is only poorly understood; although some encouraging data has recently been obtained with *M. persicae* receptors (Huang et al. 1999).

An extremely important question in relation to our basic knowledge as well as our search for new insecticides is, "are there differences between the nAChRs of different insects?" Binding experiments (Lind et al. 1998; Nauen et al. 1998a) and electrophysiology (see Sect. 4.) have revealed some differences, but from our point of view these are not decisive, and cannot explain the big differences in efficacy observed in whole animal bioassays, e.g., aphids vs. noctuid larvae. Differences in the uptake and metabolism of neonicotinoids, spinosyns and other compounds may contribute greatly to the observed heterogeneity in biological activity, but this has not yet been studied in detail.

3.2
Neonicotinoids

3.2.1
Imidacloprid and Related Structures

The discovery of imidacloprid was the result of a search for improved activity by altering the structure of the originally described six-membered tetrahydrothiazine, nithiazine (Soloway et al. 1979). Compared with nithiazine, the biological efficacy of imidacloprid against the green rice leafhopper could be enhanced by over a factor of 125 by the incorporation of a 6-chloro-3-pyridylmethyl moiety, an essential heterocyclic structural element (Kagabu 1996). Furthermore, YRC 2894, a novel insecticide with activity against sucking pests currently under worldwide development by Bayer Crop Protection, also contains the 6-chloro-3-pyridylmethyl moiety. Thiacloprid (ISO draft proposal) has been proposed as the common name for this compound. It is not surprising that two of the most important open-chained neonicotinoids, nitenpyram (Minamida et al. 1993b) from Takeda and acetamiprid (Takahashi et al. 1992) from Nippon Soda, both have the excellent 6-chloro-3-pyridylmethyl moiety.

Due to Bayer's success with imidacloprid, several different companies such as Takeda, Nippon Soda, Agro Kanesho, Mitsui Toatsu and Novartis introduced their own neonicotinoid insecticides. These chemical structures only differ regarding the heterocyclic group from the above-mentioned three chloronicotinyl structures, which are already on the market. In other structural types of neonicotinoids, thiamethoxam from Novartis (Maienfisch and Sell 1992), clothianidin from Takeda/Bayer (Uneme et al. 1998; Anonymous 1999) and the not commercialized AKD 1022 from Agro Kanesho (Maienfisch et al. 1999), the chloropyridylmethyl moiety is replaced by a 2-chloro-5-thiazolyl group. In the open-chained dinotefuran or nidinotefuran, a novel insecticide of the nitroguanidine type discovered by Mitsui Chemicals, Inc., the aromatic heterocyclic chloropyridylmethyl moiety is replaced by a non-aromatic tetrahydrofuran ring (Kodaka et al. 1998). An overview of all the different cyclic and open-chain neonicotinoid/chloronicotinyl structures discussed is given in Table 4.

In general, structural requirements for the cyclic and/or open-chain active neonicotinoids/chloronicotinyls consist of different segments: the heterocyclic group [Het] (i), the bridging chain [—CHR—] (ii), the functional group [=X—Y] as part of the pharmacophore [—N—C(Z)=X—Y] (iii) and for the cyclic type compounds the bridging chain $[R^1—R^2]$ (iv) and for the open-chain type compounds the separate substituents $[R^1, R^2]$ (iv), as shown in Fig. 4. The influence of these different structural segments on the insecticidal activity of neonicotinoids/chloronicotinyls has already been described (Kagabu et al. 1992;

Table 4. Chemical structures of cyclic and open-chain neonicotinoids (chloronicotinyls) displaying the different types of pharmacophores that provide insecticidal activity

Pharmacophores [—N—C(Z)=X—Y]	Structure type	
	Cyclic (R^1—R^2)	Open-chain (R^1, R^2)
Nitroenamines (Z = S, N) (Nitromethylenes) [—N—C(Z)=CH—NO$_2$]	Nithiazine	Nitenpyram
Nitroguanidines (Z = N) [—N—C(N)=N—NO$_2$]	Imidacloprid	Clothianidin[a]
	AKD 1022	Dinotefuran[a]
	Thiamethoxam[a]	
Cyanoamidines (Z = S, —Me) [—N—C(Z)=N—CN]	Thiacloprid[a]	Acetamiprid

[a] Proposed common name

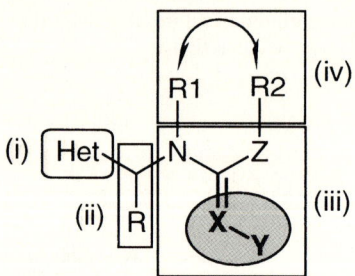

Fig. 4. Structural segments of neonicotinoids

Minamida et al. 1993a,b; Moriya et al. 1993; Tabuchi et al. 1994; Maienfisch et al. 1999).

The pharmacophore (iii) can be represented by the group [—N—C(Z) = X—Y], where = X—Y is an electron-withdrawing group and Z is an NH, NR', CH_2, O or S moiety. It is well known that the pharmacophore type has an influence on the insecticidal activity of the compound. For example, in a series of imidacloprid analogs, insecticidal activity against the green rice leafhopper, *Nephotettix cincticeps*, decreases in the order of the following functional groups [=X—Y]: nitromethylene [=CH—NO$_2$], nitroimine [=N—NO$_2$], >cyanoimine [=N—CN] >dicyanomethylene [=C(CN)$_2$] >>cyanomethylene [=CH—CN], carbonyl [=O], imine [=N—H] (Shiokawa et al. 1995). Other imino derivatives, like alkoxycarbonylimine [=N—CO—OR'] and sulfonylimine [=N—SO$_2$—R'], showed weaker activity.

In general, the greatest insecticidal activity is observed for the compounds containing the nitroenamine (or nitromethylene), nitroguanidine, or cyanoamidine pharmacophore (see also different types of pharmacophores in Table 4). Besides its influence on biological activity, the pharmacophore is also responsible for some specific properties such as photolytic stability, degradation in soil, metabolism in plants, and toxicity to different animals and beneficials.

From this perspective, relatively little work has been performed to introduce bulky substituents on the heterocyclic chloropyridylmethyl moiety (i), the bridging chain (ii) and for the substitution of the imidazolidine ring system (=>iv) of imidacloprid (Fig. 5). If the chlorine atom is required for activity at the nicotine receptor, then its replacement by heterocycles or bulky substituents should result in an unfavorable steric interaction with the nAChR giving reduced affinity or pI$_{50}$-values for housefly head membrane nAChR, using [³H]imidacloprid. Exchange of chlorine by substituted benzyloxy at the 6-position on the pyridine ring decreases the affinity and insecticidal activity against *M. persicae* and *N. cincticeps* significantly (Table 5).

In parallel to these studies, we have examined the activity of lipophilic 5-aryl(hetaryl)-substituted 6-chloronicotinyl derivatives such as 5-(4-

Fig. 5. Derivatization areas in imidacloprid

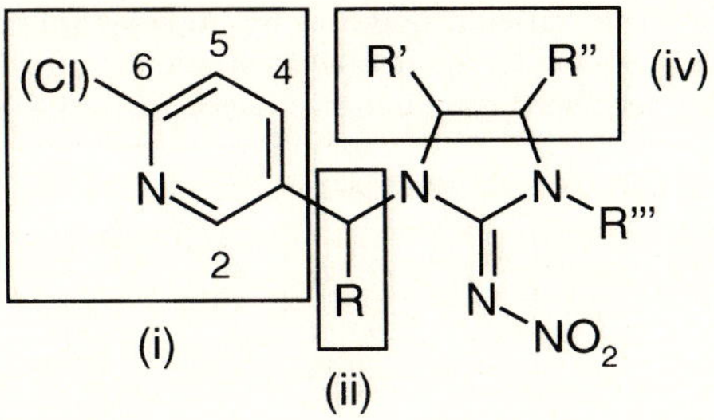

Table 5. Variation of the pyridine ring of 1-nicotinoyl-2-nitroimino-imidazolidine and resulting insecticidal activity (concentration in ppm giving 100% mortality) and affinity to nAChR in housefly head membranes expressed as a pI_{50}-value (for explanation, see Fig. 8)

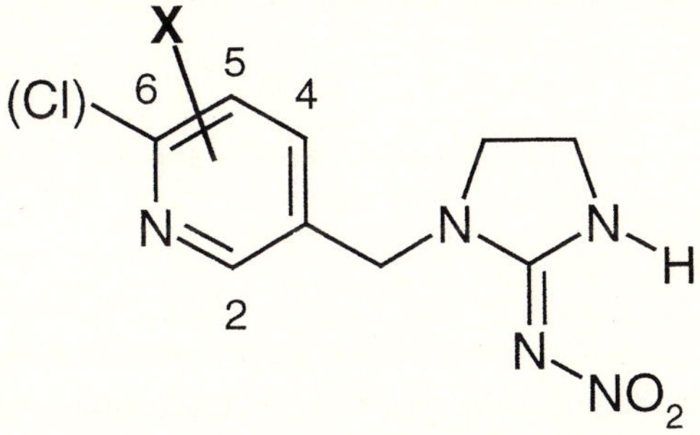

6-X:	6-Cl	6-(4-Me-benzyloxy)	6-(4-CF₃O-Ph)	6-(N¹-1,2,4-triazolyl)
M. persicae	1.6	100	>1000	>1000
N. cincticeps	0.32	100	>1000	>1000
pI_{50}-value	9.3	7.0	<5	<5

trifluoromethoxyphenyl) (pI_{50}-value 6.1) or 5-(2,4-dichlorophenyl) (pI_{50}-value 6.8) and found only low affinity levels except for the 6-(4-methylbenzyloxy) derivative. Other bulky substituents in this specific ring position also failed to induce remarkable effects.

The fact that the bulky substituents show low activity prompted the first assumption that the smaller halogen atom as substituent at the pyridine ring exerts generally a more favorable influence than the aryl, hetaryl or benzyloxy groups, respectively.

From 1-nicotinyl-2-nitromethylene imidazolines it was known that the distance between the pyridyl and the imidazoline nitrogen is important for the activity. Concerning the bond length between the pyridine ring and the imidazolidine moiety, methylene is the optimal spacer group. A study with the derivatives of imidacloprid, where the methylene bond is substituted by larger

Table 6. Variation of the methylene bond of 1-nicotinoyl-2-nitroimino-imidazolidine and resulting insecticidal activity (concentration in ppm giving 100% mortality) and affinity to nAChR in housefly head membranes expressed as a pI_{50}-value (for explanation, see Fig. 8)

R:	-H	-Ph	4-Me-Ph	4-Cl-Ph	4-CF$_3$O-Ph
M. persicae	1.6	>100	>1000	>1000	>1000
N. cincticeps	0.32	>1000	>1000	>1000	>1000
pI_{50}-value	9.3	n. d.	<5	<5	<5

substituents than hydrogen, indicated that in general the longer chain reduces the insecticidal activity against *M. persicae* and *N. cincticeps,* while methylation at the methylene bond reduces the activity only slightly.

We prepared analogs including the phenyl or 4-methyl-, phenyl, 4-chloro- and 4-trifluoromethoxyphenyl as bulky substituents at the methylene bond of imidacloprid. All these aryl-substituted derivatives also displayed lower affinity and insecticidal activity than imidacloprid, suggesting that increasing size is not tolerated at the methylene bond position (Table 6).

Hydroxylation of the imidazolidine ring of imidacloprid showed that in general the mono- (R′: —H; R″: —OH) and bishydroxylated (R′, R″: —OH) derivatives reduce affinity, as seen in Table 7. On the other hand, the mono (R′: —OH; R″: —H) derivative shows interestingly a higher level of efficacy (pI_{50}-value: 8.5). Another remarkable structure-activity relationship is that even if a double bond was introduced into the central heterocyclic ring, the insecticidal efficacies remained at a high level. It is already known from the metabolism of imidacloprid that, compared with imidacloprid, the imidazoline derivative (olefin, —CH=CH—) provides superior toxicity to some homopterans after oral ingestion (Nauen et al. 1998b, 1999a). This suggests that for the central ring system the exact arrangement of the ring atoms and not the electronic effect of the ring system seems to be necessary for activity.

A similar phenomenon has been described for the conjugated pyridone derivatives (Kagabu 1999).

Table 7. Variation of the bridging chain of 1-nicotinoyl-2-nitroimino-imidazolidine and pI_{50}-values (for explanation, see Fig. 8) for nAChR from housefly head membranes

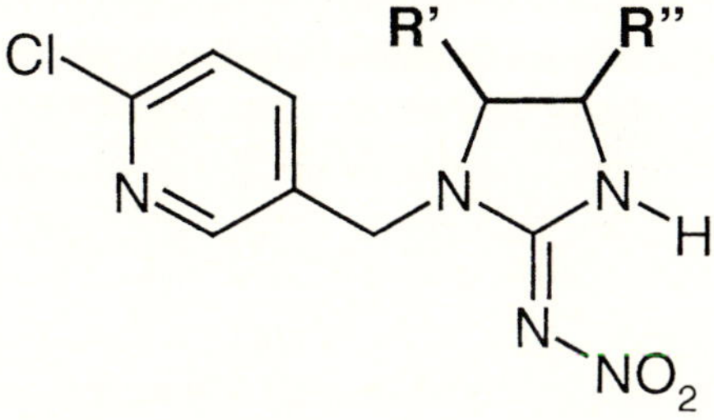

−R′HC−CHR″−	−H₂C−CH₂−	−HC=CH−	−(HO)HC−CH₂−	−CH₂−CH(OH)−	−(HO)HC−CH(OH)−
pI_{50}-value	9.3	9.6	8.5	7.6	5.5

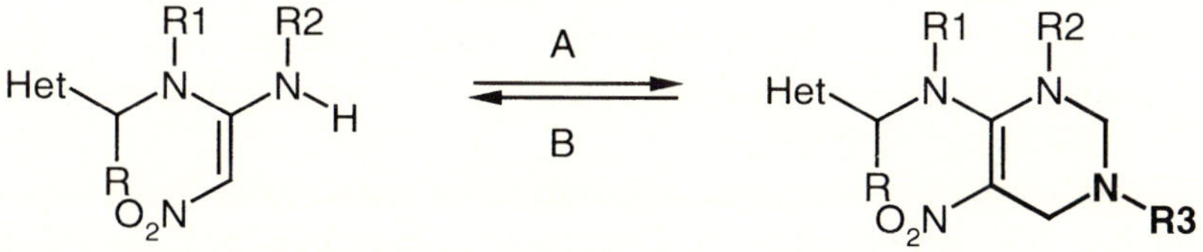

Fig. 6. Formation and hydrolysis of Mannich adducts *1*

3.2.2
Mannich Adducts as Experimental Pro-Neonicotinoids

First patent applications covering a number of 5-nitrotetrahydropyrimidines, the so-called *Mannich adducts* of certain insecticidally active nitromethylene compounds, were published in 1988 (e.g., compound Bay T 9992, Bayer). Due to the potent insecticidal and arthropodicidal properties of Mannich adducts (Kishida et al. 1992), represented by the general formula *1* in Fig. 6, different companies such as DuPont, Mitsui Toatsu Chem., Bayer, Agro-Kanesho, Nippon Soda, Ciba-Geigy (since 1996 Novartis), Takeda and others produced compounds in this class. Mannich adducts can be prepared via an aminomethylation reaction by the treatment of the CH-acidic nitromethylenes *2* (=X—Y: =CH—NO₂; Z: —NH—R₂), which react as nitroenamines (Rajappa 1981), with one or more equivalents of amine (*3*, R³—NH₂) and at least two molar equivalents of the electrophile formaldehyde (HCHO) in a suitable solvent, as shown in Fig. 6 (Mannich adduct formation, pathway A).

These reactions are typically carried out at room temperature or by heating if necessary. Suitable solvents include alcohols, water, and polar aprotic solvents such as tetrahydrofuran and dimethylformamide, or solid paraformaldehyde can be used. In some cases, a small amount of a strong, nonoxidizing acid, such as hydrochloric acid, can be used as a catalyst. The Mannich adducts have a wide range of substituents R^3 which inhibit [^{3}H]imidacloprid binding to the *Drosophila* or *Musca* nAChR by 50% at 0.7–24.0 nM (Latli et al. 1997).

However, at different pH-values, hydrolysis of the Mannich adducts can be observed as shown in Fig. 6 (Mannich adduct cleavage, pathway B). Utilizing this cleavage assay, the pH-dependence of the Mannich adduct cleavage reaction was investigated in order to identify the formed cleavage product. Table 8 shows the half-times obtained for the cleavage of the synthesized model compounds *1a-g* in acetonitrile, buffered at pH-values of 4, 7 and 9. For all derivatives the maximal splitting rates were measured at pH 4; however, for compounds *1e-g* a moderate cleavage and for compounds *1d-g* a very low cleavage behavior was observed at pH 7 and pH 9, respectively. The analytical data (high performance liquid chromatography, nuclear magnetic resonance) clearly show that it is the nitromethylene compound *2a* that is formed and that is obviously required for activity. The above data are in full agreement with the observed insecticidal activity.

Another factor that must be considered is the half-life of the Mannich adducts. It seems that the compounds which have a heterocyclic substituent at side chain R^3, e.g. the derivatives *1d-g*, would possess much longer half-lives (13–35 h) concerning hydrolysis at pH 7 to 9 than the alkyl-type substituents in *1a-c*. However, it was discovered that all the Mannich adducts exhibit the same insecticidal activity against *M. persicae* and *N. cincticeps* (at 40 pm and 8 ppm), similar to the parent methylene derivative *2a*. This data, coupled with other experimental evidence, suggest that Mannich adducts are pro-neonicotinoids/chloronicotinyls.

4
Electrophysiological Considerations

Electrophysiological measurements reported in numerous studies have revealed that nAChRs are widely expressed in the insect nervous system on both postsynaptic and presynaptic nerve terminals, on the cell bodies of interneurons, motor neurons and sensory neurons (Goodman and Spitzer 1980; Harrow and Sattelle 1983; Sattelle et al. 1983; David and Sattelle 1984; Breer and Sattelle 1987; Breer 1988; Restifo and White 1990). Schröder and Flattum (1984) using extracellular electrophysiological recordings were the first to identify that the site of action of the nitromethylene nithiazine was on the cholinergic synapse. A number of subsequent electrophysiological and biochemical binding studies revealed that the primary target of the neonicoti-

Table 8. Hydrolytic stability and log P_{OW}-values at different pH-values of 5-nitrotetrahydropy-rimidines (Mannich adducts)

Substituent R[3]	Half-life (h), acetonitrile/buffer (1:1)			log P_{OW}-values[a]	
	pH 4	pH 7	pH 9	pH 2	pH 7.5
1a	<<1	<1	≈1–2	n.d.	3.22 / 3.34
1b	<<1	<1	≈2	n. d.	3.37
1c	<1	<1	≈9	1.82	1.52
1d	<1	<1	≈13	0.11	1.26
1e	1.6	≈7	≈23	1.35	1.47
1f	<1	≈3	≈30	0.56	1.54
1g	<1	≈2	≈35	2.15	1.95

[a] Octanol/water partition coefficient

noids were the nAChRs (Benson 1989; Sattelle et al. 1989; Bai et al. 1991; Leech et al. 1991; Cheung et al. 1992; Tomizawa and Yamamoto 1992, 1993; Zwart et al. 1992; Liu and Casida 1993a; Tomizawa et al. 1996). Recent electrophysiological studies indicate that imidacloprid acts as an agonist on two distinct nAChR subtypes on cultured cockroach DUM neurons (Buckingham et al. 1997), an α-bungarotoxin (α-BGTx) sensitive nAChR with "mixed" nicotinic/muscarinic pharmacology and an α-BGTx insensitive nAChR. Such electrophysiological observations were supported by binding studies with [³H]imidacloprid in membrane preparations from the peach-potato aphid, *M. persicae*. Ligand competition studies revealed the presence of high- and low-affinity nAChR binding sites for imidacloprid in *M. persicae* (Lind et al. 1998).

The identification of multiple putative nAChR subunits by molecular cloning is consistent with the substantial diversity of insect nAChRs (Gundelfinger 1992). At present, at least five different subunits have been cloned from *D. melanogaster* (Schulz et al. 1998), from the locust *Locusta migratoria* (Hermsen et al. 1998) and from the aphid *M. persicae* (Huang et al. 1999). Despite the considerable number of subunits identified, only a few functional receptors were obtained after expression of different subunit combinations in *Xenopus* oocytes or cell lines. Initial work suggested that some subunits can form homo-oligomeric functional receptors when expressed in *Xenopus* oocytes. This was shown for Lα1 from *Schistocerca gregaria* (Marshall et al. 1990; Amar et al. 1995) and for the Mpα1 and Mpα2 from *M. persicae* (Sgard et al. 1998). However, the expression of these subunits was not very effective and generated only small inward currents (5–50 nA) following application of nicotine or acetylcholine. On the other hand, all three *Drosophila* α-subunits (ALS, SAD and Dα2) can form functional receptors in *Xenopus* oocytes when co-expressed with a chicken neuronal β2-subunit (Bertrand et al. 1994; Matsuda et al. 1998; Schulz et al. 1998), suggesting that additional insect nAChR subunits remain to be cloned. Radioligand binding studies using several *M. persicae* α-subunits co-expressed with a rat β2-subunit in the *Drosophila* S2 cell line also indicate pharmacological diversity in *M. persicae* (Huang et al. 1999). In these binding studies it was shown that imidacloprid-selective targets were formed by Mpα2 and Mpα3 but not Mpα1 subunits. These examples indicate that our understanding of the complexity of insect nAChR is still embryonic and that electrophysiology will play an essential role in determining the significance of certain subunit-combinations in the mode of action of neonicotinoid and other insecticidally active ligands.

4.1
Whole Cell Voltage Clamp of Native Neuron Preparations

The use of isolated neurons from insect CNS for electrophysiological studies is a suitable tool to investigate the mode of action of new insecticidal compounds which act on a range of neuronal target sites. Neuronal primary cell cultures from *H. virescens* (tobacco budworm) larvae, one of the most important lepidopteran pest species, is one of these suitable tools. The *Heliothis* neurons respond to the application of ACh with a fast inward current of up to

5 nA at a holding potential of –70 mV. The current reversed at a holding potential close to 0 mV, indicating the activation of nonspecific cation channels, i.e. nAChRs.

Figure 7 illustrates the whole cell currents elicited by application of 1 μM nithiazine and the second generation neonicotinoids acetamiprid, nitenpyram and clothianidin. All of these compounds act as agonists on the nAChR, but the potency and agonistic efficacy of these compounds were quite different. Imidacloprid and clothianidin were the most potent compounds in this *Heliothis* preparation with an EC$_{50}$ of 0.3 μM (Table 9). In the case of imidacloprid there was good agreement with electrophysiological measurements recorded from isolated cockroach neurons, where imidacloprid exhibited an EC$_{50}$ of 0.36 μM (Orr et al. 1997). Acetamiprid, nitenpyram and the natural toxin epibatidine exhibited an EC$_{50}$ of between 1 and 2 μM (Table 9).

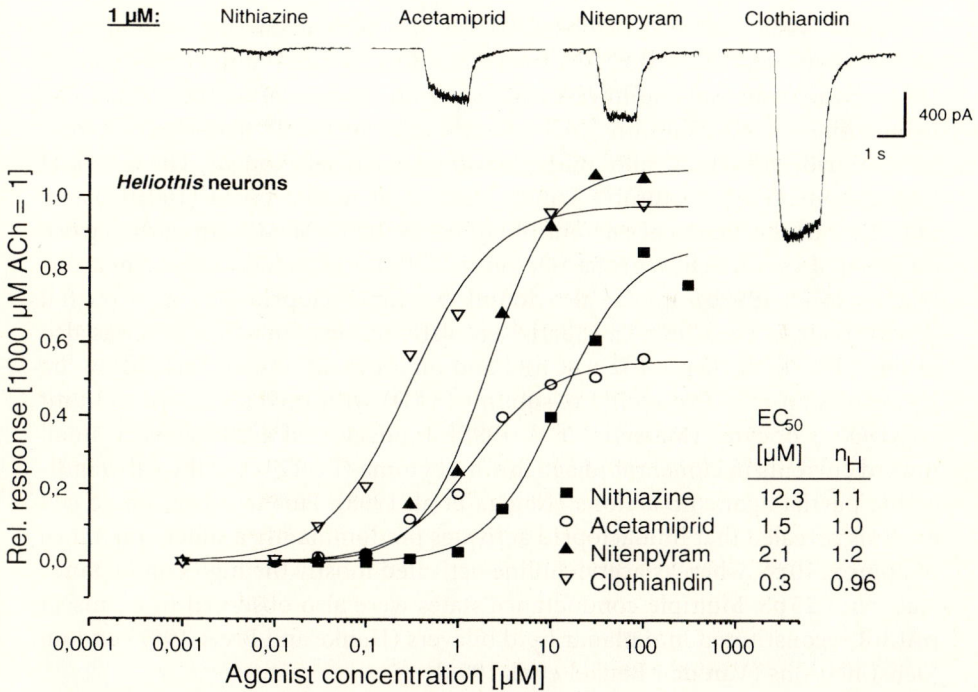

Fig. 7. Whole cell current responses of a neuron isolated from the CNS of *Heliothis virescens* after application of different neonicotinoids. The dose response curve was fitted by the Hill equation. All currents were first normalized to mean amplitudes elicited by 10 μM ACh before and after each test concentration was applied and than normalized to the relative amplitude elicited by 1000 μM ACh. EC$_{50}$-values given correspond to the half maximal activation of nAChR by each agonist. The Hill coefficient (n$_H$) of all tested compounds was close to 1. The upper *inset* shows the corresponding responses for the neonicotinoids at 1 μM (holding potential –70 mV). All currents were obtained from the very same neuron

Table 9. Comparison between electrophysiological and [³H]imidacloprid displacement potencies for different neonicotinoids and epibatidine on insect nAChRs. Electrophysiological data (EC_{50} and relative [agonist] efficacy) were obtained from neuron cell bodies isolated from the CNS of *Heliothis virescens*. EC_{50} and relative efficacy values represent the mean of n separate experiments on different neurons (±SD). The inhibition of [³H]imidacloprid binding to nAChR in housefly head membrane preparations by the compounds is expressed as a pI_{50}-value (for explanation of EC_{50} and pI_{50} refer to Fig. 8)

Compound	(*n*)	EC_{50} (μM)	Relative efficacy (1 mM ACh = 1)	[³H]imidacloprid (pI_{50})
Imidacloprid	4	0.31 ± 0.15	0.14 ± 0.02	9.3
Clothianidin	3	0.33 ± 0.03	0.99 ± 0.08	9.2
Acetamiprid	3	1.07 ± 0.37	0.56 ± 0.05	8.7
Nitenpyram	3	1.66 ± 0.38	0.98 ± 0.07	8.6
(±)-Epibatidine	3	1.69 ± 0.79	0.20 ± 0.05	6.2
Nithiazine	4	9.60 ± 3.20	0.79 ± 0.06	6.8

Similar values were also observed for the cockroach preparation with an EC_{50} between 0.5 and 0.7 μM for epibatidine and acetamiprid (Orr et al. 1997). Nithiazine had the lowest potency with an EC_{50} of about 10 μM. The neonicotinoids clothianidin and nitenpyram were full agonists, whereas acetamiprid, epibatidine and imidacloprid were partial agonists. The maximal response elicited by saturable concentrations of imidacloprid (100 μM) was only 15% of the maximal current obtained by 1000 μM ACh. In earlier work on isolated cockroach neurons (Orr et al. 1997) and isolated locust neurons (Nauen et al. 1999b), it was also found that imidacloprid acts as a partial agonist on insect nAChRs. The partial agonistic action of imidacloprid was also observed with chicken $\alpha 4\beta 2$ nAChRs and on a hybrid nAChR formed by the co-expression of a *Drosophila* α-subunit (SAD) with the chicken $\beta 2$-subunit in *Xenopus* oocytes (Matsuda et al. 1998). Imidacloprid activates very small inward currents in clonal rat phaeochromocytoma (PC 12) cells thus also indicating partial agonistic actions (Nagata et al. 1998). Furthermore, single cell analysis revealed that imidacloprid activates predominantly a subconductance of approx. 10 pS, whereas acetylcholine activated mostly the high conductance state with 25 pS. Multiple conductance states were also observed in an insect nAChR reconstituted into planar lipid bilayers (Hanke and Breer 1987) and on locust neurons (Van den Beukel et al. 1998).

4.1.1
Correlation Between Electrophysiology and Radioligand Binding Studies

There is a good correlation between electrophysiological measurements using isolated *Heliothis* neurons and radioligand binding studies on housefly head membranes regarding the affinity of different sets of ligands to nAChRs

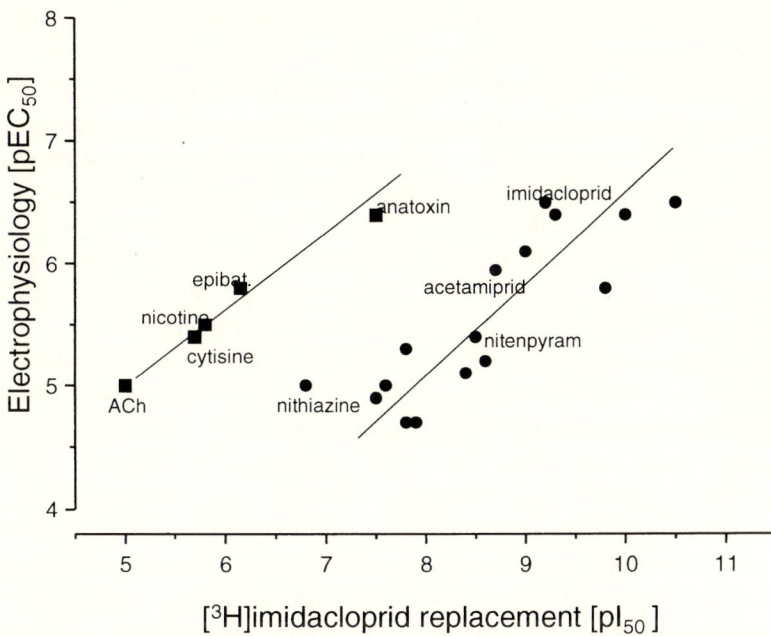

Fig. 8. Comparison between electrophysiological and binding potencies of different neonicotinoids and nicotinoids. Electrophysiological data were obtained from neuron cell bodies isolated from the CNS of *Heliothis virescens*. pEC_{50}-values ($= -\log M$) correspond to the half maximal activation of nAChR by each agonist. Binding data: pI_{50}-values ($= -\log M$) correspond to the concentration of cold ligand displacing 50% of bound [³H]imidacloprid from housefly head membranes

(Fig. 8). The good correlation for the neonicotinoids may reflect that houseflies (binding data) and tobacco budworms (electrophysiology) have similar binding sites for imidacloprid and related compounds. Biochemical investigations using [³H]imidacloprid as a radioligand in a number of different insect membrane preparations, e.g. from *Periplaneta americana*, *Lucilia sericata*, *D. melanogaster*, *Manduca sexta*, *H. virescens*, *Ctenocephalides felis*, *M. persicae* and *N. cincticeps*, indicate that many (if not all) insects have high specific imidacloprid binding sites with nearly identical Kd values of approx. 1–10 nM (Lind et al. 1998). However, it was also found that only the homopteran species seems to have an additional very high affinity binding site.

In general, the pI_{50}-values obtained by the displacement of specifically bound [³H]imidacloprid from housefly head membranes were 2–4 orders of magnitude higher than the electrophysiologically determined pEC_{50}-values obtained from isolated *Heliothis* neurons. Similar differences in biochemical binding and functional assay studies were also observed for different vertebrate nAChRs (for review see Holladay et al. 1997). A possible explanation for this phenomenon might be provided by considering the following: It is generally accepted that each nAChR can exist in multiple states, i.e. a resting state, active (open) state and one or more desensitized state(s), which have different affinities for ligands. The active state has a low affinity for ACh (Kd ranging from about 10 to 1000 μM) whereas the desensitized state(s) shows a higher affinity (Kd ranging from about 10 nM to 1 μM) for nicotinic ligands (Léna and Changeux 1993). The kinetics of the transitions between these states have been resolved for *Torpedo* nAChR in vitro. The rate of isomerization between the resting and active state lies on the μs- to ms-time scale and within the desensitized state over a time frame of seconds to minutes (see Changeux 1990 cited in Léna and Changeux 1993). Because binding studies are conducted over a time scale of minutes to hours they may reflect interaction with the desensitized state(s), whereas electrophysiological studies measure the interaction of ligands with the active state. Considering this, it is surprising that we found a direct correlation between electrophysiological and biochemical binding studies for natural alkaloids such as (–)-nicotine, cytisine, (±)-epibatidine and anatoxin A (Fig. 8). For these compounds the pI_{50}- and pEC_{50}-values were in the same range, with a good correlation. It is a general observation that natural alkaloids such as nicotine and epibatidine exhibit an agonistic potency in electrophysiological assays on isolated cockroach neurons (Bai et al. 1991; Buckingham et al. 1997; Orr et al. 1997), *Heliothis* neurons (Fig. 8) and locust neurons (Nauen et al. 1999b, see below) comparable to highly insecticidal neonicotinoids like imidacloprid. Matsuda et al. (1998), using a hybrid receptor formed by the co-expression of the *Drosophila* α-subunit SAD with the chicken β2, observed a comparable agonistic potency for both (–)-nicotine and imidacloprid; however, the agonistic potency of (+)-epibatidine was about two orders of magnitude greater. In contrast, all binding studies using [³H]imidacloprid on housefly head membranes (Liu and Casida 1993a,b; Liu et al. 1995; Wollweber and Tietjen 1999), whitefly preparations (Chao et al. 1997; Nauen and Tietjen, unpubl.) and *Myzus* preparations (Nauen et al. 1996, 1998a; Lind et al. 1998) indicate that imidacloprid has a considerably higher potency in replacing specifically bound [³H]imidacloprid than (–)-nicotine.

4.2
Agonists vs. Antagonists

The advantage of electrophysiological measurements compared with biochemical binding assays is their ability to distinguish between agonists and antagonists of the nAChR. This functional difference in the mode of action of compounds with high specificity for the nAChR is very important for insecticidal potency (Nauen et al. 1999b). Several natural compounds that act on the nAChR, e.g. dihydro-β-erythroidine, methyllycaconitine and paraherquamide, show considerably lower potential as insecticides, though their binding affinities to housefly receptors do not differ widely from those of the neonicotinoids (Fig. 3). Electrophysiological measurements from isolated housefly neurons revealed that compounds acting agonistically on the nAChR were in general insecticidal, whereas antagonistic compounds were mostly non-active (Nauen et al. 1999b; Wollweber and Tietjen 1999). In addition, fully antagonistic neonicotinoids have a very limited insecticidal efficacy (Ebbinghaus and Nauen, unpubl.). As described above, all of the neonicotinoids already introduced onto the market and those likely to be launched within the next few years are potent agonists on nAChRs. This general observation is supported by other studies which show a positive correlation between nerve activity induced in cockroach preparations and insecticidal activity against the green rice leafhopper for 19 neonicotinoids (Nishimura et al. 1994, 1998). Agonistic action on nAChRs causes first hyperexcitation and then paralysis as shown in many symptomology studies with imidacloprid on different insect species (Sone et al. 1994; Nauen 1995; Mehlhorn et al. 1999). Some antagonists, such as dihydro-β-erythroidine, alter the behavior of aphids and show a clear antifeedant effect in oral bioassays. Mortality observed in such short-term artificial feeding experiments (4 h) was high (>50%) for imidacloprid (1 mg/l), low for dihydro-β-erythroidine (100 mg/l, 4%) and also quite low for a mixture of the two (16%), suggesting that dihydro-β-erythroidine prevents the agonistic effects of imidacloprid necessary for irreversible symptoms of poisoning (Nauen et al. 1999b). Narahashi et al. (1998) concluded that in general most insecticides are neurotoxicants and cause various forms of hyperexcitation and paralysis leading to death of animals. Pyrethroids keep the voltage-gated sodium channel open causing hyperexcitation, whereas the insecticides dieldrin and hexachlorocyclohexane cause hyperexcitation by blocking the inhibitory effect of GABA receptors.

4.3
Receptor Subtypes in *Locusta migratoria*

As outlined above, and in accordance with observations on vertebrate nAChRs, a large body of evidence exists to support the remarkable heterogeneity of insect nAChRs. However, no detailed biophysical and pharmacological characterization yet exists of different nAChRs in native preparations. Such investigations were made using isolated locust neurons, because in earlier work it was observed that ACh responses varied between cells. Variation was observed not only in relation to peak amplitude, but also with respect to the time course, indicating the activation of different sub-populations of nAChRs in these preparations (Van den Beukel et al. 1998). Figure 9 clearly shows different time courses of ACh responses in these preparations. In addition to the rapidly desensitizing component present in most cells, some cells also exhibit a non-desensitizing response to ACh. Measurements on cells in which one or other type of ACh response predominated revealed considerable differences in the efficacy as well as in the potency of some agonists.

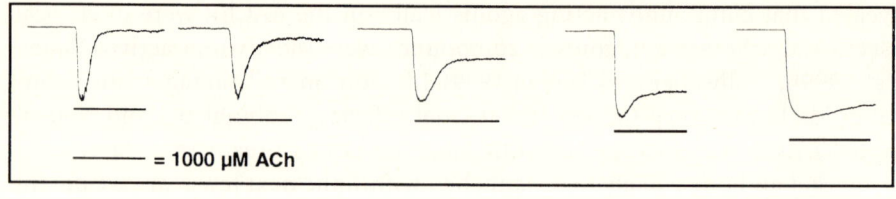

fast desensitizing component (A)

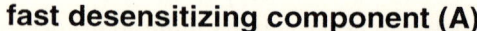

non-desensitizing component (B)

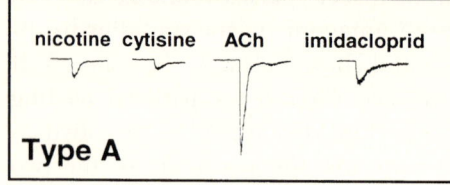

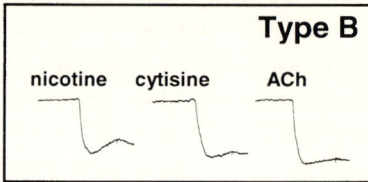

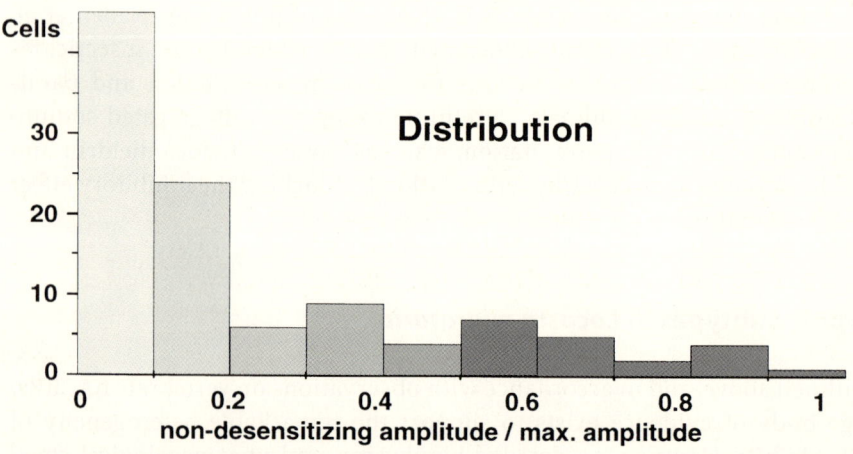

Fig. 9. Diversity of nicotinic whole cell currents on isolated locust neurons induced by ACh and some other ligands. *Upper trace* Examples of ACh induced currents on different cells. Most of the cells (see frequency distribution) respond to ACh with a transient inward current followed by a rapid and complete current decay still during the ACh application (type A currents). Those cells exhibiting a non-desensitizing response to ACh were defined as type B. *Lower trace* Maximal responses elicited by saturable concentrations of different agonists on cells with type A ACh responses (*left*) and on those with type B ACh responses (*right*)

On cells which had predominantly the fast desensitizing component (type A >90%), only ACh and the neonicotinoid clothianidin acted as full agonists, whereas anatoxin A, nicotine, cytisine and the neonicotinoid imidacloprid were all partial agonists. The agonist efficacy rank order was ACh>>anatoxin

Table 10. Efficacy of some nAChR agonists on isolated locust neurons in which ACh induced primarily a fast desensitizing component (nAChR subtype A) and a non-desensitizing component (nAChR subtype B), respectively

Compound	n	Rel. efficacy[a] Subtype A	n	Rel. efficacy[a] Subtype B
Cytisine	12	0.14 ± 0.08	2	1.01
(−)-Nicotine	7	0.23 ± 0.08	4	0.93 ± 0.10
Imidacloprid	6	0.27 ± 0.08	2	0.79
Anatoxin	4	0.57 ± 0.03	2	0.96

[a] Relative efficacy describes the maximal agonistic response (±SD) evoked by saturable concentrations of the test compound compared with ACh (1000 µM) on the very same neuron

A≫imidacloprid = nicotine (Table 10). Cytisine was the least effective agonist on these cells, eliciting a maximal response about ten times less than that observed for 1 mM ACh. Anatoxin A was the most potent agonist, with an EC_{50} of 0.4 µM.

On cells which had no fast desensitizing component (type B) ACh, anatoxin A, cytisine, nicotine and imidacloprid elicited nearly the same maximal responses at saturation concentrations. Anatoxin A was the most potent agonist with an EC_{50} <0.1 µM. On the very same cell, nicotine and cytisine exhibited practically the same EC_{50} of ~0.5 µM. ACh had the lowest potency with an EC_{50} of approx. 5 µM, and imidacloprid showed an EC_{50} of 3 µM. Taken together, these results gave a rank order of potency of anatoxin A>nicotine = cytisine>imidacloprid>ACh. With the exception of imidacloprid, the EC_{50} values for all agonists tested were lower than the corresponding EC_{50} values obtained from cells with mostly rapidly desensitizing ACh-induced currents. However, the most remarkable pharmacological difference in which one or the other type of ACh response predominated is the clearly distinct affinity for nicotine, which can differ by a factor of 100. In summary, our data (Nauen et al. 1999b) and those from others (Van den Beukel et al. 1998) confirm the existence of functional nAChRs in locust neurons with different ion channel properties and distinct affinity for ligands, suggesting at least two independent populations of nAChRs in the insect nervous system. Distinct nAChR receptor populations might be involved in different (neuro)physiological processes and can therefore be considered as different target sites when it comes to highly specialized insecticidal ligands binding to one population but not to the other.

References

Amar M, Thomas P, Wonnacott S, Lunt GG (1995) A nicotinic acetylcholine receptor subunit from insect brain forms a non-desensitizing homo-oligomeric nicotinic acetylcholine receptor when expressed in *Xenopus* oocytes. Neurosci Lett 199:107–110

Anonymous (1999) Agrow no 281, p 22

Bai D, Lummis SCR, Leicht W, Breer H, Sattelle DB (1991) Actions of imidacloprid and a related nitromethylene on cholinergic receptors of an identified insect motor neurone. Pestic Sci 33:197–204

Benson JA (1989) Insect nicotinic acetylcholine receptors as targets for insecticides. In: Mc Farlane NR, Farnham AW (eds) Progress and prospects in insect control. BCPC Monograph 43. UK British Crop Protection Council, pp 59–70

Bertrand D, Ballivet M, Gomez M, Bertrand S, Phannavong B, Gundelfinger ED (1994) Physiological properties of neuronal nicotinic receptors reconstituted from the vertebrate β2 subunit and *Drosophila* α subunits. Eur J Neurosci 6:869–875

Breer H (1988) Receptors for acetylcholine in the nervous system of insects. In: Lunt GG (ed) Neurotox 88 Molecular basis of drug and pesticide action. Excerpta Medica, Amsterdam, pp 301–309

Breer H, Sattelle DB (1987) Molecular properties and functions of insect acetylcholine receptors. J Insect Physiol 33:771–790

Buckingham SD, Lapied B, Le Corronc H, Grolleau F, Sattelle DB (1997) Imidacloprid actions on insect neuronal acetylcholine receptors. J Exp Biol 200:2685–2692

Chao SL, Dennehy TJ, Casida JE (1997) Whitefly (Hemiptera: Aleyrodidae) binding site for imidacloprid and related insecticides: a putative nicotinic acetylcholine receptor. J Econ Entomol 90(4):879–882

Cheung H, Clarke BS, Beadle DJ (1992) A patch-clamp study of the action of a nitromethylene heterocycle insecticide on cockroach neurones growing *in vitro*. Pestic Sci 34:187–193

David JA, Sattelle DB (1984) Actions of cholinergic pharmacological agents on the cell body membrane of the fast coxal depressor motoneurone of the cockroach, *Periplaneta americana*. J Exp Biol 108:119–136

Dunbar SJ, Goodchild JA, Cutler PM (1998) Actions of natural products on insect nicotinic receptors. 9th International Congress of Pesticide chemistry, Book of Abstracts 1:4B-040

Eastham HM, Lind RJ, Eastlake JL, Clarke BS, Towner CP, Reynolds SE, Wolstenholme AJ, Wonnacott S (1998) Characterisation of a nicotinic acetylcholine receptor from the insect *Manduca sexta*. Eur J Neurosci 10:879–889

Elbert A, Nauen R, Leicht W (1998) Imidacloprid, a novel chloronicotinyl insecticide: biological activity and agricultural importance. In: Ishaaya I, Degheele D (eds) Insecticides with novel modes of action: mechanism and application. Springer, Berlin Heidelberg New York, pp 50–74

Elbert A, Becker B, Hartwig J, Erdelen C (1991) Imidacloprid – a new systemic insecticide. Pflanzenschutz-Nachr Bayer 44:113–136

Elbert A, Nauen R, Cahill M, Devonshire AL, Scarr A, Sone S, Steffens R (1996) Resistance management for chloronicotinyl insecticides using imidacloprid as an example. Pflanzenschutz Nachr Bayer 49:5–54

Goodman CS, Spitzer NC (1980) Embryonic development of neurotransmitter receptors in grasshoppers. In: Sattelle DB, Hall LM, Hidebrand JG (eds) Receptors for neurotransmitters, hormones and pheromones in insects. Elsevier, Amsterdam, pp 195–307

Gundelfinger ED (1992) How complex is the nicotinic receptor system of insects? TINS 15(6):206–211

Hanke W, Breer H (1986) Channel properties of an insect neuronal acetylcholine receptor protein reconstituted in planar lipid bilayers. Nature 321:171–174

Harrow ID, Sattelle DB (1983) Acetylcholine receptors on the cell body membrane of giant interneurone 2 in the cockroach *Periplaneta americana*. J Exp Biol 105:339–350

Hermsen B, Stetzer E, Thees R, Heiermann R, Schrattenholz A, Ebbinghaus U, Kretschmer A, Methfessel C, Reinhardt R, Maelicke A (1998) Neuronal nicotinic receptors in the locust *Locusta migratoria*. J Biol Chem 273(29):18394–18404

Holladay MW, Dart MJ, Lynch JK (1997) Neuronal nicotinic acetylcholine receptors as targets for drug discovery. J Med Chem 40:4169–4194

Huang Y, Williamson MS, Devonshire AL, Windass JD, Lansdell SJ, Millar NS (1999) Molecular characterisation and imidacloprid selectivity of nicotinic acetylcholine receptor subunits from the peach-potato aphid *Myzus persicae*. J Neurochem 73:380–389

Kagabu S (1996) Studies on the synthesis and insecticidal activity of neonicotinoid compounds. Pestic Sci 46:231–239

Kagabu S (1999) Discovery of chloronicotinyl insecticides. In: Yamamoto I, Casida JE (eds) Nicotinoid insecticides and nicotinic acetylcholine receptor. Springer, Berlin Heidelberg New York, pp 91–106

Kagabu S, Moriya K, Shibuya K, Hattori Y, Tsuboi S, Shiokawa K (1992) 1-(6-Halonicotinyl)-2-nitromethylene-imidazolidines as potential new insecticides. Biosci Biotechnol Biochem 56:326–363

Kishida H, Sakamoto N, Umeda K, Fujimoto H (1992) Preparation of nitropyrimidine derivatives as insecticides. Jpn Kokai Tokkyo Koho JP04,173,788, Chem Abstr 118:22251q

Kodaka K, Kinoshita K, Wakita T, Kawahara N, Yasui N (1998) MTI 446: a novel systemic insect control compound. Proc Brighton Crop Protection Conf-Pests and Diseases, pp 21–26

Kollmeyer WD, Flattum RF, Foster JP, Powell JE, Schroeder ME, Soloway SB (1999) Discovery of the nitromethylene heterocycle insecticides. In: Yamamoto I, Casida JE (eds) Nicotinoid insecticides and nicotinic acetylcholine receptor. Springer, Berlin Heidelberg New York, pp 71–91

Krause RM, Buisson B, Bertrand S, Corringer PJ, Galzi JL, Changeux JP, Bertrand D (1998) Ivermectin: a positive allosteric effector of the α7 neuronal nicotinic acetylcholine receptor. Mol Pharmacol 53:283–294

Latli B, Tomizawa M, Casida JE (1997) Synthesis of a novel [^{125}I]neonicotinoid photoaffinity probe for the *Drosophila* nicotinic acetylcholine receptor. Bioconjugate Chem 8:7–14

Leech CA, Jewess P, Marshall J, Sattelle DB (1991) Nitromethylene actions on in situ and expressed insect nicotinic acetylcholine receptors. FEBS Lett 290:90–94

Léna C, Changeux J-P (1993) Allosteric modulations of the nicotinic acetylcholine receptor. TINS 16(5):181–186

Leicht W (1996) Imidacloprid – a chloronicotinyl insecticide: biological activity and agricultural significance. Pflanzenschutz-Nachr Bayer 49:71–84

Lind RJ, Clough MS, Reynolds SE, Earley FGP (1998) [^{3}H]imidaclorid labels high- and low-affinity nicotinic acetylcholine receptor-like binding sites in the aphid *Myzus persicae* (Hemiptera: Aphididae). Pestic Biochem Physiol 62:3–14

Lind RJ, Clough MS, Earley FGP, Wonnacott S, Reynolds SE (1999) Characterisation of multiple α-bungarotoxin binding sites in the aphid *Myzus persicae* (Hemiptera: Aphididae). Insect Biochem Mol Biol 29:979–988

Liu M-Y, Casida JE (1993a) Relevance of [^{3}H]imidaclorid binding site in housefly head acetylcholine receptor to insecticidal activity of 2-nitromethylene- and 2-nitroimino-imidazolidines. Pestic Biochem Physiol 46:200–206

Liu M-Y, Casida JE (1993b) High affinity binding of [^{3}H]imidaclorid in the insect acetylcholine receptor. Pestic Biochem Physiol 46:40–46

Liu M-Y, Latli B, Casida JE (1995) Imidacloprid binding site in *Musca* nicotinic acetylcholine receptor: interactions with physostigmine and a variety of nicotinic agonists with chloropyridyl and chlorothiazolyl substituents. Pestic Biochem Physiol 52:170–181

Maienfisch P, Sell L (1992) Preparation of 3-(heterocyclylmethyl)-4-iminoperhydro-1,3,5-oxadiazine derivatives as pesticides. Eur Pat Appl EP 580553 A2 940126

Maienfisch P, Brandl F, Kobel W, Rindlisbacher A, Senn R (1999) CGA 293'343: a novel, broad-spectrum neonicotinoid insecticide. In: Yamamoto I, Casida JE (eds) Nicotinoid insecticides and the nicotinic acetylcholine receptor. Springer, Berlin Heidelberg New York, pp 177–213

Marshall J, Buckingham SD, Shingai R, Lunt GG, Goosey MW, Darlison MG, Sattelle DB, Barnard EA (1990) Sequence and functional expression of a single α subunit of an insect nicotinic acetylcholine receptor. EMBO J 9:4391–4398

Matsuda K, Buckingham SD, Freeman JC, Squire MD, Baylis HA, Sattelle DB (1998) Effects of the α subunit on imidacloprid sensitivity of recombinant nicotinic acetylcholine receptors. Br J Pharmacol 123:518–524

Mehlhorn H, Mencke N, Hansen O (1999) Effects of imidacloprid on adult and larval stages of the flea *Ctenocephalides felis* after in vivo and in vitro application: a light- and electron-microscopy study. Parasitol Res 85:625–637

Minamida I, Iwanaga K, Tabuchi T (1993a) Synthesis and insecticidal activity of acyclic nitroethene compounds containing heteroarylmethylamino groups. J Pestic Sci 18:41–48

Minamida I, Iwanaga K, Tabuchi T (1993b) Synthesis and insecticidal activity of acyclic nitroethene compounds containing a 3-pyridylmethylamino group. J Pestic Sci 18:31–40

Miyazawa A, Fujiyoshi Y, Stowell M, Unwin N (1999) Nicotinic acetylcholine receptor at 4.6 ang resolution: transverse tunnels in the channel wall. J Mol Biol 288:765–786

Mongan NP, Baylis HA, Adcock C, Smith GR, Sansom MSP, Sattelle DB (1998) An extensive and diverse gene family of nicotinic acetylcholine receptor α subunits in Caenorhabditis elegans. Recept Channels 6:213–228

Moriya K, Shibuya K, Hattori Y, Tsuboi S, Shiokawa K, Kagabu S (1993) 1-Diazinylmethyl-2-nitromethylene-imidazolidines and 2-nitroimino-imidazolidines as new potential insecticides. J Pestic Sci 18:119–123

Mullins JW (1993) Imidacloprid – a new nitroguanidine insecticide. ACS Symp Ser 254:183–198

Nagata K, Aistrup GL, Song J-H, Narahashi T (1996) Subconductance-state currents generated by imidacloprid at the nicotinic acetylcholine receptor in PC 12 cells. NeuroReport 7(5):1025–1028

Nagata K, Song J-H, Shono T, Narahashi T (1998) Modulation of the neuronal nicotinic acetylcholine receptor-channel by the nitromethylene heterocycle imidacloprid. J Pharmacol Exp Ther 285:731–738

Nauen R (1995) Behaviour modifying effects of low systemic concentrations of imidacloprid on Myzus persicae with special reference to an antifeeding response. Pestic Sci 44:145–153

Nauen R, Strobel J, Tietjen K, Otsu Y, Erdelen C, Elbert A (1996) Aphicidal activity of imidacloprid against a tobacco feeding strain of Myzus persicae (Homoptera: Aphididae) from Japan closely related to Myzus nicotianae and highly resistant to carbamates and organophosphates. Bull Entomol Res 86:165–171

Nauen R, Tollo B, Tietjen K, Elbert A (1998a) Antifeedant effect, biological efficacy and high affinity binding of imidacloprid to acetylcholine receptors in Myzus persicae and Myzus nicotianae. Pestic Sci 51:52–56

Nauen R, Tietjen K, Wagner K, Elbert A (1998b) Efficacy of plant metabolites of imidacloprid against Myzus persicae and Aphis gossypii (Homoptera: Aphididae). Pestic Sci 52:53–57

Nauen R, Reckmann U, Armborst S, Stupp HP, Elbert A (1999a) Whitefly-active metabolites of imidacloprid: biological efficacy and translocation in cotton plants. Pestic Sci 55:265–271

Nauen R, Ebbinghaus U, Tietjen K (1999b) Ligands of the nicotinic acetylcholine receptor as insecticides. Pestic Sci 55:608–610

Narahashi T, Ginsburg KS, Nagata K, Song JH, Tatebayashi H (1998) Ion channels as targets for insecticides. NeuroTox 19:581–590

Nishimura K, Kanda Y, Okazawa A, Ueno T (1994) Relationship between insecticidal and neurophysiological activities of imidacloprid and related compounds. Pestic Biochem Physiol 50:51–59

Nishimura K, Tanaka M, Iwaya K, Kagabu S (1998) Relationship between insecticidal and nerve-excitatory activities of imidacloprid and its alkylated congeners at the imidazoline NH site. Pestic Biochem Physiol 62:172–178

Orr N, Shaffner J, Watson GB (1997) Pharmacological characterization of an epibatidine binding site in the nerve cord of Periplaneta americana. Pestic Biochem Physiol 58:183–192

Pennisi E (1999) Fruit fly researchers sign pact with Celera. Science 283:767

Pitman RM (1971) Transmitter substances in insects: a review. Comp Gen Pharmacol 2:347–371

Rajappa S (1981) Nitroenamines: preparation, structure and synthetic potential. Tetrahedron Lett 37:1453–1480

Restifo LL, White K (1990) Molecular and genetic approaches to neurotransmitter and neuromodulator systems in Drosophila. Adv Insect Physiol 22:115–219

Salgado VL (1998) Studies on the mode of action of spinosad: insect symptoms and physiological correlates. Pestic Biochem Physiol 60:91–102

Salgado VL, Watson GB, Sheets JJ (1997) Studies on the mode of action of spinosad, the active ingredient in tracer insect control. Proc Beltwide Cotton Conf 2:1082–1084

Sattelle DB, Harrow ID, Hue B, Pelhate M, Gepner JI, Hall LM (1983) α-Bungarotoxin blocks excitatory synaptic transmission between cercal sensory neurones and giant interneurones. J Exp
Biol 107:473–489

Sattelle DB, Buckingham SD, Wafford KA, Sherby SM, Barkry NM, Eldefrawi AT, Eldefrawi ME,
May TE (1989) Actions of the insecticide 2(nitromethylene)tetrahydro-1,3-thiazine on insect
and vertebrate nicotinic acetylcholine receptors. Proc R Soc Lond B 237:501–514

Schröder ME, Flattum RF (1984) The mode of action and neurotoxic properties of the nitromethylene heterocycle insecticides. Pestic Biochem Physiol 22:148–160

Schulz R, Sawruk E, Mülhardt C, Bertrand S, Baumann A, Phannavong B, Betz H, Bertrand D,
Gundelfinger ED, Schmitt B (1998) Da3, a new functional α subunit of nicotinic acetylcholine
receptors from Drosophila. J Neurochem 71:853–862

Schuster R, Phannavong B, Schroeder C, Gundelfinger ED (1993) Immunohistochemical localization of a ligand-binding and a structural subunit of nicotinic acetylcholine receptors in the
central nervous system of Drosophila melanogaster. J Comp Neurol 335:149–162

Sgard F, Fraser SP, Katkowska MJ, Djamgoz MBA, Dunbar SJ, Winddass JD (1998) Cloning and
functional characterisation of two novel nicotinic acetylcholine receptor α subunits from the
insect pest Myzus persicae. J Neurochem 71:903–912

Shiokawa K, Tsuboi S, Moriya K, Kagabu S (1995) Chloronicotinyl insecticides: development of
imidacloprid. In: Ragsdale NN, Kearney PC, Plimmer JR (eds) 8th international congress of
pesticide chemistry – option 2000. ACS Conf Proc Ser, pp 49–59

Soloway SB, Henry AC, Kollmeyer WD, Padgett WM, Powell JE, Roman SA, Tieman CH, Corey RA,
Horne CA (1979) Nitromethylene insecticides. In: Geissbühler H, Brooks GT, Kearney PC (eds)
Advances in pesicide science, part 2. Pergamon Press, Oxford, pp 206–217

Sone S, Nagata K, Tsuboi S, Shono T (1994) Toxic symptoms and neural effect of a new class of
insecticide, imidacloprid, on the American cockroach, Periplaneta americana (L.). J Pestic Sci
19:69–72

Tabuchi T, Fusaka T, Iwanaga K (1994) Synthesis and insecticidal activity of acyclic nitroethene
compounds containing a (6-substituted)-3-pyridylamino group. J Pestic Sci 19:119–125

Takahashi H, Mitsui J, Takakusa N (1992) NI-25, a new type of systemic and broad spectrum insecticide. Proc Brighton Crop Protection Conf Pest Diseases 1:89–96

Thompson G, Hutchins S (1999) Spinosad. Pesticide Outlook 4:78–81

Tomizawa M, Yamamoto I (1992) Binding of nicotinoids and the related compounds to the insect
nicotinic acetylcholine receptor. J Pestic Sci 17:231–236

Tomizawa M, Yamamoto I (1993) Structure-activity relationships of nicotinoids and imidacloprid
analogs. J Pestic Sci 18:91–98

Tomizawa M, Latli B, Casida JE (1996) Novel neonicotinoid-agarose affinity column for Drosophila
and Musca nicotinic acetylcholine receptors. J Neurochem 67:1669–1676

Tomizawa M, Latli B, Casida JE (1999) Structure and function of insect nicotinic acetylcholine
receptors studied with nicotinoid insecticide affinity probes. In: Yamamoto I, Casida JE (eds)
Nicotinoid insecticides and the nicotinic acetylcholine receptor. Springer, Berlin Heidelberg
New York, pp 271–292

Uneme H, Iwanaga K, Higuchi N, Kando Y, Okauchi T, Akayama A, Minamida I (1998) Synthesis
and insecticidal activity of nitroguanidine derivatives. 9th IUPAC Congress on Pesticide
chemistry, Book of Abstracts 1:1D-009

van den Beukel I, van Kleef RGDM, Zwart R, Oortgiesen M (1998) Physiostigmine and acetylcholine differentially activate nicotinic receptor subpopulations in Locusta migratoria
neurons. Brain Res 789:263–273

Wollweber D, Tietjen K (1999) Chloronicotinyl insecticides: a success of the new chemistry. In:
Yamamoto I, Casida JE (eds) Nicotinoid insecticides and the nicotinic acetylcholine receptor.
Springer, Berlin Heidelberg New York, pp 109–125

Zwart R, Oortgiesen M, Vijverberg HP (1992) The nitromethylene heterocycle 1-(pyridin-3-
ylmethyl)-2-nitromethylene-imidazolidine distinguishes mammalian from insect nicotinic
receptor subtypes. Eur J Pharmacol 228:165–169

Ecdysteroid and Juvenile Hormone Receptors: Properties and Importance in Developing Novel Insecticides

S. R. Palli[1] and A. Retnakaran[2]

1
Introduction

From time immemorial man has been in a state of constant confrontation with insects, his main competitor for food and fiber. The first recorded use of a chemical, sulfur, to control insects can be found in the writings of Homer (before 1000 B.C.). Later, in A.D. 79, Pliny recommended arsenic (Cremlyn 1978). These first-generation inorganic insecticides were later replaced by powerful second-generation organic insecticides, the most notable one being DDT, which was developed by Paul Müller in 1940 (West and Campbell 1950). While the organochlorines enjoyed tremendous success in insect control, their impact on the environment was dramatic, a fact brought to light by Rachel Carson (1962). An intense search for alternatives less harmful to the environment was initiated in laboratories around the world. Carol Williams (1967), the well-known insect endocrinologist, coined the term "third-generation pesticides" for insect hormones such as juvenile hormone and suggested that they could be insect specific and cause little or no harm to the environment. Since then there has been steady progress towards the development of narrow spectrum insecticides that are target specific and environmentally safe. There has also been a surge of interest in biological control agents, led by *Bacillus thuringiensis* (Bt). The underlying theme has been to search for unique biochemical sites in insect pests that can be selectively targeted. Various approaches are currently being examined; one of these is the hormonal regulation of key developmental processes in insects.

2
Ecdysteroids

Endocrine approaches to insect control have focused largely on adversely interfering with molting and metamorphosis. While there are many hormones involved in regulating growth and metamorphosis in insects, the two princi-

[1] Rohm and Haas Research Laboratories, 727 Norristown Road, Spring House, PA, 19477, USA
[2] Canadian Forest Service, Great Lakes Forestry Centre, 1219 Queen Street East, Sault Ste. Marie, Ontario, Canada, P6A 5M7

Isaac Ishaaya (Ed.): Biochemical Sites of
Insecticide Action and Resistance
© Springer-Verlag Berlin, Heidelberg 2001

pal players are the steroid, molting hormone, ecdysone, and the terpenoid "status-quo hormone", juvenile hormone (JH). The physiologically active molting hormone in most insects is 20-hydroxyecdysone (20 E). The generic term "ecdysteroid", which denotes compounds that have molting hormone activity (Goodwin et al. 1978), includes more than 250 polyhydroxysterols that occur both in plants and animals (Lafont 1997; Dinan et al. 1999).

2.1
Biology, Endocrinology and Molecular Biology

The importance of the thoracic center for inducing the molting process in insects was discovered in 1931 (Hachlow), and was later traced to the paired prothoracic glands located in the prothoracic segment (Fukuda 1940). The molting hormone secreted by these glands was isolated from 500 kg of silk worm, *Bombyx mori*, pupae in 1954 by Butenandt and Karlson, and its structure was later elucidated as a polyhydroxylated steroid by Karlson in 1967. During development, when a larval insect reaches a critical size, proprioreceptors on the body wall transmit the "stretch" signal to the brain via the nervous system. In response to this signal the lateral neurosecretory cells in the brain secrete a neuropeptide, the prothoracicotropic hormone (PTTH), which is axonally translocated and released from the corpus allatum into the hemolymph. The PTTH triggers the prothoracic glands to secrete ecdysone (see review by Steel and Davey 1985) (Fig. 1).

The prothoracic glands are a pair of translucent glands that lie on the dorsal surface of the lateral tracheal trunks in the prothoracic segment. They resemble a bunch of grapes, made up of large cells that have deeply infolded plasma membrane, dense endoplasmic reticulum and elaborate mitochondria (Nijhout 1994). Upon stimulation by PTTH, the prothoracic glands secrete ecdysone into the hemolymph, which is hydroxylated to the active hormone, 20-hydroxyecdysone (20 E), and this conversion, although it occurs primarily in the epidermis can also occur in other tissues such as the salivary glands, ovaries and imaginal discs (O'Connor 1985). Ecdysteroids have also been shown to be synthesized outside the prothoracic glands in other tissues such as the ovaries and abdominal integument (Hoffmann and Gerstenlauer 1997). Ecdysteroids act through DNA binding nuclear receptors and act at the genetic level regulating the sequential expression of genes, which will be addressed in greater detail later (Riddiford and Truman 1993). Ecdysteroids play a critical role in regulating growth, development, metamorphosis, and reproduction.

A series of pioneering experiments by Clever and Karlson (1960) and Ashburner et al. (1974) showed that in response to 20 E, a sequence of specific puffs are induced in the polytene chromosomes of the salivary glands in *Chironomus tentans* and *Drosophila melanogaster* and this sequence is a mirror image of the events that occur during the molting process. Ashburner et al. (1974)

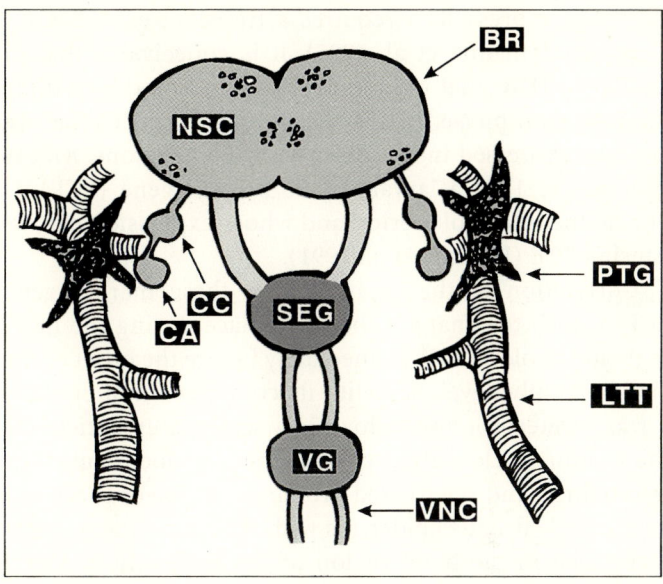

Fig. 1. Generalized view of the neuroendocrine system in insects based on our observations in the spruce budworm, *Choristoneura fumiferana*. *BR* Brain; *CA* corpus allatum; *CC* corpus cardiacum; *LTT* lateral tracheal trunk; *NSC* neuroscretory cells; *PTG* prothoracic gland; *SEG* subesophageal ganglion; *VNC* ventral nerve cord

hypothesized that the ecdysone-receptor complex induces the transcription of "early genes" whose protein products in turn induce the transcription of "late genes" as well as suppress the expression of "early genes". During the past two decades, Ashburner's hypothesis has become the central paradigm for explaining the mode of action of ecdysteroids.

During each larval stage, PTTH appears as a sharp peak in the hemolymph and, immediately after this peak, a wider ecdysteroid peak which extends over a few days is observed. The whole series of developmental changes that occur during the molting process extends from the start of one ecdysteroid peak in one larval instar to the beginning of the ecdysteroid peak in the subsequent instar. The entire molting cycle can be divided into three phases: an early phase, which marks the beginning of the molting cycle at the onset of ecdysteroid secretion; a late phase which occurs at the end of the ecdysteroid peak and extends to the ecdysis of the old cuticle; and finally the intermolt phase, which starts from the newly molted larva and ends immediately before the rise of the ecdysteroid peak. During the molting cycle, a sequential cascade of 20 E-regulated gene expression and repression takes place. In the spruce budworm, *Choristoneura fumiferana*, at 20 E rise, genes such as *Choristoneura* hormone receptor 75 (CHR75, Palli et al. 1997) and *Choristoneura* hormone receptor 3 (CHR3, Palli et al. 1996) are expressed along with perhaps a few other early genes such as *broad complex* and E74. At 20 E fall, dopadecarboxylase

(DDC) is expressed; it requires a transient exposure to 20 E followed by its clearance (Hiruma et al. 1995). It is conceivable that there are other genes similar to DDC that require exposure followed by withdrawal of 20 E. During the intermolt phase, 20 E is absent and the genes that were repressed by 20 E are now expressed in the absence of the hormone. An example of this type of gene is the 14 kDa larval cuticle protein gene (LCP14), which is expressed during the intermolt period and whose expression can be inhibited by administering 20 E (Hiruma et al. 1991).

Correlation of the expression of these marker genes, CHR3, DDC and LCP14, with the changes that take place during the molting cycle reveals the regulatory role of 20 E. Immediately before the onset of the 20 E peak, the larval cuticle is fully developed with its complement of cuticulin layer covered with a waxy waterproofing followed by a dense epicuticle. Below this layer lies a thick endocuticle with its characteristic endocuticular lamellae made of chitin microfibrils and sclerotized proteins. This layer is often melanized. Under the endocuticle lies the epidermis with the apical microvilli of the epidermal cells embedded in the basal region of the endocuticle. Muscles of the body wall below the epidermis are anchored into the endocuticle. For a detailed description, see Locke (1998). At the rise of the ecdysone peak, the dissolution of the old cuticle is initiated. Ecdysial droplets are secreted by the epidermis into the base of the endocuticle and a new cuticulin layer is formed to protect the epidermal cells. The ecdysial droplets contain chitinases and proteases and by their combined action progressively dissolve the endocuticle and create an endocuticular space between the old cuticle and the newly formed cuticulin layer. This separation of the old partially digested cuticle from the newly formed cuticle is known as apolysis. The epicuticle layer immediately below the cuticulin layer is then laid down by the epidermis. The epidermal cells also divide during this time, expanding the epidermis. The dissolution of the old cuticle results in the release of the muscles anchored in the endocuticle. The larva lies quiescent until the muscles are re-anchored into the new cuticle. The overt phenotypic effect characteristic of this stage is the head capsule slippage. The formation of the new head capsule pushes the old head capsule away leaving a clear band between the two. At the end of the fall of the ecdystroid peak, the DDC gene is expressed whose enzyme product is necessary for the synthesis of tanning precursors (Hiruma et al. 1995).

There may be other genes that are expressed in a similar fashion. The pherate larva inside the shell of the old cuticle is ready to ecdyse. The eclosion hormone, which is necessary to provide the nervous stimulus to initiate the peristaltic eclosion response, is stored in the paired protodeal nerves and is released in the absence of 20 E (Truman 1990, 1992). At the time of ecdysis, the cuticle is thin and untanned. The hormone bursicon, which acts on the epidermal cells to secrete a variety of phenolic compounds necessary for sclerotization, is released (Sugumaran 1988; Hopkins and Kramer 1992). The cuticle

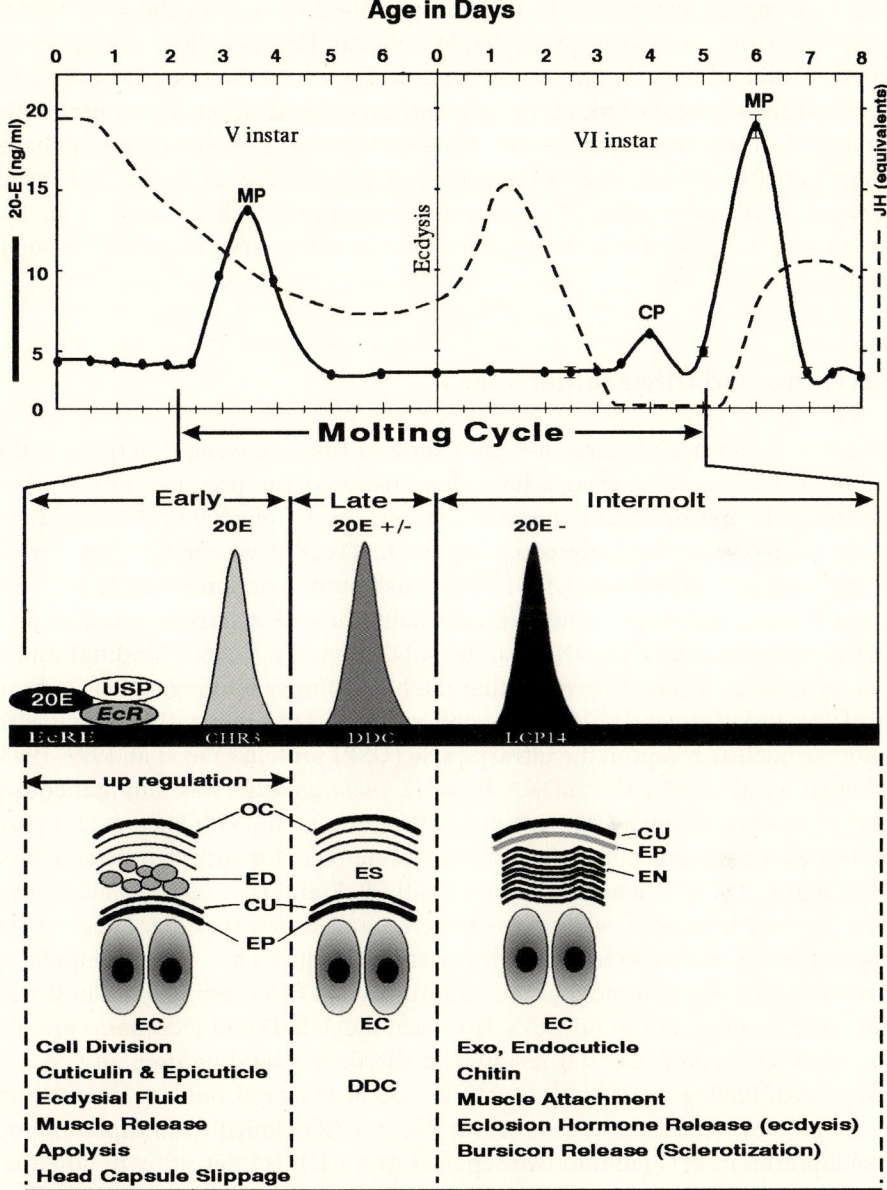

Fig. 2. Diagrammatic representation of the major events that occur during a typical molting cycle in the spruce budworm, *Choristoneura fumiferana*. *CHR3* Choristoneura hormone receptor 3; *CP* commitment peak of 20 E; *CU* cuticulin; *DDC* dopadecarboxylase; *20 E* 20-hydroxyecdysone; *EC* epidermal cells; *EcR* ecdysone recptor; *EcRE* ecdysone response element; *ED* ecdysial droplets; *EN* endocuticle; *EP* epicuticle; *ES* ecdysial space; *Exo* exocuticle; *JH* juvenile hormone; *LCP14* larval cuticular protein 14 kDa; *MP* molt peak of 20 E; *USP* ultraspiracle; *OC* old cuticle

becomes rapidly elaborated in the newly molted larva with the synthesis of chitin and the formation of the endocuticular lamellae (Retnakaran et al. 1997b; Palli and Retnakaran 1999). Structural proteins such as LCP14 are synthesized for the cross-linking process during sclerotization. The muscles are anchored in the new cuticle by this time and the larva enters a growth phase. Soon the 20 E is released and the cycle is repeated (Fig. 2). In the last instar larva, a small 20 E pulse, the programming peak which is secreted in the absence of JH prior to the molt peak, programs the larva to go into a pupal molt (Riddiford 1991).

2.2
Receptors and Other Target Sites

Ever since Clever and Karlson (1960) showed that ecdysone functions at the genomic level, several groups have demonstrated the presence of ecdysone receptors in cytosolic and nuclear extracts using biochemical techniques. Three decades later the ecdysone receptor (EcR) cDNA was finally cloned from *D. melanogaster* (Koelle et al. 1991). EcR was found to be a member of a steroid hormone receptor superfamily and contains the characteristic five domains, A/B (transactivation), C (DNA binding), D (hinge), E (ligand binding) and F domains. Soon it was discovered that the high affinity binding of EcR to both the DNA and the ligand (20 E) depended on the heterodimerization of EcR with another nuclear receptor, the ultraspiracle (USP) protein (Yao et al. 1992, 1993; Thomas et al. 1993). USP cDNA from *D. melanogaster* was simultaneously cloned by three groups (Henrich et al. 1990; Oro et al. 1990; Shea et al. 1990). USP is also a member of the steroid hormone receptor superfamily and contains A/B, C, D and E domains. Subsequently, EcR and USP cDNAs were cloned from several insects as well as from a crab and a tick (see Fig. 3 legend for species names and references). Comparison of deduced amino acid sequences from these cDNAs showed that the 66 amino acid DNA binding domain is well conserved among EcRs and USPs. However, the A/B, D and F domains are not very well conserved. Critical residues in the ligand binding domain are well conserved. Phylogenetic trees of 222 amino acid ligand binding domains of EcR and USP sequences are shown in Fig. 3. EcRs cloned from dipteran and lepidopteran insects fall into two separate groups. It is interesting to note that orthopteran, coleopteran and homopteran EcRs fall into one group. LXR and FXR are the two most closely related mammalian receptors to EcR. Similar to the EcRs, the USPs cloned from dipteran and lepidopteran insects also fall into two separate groups. It is interesting to note that the USP homologues cloned from the locust, tick and crab fall into a group which includes *Xenopus laevis* and *Homo sapiens* RXRs. These differences in the ligand binding domains of EcRs and USPs from insects belonging to different orders can be exploited to discover order-specific insecticides.

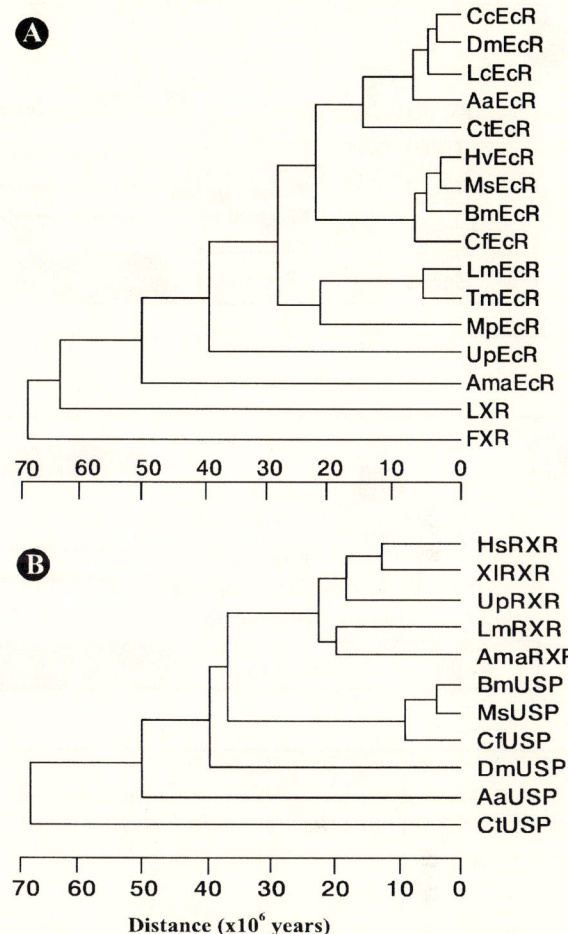

Fig. 3A,B. Phylogenetic tree of EcR and USP/RXR ligand binding domain sequences. Phylogenetic tree was prepared using DNA star (DNA star Inc, Madison, WI, USA). The sequences used are CcEcR (*Ceratitis capitata* EcR, Verras et al. 1999), DmEcR (*Drosophila melanogaster* EcR, Koelle et al. 1991), *LcEcR (Lucilia cuprina EcR,* Hannan and Hill 1997), AaEcR (*Aedes aegypti* EcR, Cho et al. 1995), CtEcR (*Chironomus tentans,* EcR, Imhof et al. 1993), HvEcR (*Heliothis virescens* EcR, Martinez et al. 1999a), MsEcR (*Manduca sexta* EcR, Fujiwara et al. 1995), BmEcR (*Bombyx mori* EcR, Swevers et al. 1995; Kamimura et al. 1996), CfEcR (*Choristoneura fumiferana* EcR, Kothapalli et al. 1995; Perera et al. 1999), LmEcR (*Locusta migratoria* EcR, Saleh et al. 1998), TmEcR (*Tenebrio molitor* EcR, Mouillet et al. 1997), MpEcR (*Myzus persicae* EcR, Hill 1999), UpEcR (fiddler *crab Uca pugilator* EcR, Chung et al. 1998), AmaEcR (ixodid tick *Amblyomma americanum* EcR, Guo et al. 1997), LXR (*Homo sapiens* Liver X receptor or nuclear oxysterol receptor, Willy et al. 1995), FXR (*Homo sapiens* farnesoid X receptor, Forman et al. 1995), HsRXR *(Homo sapiens* retinoid X receptor, Mangelsdorf et al. 1990), XlRXR (*Xenopus laevis* retinoid X receptor, Bloomberg et al. 1992), UpRXR (fiddler crab, *Uca pugilator* RXR homolog, Chung et al. 1998), LmRXR (*Locusta migratoria* RXR homolog, Hayward et al. 1999), AmaRXR (*Amblyomma americanum* RXR homolog, Guo et al. 1998), BmUSP (*Bombyx mori* USP, Tzertzinis et al. 1994), MsUSP (*Manduca sexta* USP, Jindra et al. 1997), CfUSP (*Choristoneura fumiferana* USP, Perera et al. 1998), DmUSP (*Drosophila melanogaster* USP, Henrich et al. 1990; Ora et al. 1990; Shea et al. 1990), AaUSP (*Aedes aegypti* USP, Kapitskaya et al. 1996), CtUSP (*Chironomus tentans* USP, Vogtli et al. 1999).

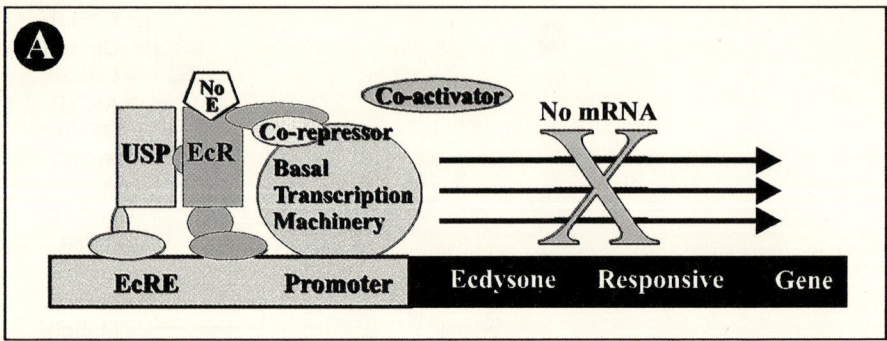

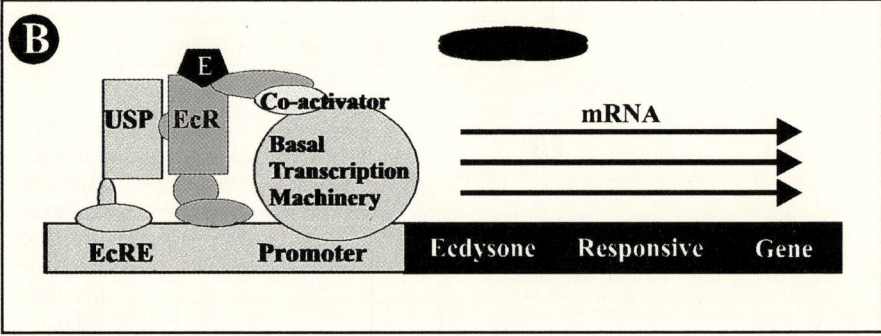

Fig. 4A,B. Mode of action of ecdysteroids. Illustration of function of ecdysteroids in cell nucleus. *EcR* Ecdysone receptor; *USP* ultraspiracle; *E* ecdysteroid; EcRE ecdysone response element

Since the cloning of EcR cDNA and its isoforms (Bender et al. 1997), an enormous amount of research has been conducted to elucidate the mode of action of 20E at the molecular level. Receptors, transcription factors (early genes), late genes, and co-repressors were identified and characterized (see Thummel 1995, 1997; Buszczak and Segraves 1998; Tsai et al. 1999). Our latest understanding of how 20E works to activate ecdysone responsive genes is depicted schematically in Fig. 4. In the absence of 20E, EcR and USP can heterodimerize and bind to EcRE present in the promoter region of the ecdysone responsive genes. However, transcription of these genes is not activated due to the presence of a group of proteins called co-repressors. When 20E is present it can bind to EcR causing a conformational change that results in the replacement of the co-repressors with co-activators leading to transcriptional activation. Most of the players depicted in the model, except for the co-activators, have been identified from insects.

Some of the enzymes involved in biosynthesis and degradation of ecdysteroids, as well as proteins that are critical for secretion and uptake of ecdysteroids into cells, can be used as targets for developing new insecticides. Some of the cDNAs can also be used to interfere with these processes by over/under expressing them either through microorganisms or transgenic insects.

2.3
Non-Steroidal Ecdysone Analogs and Their Mode of Action

Over the years several attempts have been made to use ecdysone analogs for insect control. Hundreds of phytoecdysteroids have been discovered in plants, evolved presumably inter alia to deter insect attacks. These compounds have been tested with limited success as pest control agents (Horn and Bergamasco 1985; Dinan et al. 1997). Some synthetic steroid analogs such as azasteroids showed activity in blocking ecdysone synthesis (Svoboda et al. 1972). The high cost of synthesizing steroid analogs of ecdysone, combined with their tendency to undergo oxidative degradation, have resulted in little commercial success. Some natural products such as the triterpenoid cucurbitacins have been shown to have antagonistic activity at the ecdysteroid receptor level (Dinan et al. 1997). The most dramatic discovery happened in the early 1980s when chemists at Rohm and Haas Company (Spring House, PA, USA) synthesized 1,2-diacyl-1-substituted hydrazines that had potent insecticidal activity (Hsu et al. 1997). This class of compounds induces a precocious larval molt in the susceptible species acting through the ecdysone receptor complex (Wing 1988; Wing et al. 1988), and has recently been reviewed (Dhadialla et al. 1998).

Similar to 20 E, diacyl hydrazines, or Rohm and Haas compounds (RH), bind to the receptor complex, ecdysone receptor (EcR)–ultraspiracle (USP) heterodimer (Fig. 5). The combined ligand (20 E)–receptor complex (EcR–USP) binds to the ecdysone response element (EcRE) and transactivates a cascade of genes. Both 20 E and RH compounds induce both isoforms of EcR and the early genes, CHR75 and CHR3. Therefore, the early events that occur in the presence of 20 E are similar in both cases. In the case of 20 E, after the events that occur in the presence of the hormone, the 20 E is cleared from the system so that the subsequent two phases that require the absence of 20 E can proceed. Whereas, because of the persistence of the RH compounds, the events that occur in the absence of 20 E do not proceed and the end result is an incomplete precocious molt resulting in the mortality of the larva (Smagghe and Degheele 1994; Retnakaran et al. 1995, 1997a,b; Dhadialla et al. 1998).

The effects of RH compounds on different insects can be classified into susceptible insects and resistant species. The susceptible species can be divided into insects that are (1) totally susceptible and those that are (2) partially susceptible, and which show recovery. The resistant group can be divided into three groups, those that (3) eliminate the toxicant, ones that (4) metabolize the

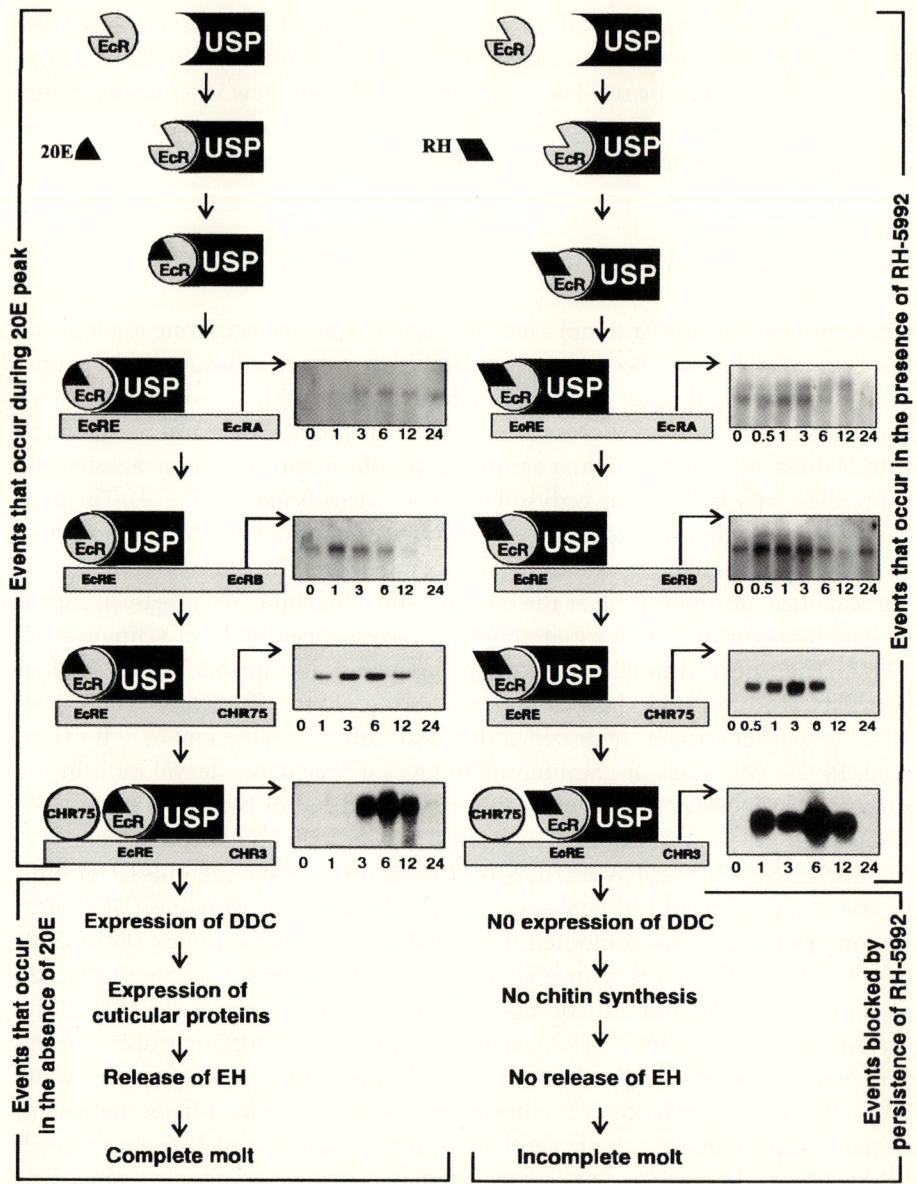

Fig. 5. Mode of action of RH-5992 compared with that of 20-hydroxyecdysone based on the data presented in Palli et al. (1996, 1997) and Perera et al. (1999). *EcR* Ecdysone receptor; *USP* ultra-spiracle; *20 E* 20-hydroxyecdysone; *RH* RH-5992; *CHR75 Choristoneura* hormone receptor 75; *CHR3 Choristoneura* hormone receptor 3; *DDC* dopadecarboxylase; *EH* eclosion hormone

compound and finally those that have a receptor structure where RH compounds do not fit well.

A typical example of the totally susceptible species is the spruce budworm, *Choristoneura fumiferana*, to tebufenozide (RH-5992). Upon ingestion of this compound, a newly molted larva goes into a precocious molt within 24 h and shows head capsule slippage within 48 h. The larva remains moribund at this stage and dies of starvation and dehydration (Retnakaran et al. 1997b).

Tebufenozide is effective only if it is ingested prior to the appearance of the ecdysteroid peak in the hemolymph. If the larva ingests the material after the peak, the effect is manifested in the succeeding larval instar (Palli et al. 1995). Once initiated the irreversible sequential cascade of gene expression and repression cannot be restarted. Tebufenozide induces the expression of early genes such as the transcription factors CHR75 and CHR3, but since the compound binds strongly to the receptor and persists in the epidermal cells, the genes that are expressed in the absence of ecdysone are not expressed (Retnakaran et al. 1995; Palli et al. 1996; Fig. 5).

Recently, we discovered in our laboratory that the white marked tussock moth, *Orgyia leucostigma*, shows partial susceptibility to tebufenozide. The first instar is very susceptible but the later instars are progressively less and less so. The last larval instar shows precocious head capsule slippage and becomes moribund. However, after a few days, the molting process is complete. The newly molted larva is totally resistant to tebufenozide (Gelbic and Retnakaran, unpublished).

Tebufenozide is very effective on most lepidopteran species but is ineffective on members of most other orders of insects. There are at least three ways in which resistance can be achieved: (1) the compound can be pumped out of the midgut cells into the lumen and eliminated without being absorbed. Using cell lines, it has been shown that there is an intracellular pump that eliminates tebufenozide in resistant cell lines (Sundaram et al. 1998a,b); (2) the compound can be metabolized and eliminated, although the differences in the metabolism of tebufenozide between susceptible and resistant species seem to be minimal (Dhadialla et al. 1998); (3) the ecdysone receptor/USP heterodimer may be conformationally incompatible to binding. The binding affinities of tebufenozide to various EcR/USP combinations indicate high affinity binding in susceptible species and little or no binding in resistant insects (Dhadialla et al. 1998). Four stable ecdysteroid agonists, tebufenozide, methoxyfenozide, halofenozide and chromafenoside (Fig. 6), are being marketed or are under commercial development. With the exception of halofenozide, they all act only on lepidopteran pests. Halofenozide has been shown to control some coleopteran pests of turf and lawn.

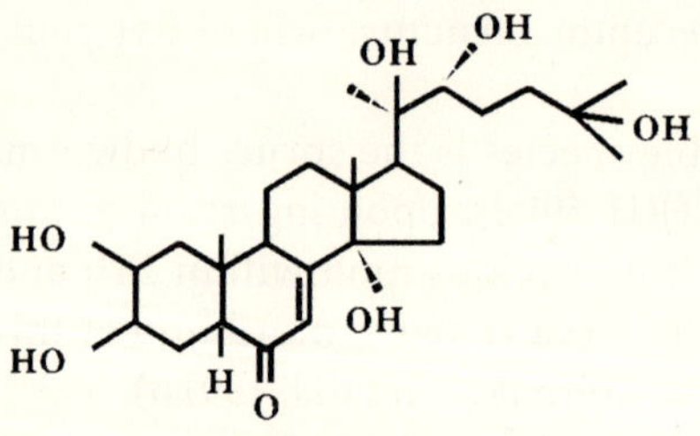

❶   20-HYDROXYECDYSONE

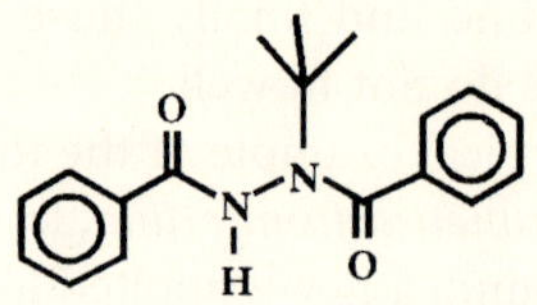

❷   RH-5849

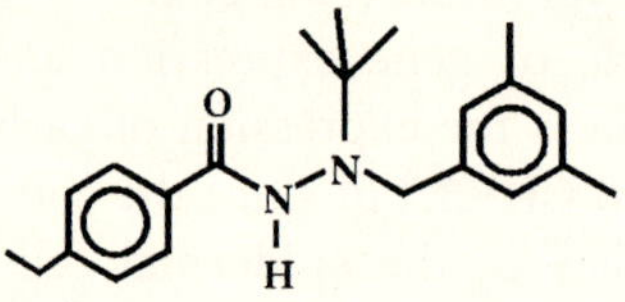

❸   RH-5992 (Tebufenozide)

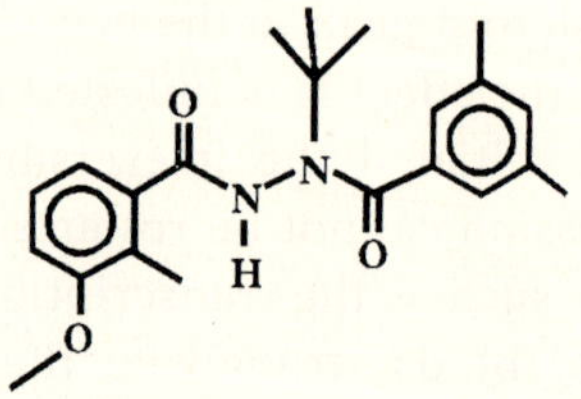

❹   RH-2485 (Methoxyfenozide)

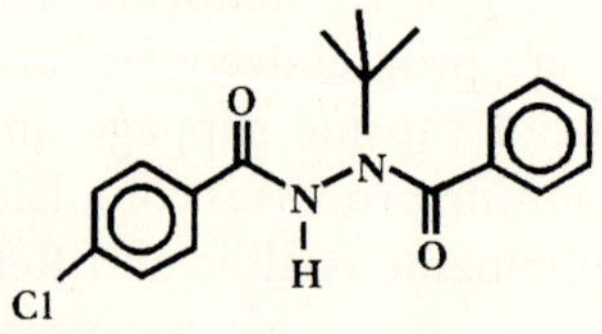

❺   RH-0345 (Halofenozide)

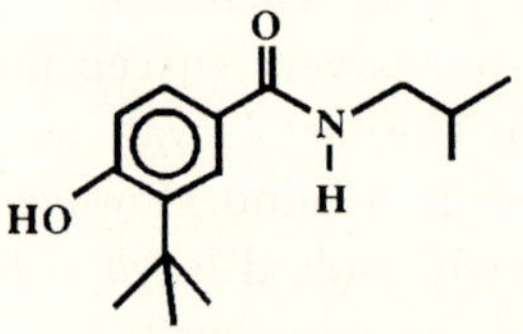

❻   DTBHIB

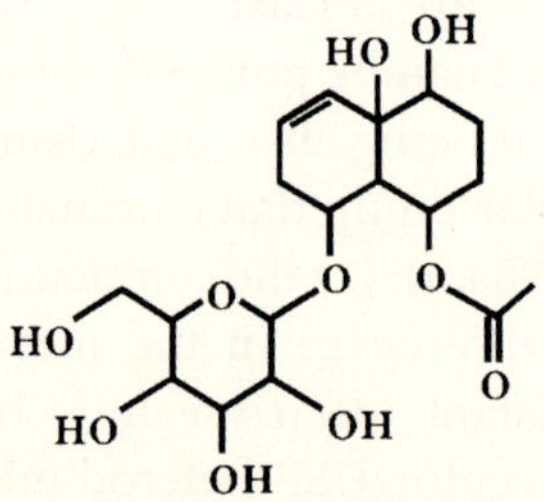

❼   8-0-Acetylharpagide

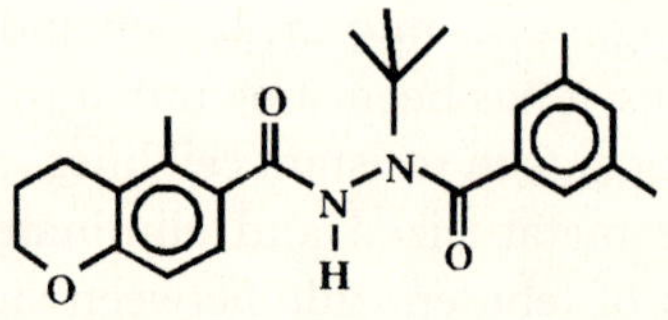

❽   Chromafenozide

Fig. 6. Chemical structures of the molting hormone, 20-hydroxyecdysone (*1*), Rohm and Haas ecdysteroid agonists (*2,3,4,5*), Sumitomo's agonist, 3,5-di-*tert*-butyl-4-hydroxy-*N*-isobutylbenzamide (*6*), Merck's iridoid glycoside isolated from the plant *Ajuga reptans* (*7*), and Nippon Kayaku's Chromafenozide (*8*)

2.4
Receptor-Based Screening Assays

The availability of cDNA clones for EcR and USP from several insect species has opened up possibilities for the development of high throughput EcR/USP based screening assays. This can be done in several ways and, depending on the screening strategy, any one of them or a combination of them can be used.

1. Binding assay using EcR/USP: EcR/USP proteins can be produced by in vitro transcription and translation of cDNAs in rabbit reticulocyte lysate or by expressing the cDNAs in bacteria, yeast, baculovirus, insect cells or mammalian cells. Either crude or purified proteins can be used in displacement assays using labeled ecdysone or ponasterone A. This kind of assay can be developed into 96- or 384-well format for high throughput screening. Several groups have successfully demonstrated the functionality of in vitro produced EcR and USP (Kothapalli et al. 1995; Kapitskaya et al. 1996; Elke et al. 1997; Dela Cruz and Mak 1997; Halling et al. 1999).
2. Yeast-based assay: it is possible to engineer EcR and USP cDNAs along with EcRE+minimal promoter+reporter cassette into yeast to create a recombinant yeast strain that can be used for high throughput screening of compounds for ecdysteroid activity. Dela Cruz and Mak (1997) attempted to develop such an assay and reported that *Drosophila* EcR functions as a constitutive activator in yeast cells. It may be possible to further modify this system and these modifications could lead to ligand-dependent activation of the reporter. Once such a system is developed, the yeast-based assay will become very useful for screening large numbers of candidate compounds.
3. Cell-based assays: another approach is to engineer EcR and USP cDNAs along with EcRE+minimal promoter+reporter cassette into cells that normally lack these molecules and use these cells in a screening assay. Insect cell lines that are deficient in EcR can be used for this purpose. L57 is one such cell line created by Drs. Peter and Lucy Cherbas by inactivating the EcR in Kc cells. We have produced an ecdysone-resistant CF-203 cell line by serial passaging of these cells in medium containing 20 E (S.S. Sohi, B. Cook, unpublished). These resistant cells no longer respond to 20 E or tebufenozide and contain very low levels of EcR when compared with normal cells. Cell lines from vertebrate species that lack EcR/USP can also be used for this purpose. Several-fold induction of reporter gene activity in a number of vertebrate cell lines engineered with EcR cDNAs has been reported (No et al. 1996; Suhr et al. 1998). Stably transformed cell lines expressing EcR/USP cDNAs and with an EcRE+minimal promoter+reporter cassette can be used for screening large numbers of compounds in high throughput mode.

4. Screening assays based on expression of early genes: several studies in our laboratory and elsewhere have shown that expression of hormone receptor 3 (HR3) can serve as a reliable indicator of 20E activity (Palli et al. 1992, 1996). The mRNAs for HR3s are induced within an hour after exposure to 20E or stable ecdysteroid analogs. Cells that are not exposed to 20E or stable ecdysteroid analogs do not have any detectable levels of HR3 mRNAs. Sundaram et al. (1998b) showed that in vivo relative toxicity of four ecdysone agonists, RH-2485, RH-5992, RH-0345 and RH-5849, correlates well with their ability to induce CHR3 mRNAs in the spruce budworm larvae. Subsequent studies showed that the relative toxicity of these four ecdysone agonists also correlates well with their ability to induce CHR3 mRNAs in ecdysone responsive CF-203 cells (M.S. Sundaram and S.R. Palli, unpubl.). Thus, measurement of mRNA or protein levels of HR3 or similar early genes can be used as a reliable end point to screen for new ecdysteroid analogs. Cell lines from insects belonging to a specific order can be used to discover compounds that will be effective against insects belonging to that order.

2.5
Future Directions

The discovery of non-steroidal ecdysone agonists, the RH compounds, has been one of the most exciting developments in ecdysteroid research in recent years. Their narrow spectrum of activity makes them one of the best fit compounds for use in integrated pest management programs. For example, tebufenozide is effective against many lepidopteran pests but has little or no effect on their hymenopteran parasites. Rohm and Haas scientists recently received a "Green Chemistry" award from the US government for the discovery of these non-steroidal ecdysone agonists. Due to the commercial success of RH compounds, several other commercial companies have initiated programs to discover new 20E agonists and antagonists. Recently, the Sumitomo Company and Merck Research Laboratories have both reported an ecdysteroid agonist (Fig. 6). We should see the commercialization of new stable ecdysteroids that are effective on pest species belonging to different orders of insects in the near future.

It was discovered recently that EcR/USP and stable ecdysteroids can be used as gene switches to regulate expression of engineered genes in mammals and plants (No et al. 1996; Martinez et al. 1999b). This provides an enormous opportunity for the utilization of EcR/USP combinations as well as stable ecdysteroid analogs in human gene therapy, production of therapeutic proteins and transgenic plants. It is likely that in the immediate future there will be an intensive research effort dedicated to both understanding the EcR/USP transactivation mechanism and to the discovery of new more active and safer stable ecdysteroid analogs.

3
Juvenile Hormone

It was Carrol Williams (1967) who suggested that juvenile hormones could be used as insect-specific control agents to which pest species may be unable to develop resistance and coined the term "third-generation pesticides". JH prevents metamorphosis and regulates reproductive maturation. The presence of JH prevents the switching of gene expression from larval to pupal or pupal to adult stages. In the ultimate larval instar, JH titers decrease and a pulse of ecdysteroids is released. This small ecdysteroid peak that is secreted in the absence of JH causes the commitment of larvae to become pupae (Fig. 2). This was elegantly demonstrated in the tobacco horn worm, *Manduca sexta*, by Riddiford (1976). The reproductive role of JH is primarily on oocyte development, yolk deposition and accessory gland development (Braun and Wyayy 1995; Wyatt and Davey 1996).

3.1
Biology, Endocrinology and Molecular Biology

Juvenile hormone (JH) is secreted by a pair of tiny glands, the corpora allata (Fig. 1), attached to the base of the brain, and is controlled by neuropeptides, allatostatins (inhibit) and allatotropins (stimulate) secreted by the neurosecretory cells in the brain (Tobe and Stay 1985; Kataoka et al. 1989). Upsetting the titer of JH in the last larval instar results in the production of larval-pupal intermediates that seldom survive.

If JH is applied on the first day of the last larval instar prior to the ecdysteroid programming peak, the larva molts into a supernumerary larva. If applied at the time of the peak, larval-pupal intermediates are formed. However, if applied after the programming is over, there is no effect (Retnakaran 1973b). Adult diapause, especially in many beetles, is characterized among other things by the lack of ovarian development and can be terminated by the application of JH (Retnakaran 1974). JH controls the synthesis of the yolk protein, vitellogenin, in the fat body and regulates its transport into the oocyte. In the locust, *Locusta migratoria*, and the blood sucking bug, *Rhodnius prolixus*, JH causes the follicle cells to shrink and, as a result, large intercellular spaces are created through which the yolk protein passes. Using this "potency" or opening of the intercellular spaces in the follicular epithelium by JH, evidence for the existence of a JH membrane receptor has been reported (Davey and Gordon 1996). The metamorphic role of JH through its "status-quo" action is thought to be through a nuclear receptor (Palli et al. 1994; Jones and Sharp 1997). Several aspects of JH action are covered in two excellent reviews (Riddiford 1996; Wyatt and Davey 1996).

3.2
Receptors and Other Target Sites

Several hemolymph, cytosolic and nuclear JH binding proteins have been identified and characterized from insects such as *Manduca sexta, Drosophila melanogaster, Locusta migratoria,* and *Leucophaea maderae* (see Goodman 1990; Palli et al. 1991 for reviews). Most, if not all, of the binding proteins identified so far have been shown to bind to JH with high affinity, implicating them in the JH signal transduction pathway. Some of these so-called JH binding proteins or carrier proteins appear to have contaminated nuclear preparations that were made for isolating receptors (see Wyatt and Davey 1996 for a discussion of this topic). It is indeed possible for hemolymph binding proteins to be present in the cytosol and the cytosol binding proteins to be present in the nucleus because they could act as carrier proteins to transport and protect JH in various cellular compartments. Therefore, the presence of JH carrier proteins in the nuclear receptor preparations may not be contamination but reflect the real situation within the cell.

Based on the assumption that a JH receptor would be a member of the steroid/thyroid hormone superfamily (Evans 1988), several groups used probes derived from various steroid/thyroid family members to clone JH receptor cDNA using low stringency hybridization techniques. Palli et al. (1991) used human retinoic acid receptor cDNA as a probe and identified a steroid/thyroid superfamily member from *Manduca sexta*. Further characterization of this cDNA revealed that this is not a JH receptor but codes for an ecdysone-induced transcription factor, which plays a critical role in ecdysone signal transduction and is related to the *Drosophila* hormone receptor 3 (Koelle et al. 1992). Therefore it is named *Manduca* hormone receptor 3 (Palli et al. 1992). The reason for using retinoic acid, a vertebrate morphogenetic hormone, is because of its structural similarity to JH. Nuclear receptors for retinoic acid have been identified and shown to be members of the steroid/thyroid hormone receptor superfamily (Leid et al. 1992 and references therein). Retinoic acids can bind to retinoic acid receptor (RAR) homodimers or RAR and retinoid X receptor (RXR) heterodimers. RXR was also shown to form heterodimers with several nuclear receptors including the farnesoid X-activated receptor (FXR). JH III but not JH acid or methoprene can bind/activate FXR and RXR heterodimer (Forman et al. 1995). Methoprene and methoprene acid but not JHIII can activate RXR (Harmon et al. 1995). These two studies suggested that RXR or its insect homologue *ultraspiracle* (USP) could play an important role in signal transduction of JH or JH-related compounds. As explained in the previous section, USP was first cloned from *D. melanogaster* and was found to be a necessary heterodimeric partner for EcR for functional DNA or ligand binding. 20 E binds to EcR in the heterodimer complex; now the question is whether USP is a non-liganded partner in this complex, or if it can bind to its own ligand?

Jones and Sharp (1997) showed that both JH III and JHB3 (Fig. 7) bind to a *D. melanogaster* USP homodimer. Although these results are preliminary and several key questions remain unanswered, this is an exciting result because competition between JH and 20 E to recruit USP/USP or EcR/USP complex would provide an elegant system for cells to decide whether to undergo molting or metamorphosis (Jones and Sharp 1997; Feyereisen 1998). Questions such as: how would JH binding to USP affect EcR function? or whether USP is the JH

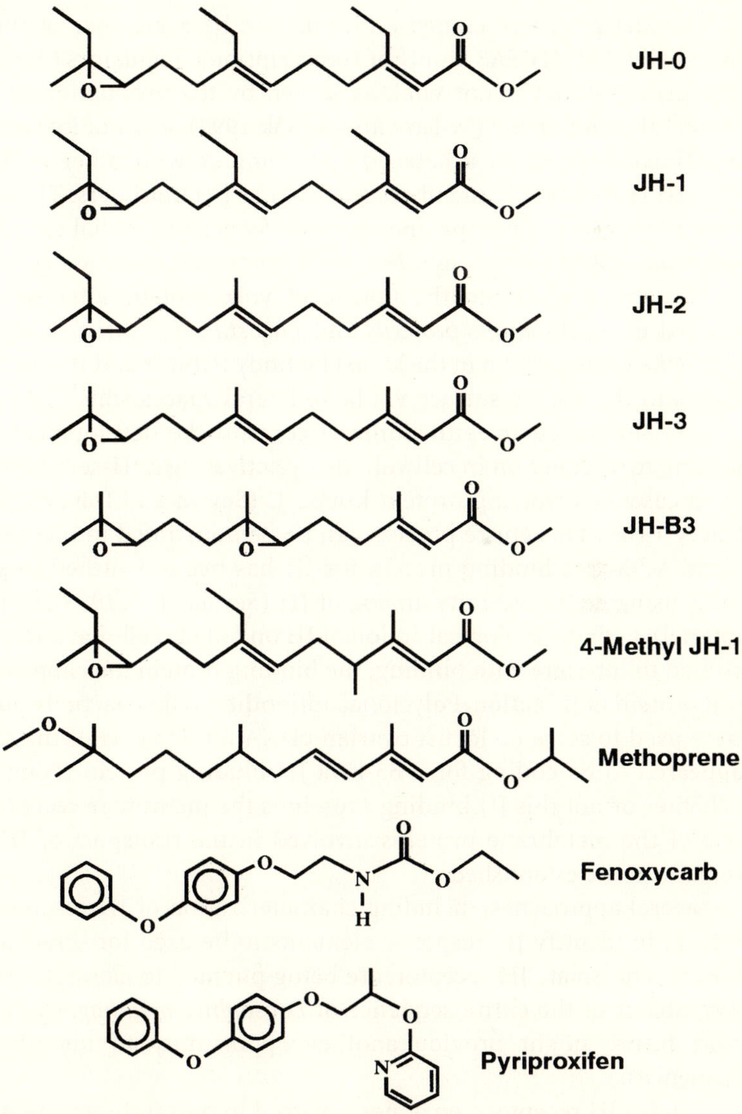

Fig. 7. Structures of juvenile hormones and some of their analogs

receptor or one of many molecules that are involved in JH signal transduction, remain enigmatic at best.

Wilson and Fabian (1986) generated *D. melanogaster* mutants that are resistant to methoprene (*Met*). *Met* flies are also resistant to JH III, JHB$_3$, and other JHAs, but not to other classes of insecticides. An 85-kDa protein that binds to JHIII with high affinity was identified in the fat body and other tissues of the wild-type *D. melanogaster*. The same 85-kDa protein isolated from *Met* flies showed a six-fold lower affinity than the wild-type protein for JHIII (Shemshedini et al. 1990).

The *Met* gene was cloned and found to be a member of the basic helix-loop-helix (bHLH)-PAS family of transcriptional regulators (Ashok et al. 1998). *Met* gene product is not vital, as shown by the production of null mutants (*Met^{27}*) that are viable (Wilson and Ashok 1998). It is not known whether *Met* functions alone or in a heterodimer complex with other proteins. Most of the bHLH-PAS family members function as heterodimers. Therefore, identification of a heterodimer partner for the *Met* gene is crucial for validation of its function as well as to design *Met*-gene-based screening assays.

The action of JH on the uptake of yolk protein, vitellogenin (Vg), was studied using *Rhodnius prolixus* and *Locusta migratoria*. JH appears to activate gene transcription in the locust fat body (Glinka and Wyatt 1996). Vg gains access to the oocyte surface via large lateral spaces which open between the cells of the follicular epithelium. JH controls the development of the spaces leading to a reduction in cell volume by activating a JH-sensitive Na/K ATPase via a cascade involving protein kinase C (Sevala and Davey 1989; Wyatt and Davey 1996). This entire process can be demonstrated in membrane preparations. A 35-kDa binding protein for JH has been identified in such preparations using a photoaffinity analog of JH (Sevala et al. 1995). Compounds that interfere with the biological action of JH on follicle cells have also been demonstrated to interfere with binding; the binding protein thus appears to be a critical protein in JH action. Polyclonal antibodies to this partially purified protein were used to screen a locust ovarian cDNA library. One of the clones isolated appeared to be coding for a 35-kDa JH binding protein (Feng et al. 1999b). Whether or not this JH binding protein is the membrane receptor for JH or is one of the membrane proteins involved in the transport of JH into the cell remains to be established.

Several approaches, including characterization of JH-responsive gene promoters to identify JH-response elements to be used for screening expression libraries to isolate JH receptor, are being pursued to clone JH receptor cDNA. Availability of the entire sequence of *Drosophila melanogaster* genome in the near future might provide another opportunity to identify JH receptor sequences.

Besides JH receptors, enzymes involved in biosynthesis and degradation of JH can be useful targets (Feyereisen and Farnsworth 1998; Feng et al. 1999a).

Several research groups are involved in synthesizing inhibitors for JH-degrading enzymes, JH esterase and JH epoxidehydralase. The cDNAs coding for JHE and JHEH have also been used to improve the efficacy of insect-specific baculoviruses (Bonning et al. 1995). With the availability of sequences for most of the JH biosynthesis and JH-degrading enzymes in the near future, these lines of research activity to discover chemical or biological means of interfering with JH biosynthesis and degradation should accelerate.

3.3
JH Analogs and Their Modes of Action

Numerous analogs of JH have been synthesized and tested for their control potential (Retnakaran et al. 1985). JH analogs (JHAs) have also been found to be naturally occurring in many plants supposedly as a defence mechanism. The JHAs can be divided into two groups, the agonists such as methoprene, fenoxy-carb and pyriproxifen (Fig. 7) and the antagonists such as precocenes. The modes of action of JHAs are varied. A few of the principal ones will be addressed. The anti-JH, precocene, acts as a "suicide substrate" by the activation of enzymes in the corpus allatum to make the compound cytotoxic to the endocrine gland (Bowers 1984). The JHA methoprene produces all the characteristic effects of JH inducing larval-pupal deformities in last instar larvae. A methoprene-tolerant gene, *Met* gene, was identified in *Drosophila*. Upon mutation of this gene, the insect shows resistance to both methoprene and JH (Wilson and Ashok 1998). It appears that methoprene acts by interfering with the function of the broad complex gene (BRC), which is an ecdysteroid-inducible transcription factor. Methoprene phenocopies BRC defects as seen in BRC mutants. In addition it also causes phenotypes or deformities not associated with known JH mutations (Restifo and Wilson 1998). Fenoxycarb is a strange JHA that has a carbamate moiety in its structure (Fig. 7). When applied to two-day-old 5th-instar larvae of the silkworm, *Bombyx mori*, it induced permanent larvae, whose prothoracic glands secreted little or no ecdysteroids. It appears that the cAMP signalling cascade in the prothoracic glands, which is necessary for ecdysteroid synthesis, was made non-functional in these fenoxycarb-treated insects (Dedos and Fugo 1999).

JHAs have lethal effects on embryonic development and this property has been used in the control of various types of lice and fleas by treating either females or eggs (Retnakaran 1970, 1973a; Enslee and Riddiford 1977). Pyriproxifen- and methoprene-treated eggs of the cat flea, *Ctenocephalides felis,* die at the blastokinesis stage (Palma et al. 1993). The mode of action of JHAs during embryonic development is not clear at present. JHA-treated insects show a progressive decline in their respiratory rate (Retnakaran 1975) and also the embryonic cells show pycnotic nuclei, clumped heterochromatin, swollen mitochondria and cellular autolysis (Marchiondo et al. 1990).

3.4
Receptor-Based Screening Assays

Since we do not know the exact composition of the JH receptor, it is difficult to design target site assays at this time. However, if the JH receptor turns out be a nuclear receptor functioning through hormone response elements, all the assays described in Section 2.3 can be readily adopted for this purpose. On the other hand, if the JH receptor turns out to be a membrane receptor that transduces the JH signal by altering intracellular signaling systems (affecting ion channels, phospholipids, second messengers and phosphorylation of transcription factors), then there are numerous well-established high throughput screening assays developed by the pharmaceutical industries that can be adapted for the screening of compounds that block or promote JH signal transduction.

It is possible to develop target site assays using cDNA coding for enzymes/proteins that regulate JH biosynthesis, secretion, uptake and metabolism. Traditionally, these types of assays have been conducted using tissue preparations, which is not only labor intensive but also can be used to screen only a limited number of compounds. The cloned cDNAs can be used to produce functional proteins in bacteria, yeast, baculoviruses, insect or mammalian cells, and these proteins can then be used to screen for large numbers of compounds that function as promoters/agonists, inhibitors/antagonists of a given enzyme or a regulatory protein.

3.5
Future Directions

The concept of using JHAs to control pests that cause damage during larval stages has not been very successful, because JHAs were mostly effective at the last larval stage by which time most of the damage had occurred. JH antagonists have either been not very potent and stable or have other side effects (Socha and Marec 1989). JHAs have achieved more success in controlling adult pests. The ovicidal effect was successfully used to control fleas and lice. On the whole, the commercial success of JH-based insecticides has been relatively disappointing. For a few compounds, such as pyriproxyfen that has had some commercial success, key pests have already developed resistance. Almost all JHAs discovered to date have been identified by assaying their effects on insects. Cloning and characterization of cDNAs for the JH receptor as well as enzymes involved in JH biosynthesis and degradation will provide tools for the development of high throughput target site assays that can be used for screening large numbers of compounds. The same cDNAs can also be used to interfere with JH action, JH biosynthesis and JH degradation by over or under expressing these proteins either through microorganisms or transgenic insects. Molecular approaches to developing JH-based control agents will probably be more successful than the time-honored screening methods using live insects.

Acknowledgements. The authors wish to thank Mark Primavera for the illustrations, Marilyn Scott for typing the manuscript, and Karen Jamieson for her editorial assistance.

References

Ashburner M, Chilhara C, Meltze P, Richards G (1974) Temporal control of puffing activity in polytene chromosomes of *Drosophila melansgaster*. Cold Spring Harbor Symp Quant Biol 38:655–662

Ashok M, Turner C, Wilson TG (1998) Insect juvenile hormone resistance gene homology with the bHLH-PAS family of transcriptional regulators. Proc Natl Acad Sci USA 95:2761–2766

Bender M, Imam FB, Talbot WS, Ganetzky B, Hogness DS (1997) *Drosophila* ecdysone receptor mutations reveal functional differences among receptor isoforms. Cell 91:777–788

Bloomberg B, Mangelsdorf DJ, Dyck JA, Bittner DA, Evans RM, De Robertis EM (1992) Multiple retinoid-responsive receptors in a single cell: families of retinoid 'X' receptors and retinoic acid receptors in the Xenopus egg. Proc Natl Acad Sci USA 89:2321–2325

Bonning BC, Hoover K, Booth TF, Duffey S, Hammock BD (1995) Development of a recombinant baculovirus expressing a modified juvenile hormone esterase with potential for insect control. Arch Insect Biochem Physiol 30:177–194

Braun RP, Wyatt GR (1995) Growth of the male accessory gland in adult locusts: roles of juvenile hormone, JH esterase, and JH binding proteins. Arch Insect Biochem Physiol 30:383–400

Bowers WS (1984) Insect-plant interactions: endocrine defences. Ciba Found Symp 102:119–137

Buszczak M, Segraves WA (1998) *Drosophila* metamorphosis: the only way is USP? Curr Biol 24:R879–82

Butenandt A, Karlson P (1954) Ueber die Isolierung eines Metamorphose-Hormons der Insekten in kristallisierter Form. Z Naturf 96:389–391

Carson R (1962) Silent spring. Houghton-Mifflin, Boston

Cho W-L, Kapitskaya MZ, Raikhel AS (1995) Mosquito ecdysteroid receptor: analysis of the cDNA and expression during vitellogenesis. Insect Biochem Mol Biol 25:19–27

Chung AC, Durica DS, Clifton SW, Roe BA, Hopkins PM (1998) Cloning of crustacean ecdysteroid receptor and retinoid-X receptor gene homologs and elevation of retinoid-X receptor mRNA by retinoic acid. Mol Cell Endocrinol 139:209–227

Clever U, Karlson P (1960) Induktion von Puff-Veränderungen in den Speicheldrüsenchromosomen von *Chironomus tentans* durch Ecdyson. Exp Cell Res 20:623–626

Cremlyn R (1978) Pesticides. Preparation and mode of action. Wiley, New York

Davey KG, Gordon DRB (1996) Fenoxycarb and thyroid hormones have JH-like effects on the follicle cells of *Locusta migratoria* in vitro. Arch Insect Biochem Physiol 32:613–622

Dedos SG, Fugo H (1999) Downregulation of the CAMP signal transduction cascade in the prothoracic glands is responsible for the fenoxycarb-mediated induction of permanent 5th instar larvae in *Bombyx mori*. Insect Biochem Mol Biol 29:723–729

Dela Cruz F, Mak P (1997) Drosophila ecdysone receptor functions as a constitutive activator in yeast. J Steroid Biochem Mol Biol 62:353–359

Dhadialla TS, Carlson GR, Le DP (1998) New insecticides with ecdysteroidal and juvenile hormone activity. Annu Rev Entomol 43:545–569

Dinan L, Whiting P, Girault JP, Lafont R, Dhadialla TS, Cress DE, Mugal B, Antomiewski C, Lepesant JA (1997) Cucurbitacins are insect steroid hormone antagonists acting at the ecdysteroid receptor. Biochem J 327:643–650

Dinan L, Sarker SD, Bourne P, Whiting P, Sik V, Rees HH (1999) Phytoecdysteroids in seeds and plants of *Rhagodia baccata* (Labill.) Mog. (Chenopodiaceae). Arch Insect Biochem Physiol 41:18–23

Elke C, Vogtli M, Rauch P, Spindler-Barth M, Lezzi M (1997) Expression of EcR and USP in *Escherichia coli*: purification and functional studies. Arch Insect Biochem Physiol 35:59–70

Enslee EC, Riddiford LM (1977) Morphological effects of juvenile hormone mimics on embryonic development in the bug, *Pyrrhocoris apterus*. Roux's Arch Dev Biol 181:163–181

Evans RM (1988) The steroid and thyroid hormone receptor superfamily. Science 240:385–397

Feng QL, Ladd TR, Tomkins BL, Sundaram M, Sohi SS, Retnakaran A, Davey KG, Palli SR (1999a) Spruce budworm (*Choristoneura fumiferana*) juvenile hormone esterase: hormonal regulation, developmental expression and cDNA cloning. Mol Cell Endocrinol 148:95–108

Feng QL, Palli SR, Davey KG, Ladd TR, Sevala VL (1999b) Cloning and characterization of cDNA for 32-kDa protein that mediates uptake of vitellogenins in ovarian follicle cells of *Locusta migratoria*. Presented at the 7th International Conference on Juvenile Hormones, Jerusalem, Israel

Feyereisen R (1998) Juvenile hormone resistance: ! no PASaran ! Proc Natl Acad Sci USA 95:2725–2726

Feyereisen R, Farnsworth DE (1988) Forced synthesis of trace amounts of juvenile hormone II from propionate by corpora allata of a juvenile hormone III producing insect. Experientia 44:47–49

Forman BM, Goode E, Chen J, Oro AE, Bradley DJ, Perlmann T, Noonan DJ, Burka LT, McMorris T, Lamph WW, Evans RM, Weinberger C (1995) Identification of a nuclear receptor that is activated by farnesol metabolites. Cell 81:687–693

Fujiwara H, Jindra M, Newitt R, Palli SR, Hiruma K, Riddiford LM (1995) Cloning of an ecdysone receptor homolog from *Manduca sexta* and the developmental profile of its mRNA in wings. Insect Biochem Mol Biol 25:845–856

Fukuda S (1940) Hormonal control of molting and pupation in the silkworm. Proc Imp Acad Jpn 16:417–420

Glinka AV, Wyatt GR (1996) Juvenile hormone activation of gene transcription in the locust fat body. Insect Biochem Mol Biol 26:13–18

Goodman WG (1990) Biosynthesis, titer, regulation and transport of juvenile hormones. In: Gupta AP (ed) Morphogenetic hormones of arthropods. Rutgers University Press, New Brunswick, pp 83–124

Goodwin TW, Horn DHS, Karlson P, Koolman J, Nakanishi K, Robbins WE, Siddall JB, Takemoto T (1978) Ecdysteroids: a new generic term. Nature 272:122

Guo X, Harmon MA, Laudet V, Mangelsdorf DJ, Palmer MJ (1997) Isolation of a functional ecdysteroid receptor from the ixodid tick *Amblyomma americanum* (L.). Insect Biochem Mol Biol 27:945–962

Guo X, Xu Q, Harmon MA, Jin X, Laudet V, Mangelsdorf DJ, Palmer MJ (1998) Isolation of two functional retinoid X receptor subtypes from the ixodid tick, *Amblyomma americanum* (L.). Genebank accession number AF035578

Hachlow V (1931) Control of pupal development. Arch Entw Mech 125:26–49

Halling BP, Yuhas DA, Eldridge RR, Gilbey SN, Deutsch VA, Herron JD (1999) Expression and purification of the hormone binding domain of the Drosophila ecdysone and ultraspiracle receptors. Protein Expr Purif 17:373–386

Hannan GN, Hill RJ (1997) Cloning and characterization of *LcEcR*: a functional ecdysone receptor from the sheep blowfly *Lucilia cuprina*. Insect Biochem Mol Biol 27:479–488

Harmon MA, Boehm MF, Heyman RA, Mangelsdorf DJ (1995) Activation of mammalian retinoid X receptors by the insect growth regulator methoprene. Proc Natl Acad Sci USA 92:6157–6160

Hayward DC, Bastiani MJ, Trueman JWH, Truman JW, Riddiford LM, Ball EE (1999) The sequence of Locusta RXR, homologous to *Drosophila* ultraspiracle, and its evolutionary implications. Dev Genes Evol 209:564–571

Henrich VC, Sliter TJ, Lubahn DB, Gilbert LI (1990) A steroid/thyroid hormone receptor superfamily member in *Drosophila melanogaster* that shares extensive sequence similarity with a mammalian homologue. Nucleic Acids Res 18:4143–4148

Hill RJ (1999) Novel genetic sequences encoding steroid and juvenile hormone receptor polypeptides and insecticidal modalities thereof. International Patent application number WO 99/36520

Hiruma K, Hardie J, Riddiford LM (1991) Hormonal regulation of epidermal metamorphosis in vitro: control of expression of a larval-specific cuticle gene. Dev Biol 144:369–379

Hiruma K, Carter MS, Riddiford LM (1995) Characterization of the dopa decarboxylase gene of *Mandula sexta* and its suppression by 20-hydroxyecdysone. Dev Biol 169:195–209

Hoffmann KH, Gerstenlauer B (1997) Effects of ovariectomy and allatectomy on ecdysteroid synthesis and ecdysteroid titers during larval-adult development of *Gryllus bimaculatus* de Geer (Ensifera: Gryllidae). Arch Insect Biochem Physiol 35:149–158

Hopkins TL, Kramer KJ (1992) Insect cuticle sclerotization. Annu Rev Entomol 37:273–302

Horn DHS, Bergamasco R (1985) Chemistry of ecdysteroids. In: Kerkut GA, Gilbert LI (eds) Comprehensive insect physiology, biochemistry and pharmacology, vol 7. Pergamon Press, Oxford, pp 185–248

Hsu AC-T, Fujimoto TT, Dhadialla TS (1997) Structure activity study and conformational analysis of RH-5992, the first commercialized non-steroidal ecdysone agonist. In: Hedin PA (ed) Phytochemicals for pest control. ACS Symposium Series 658. American Chemical Society, Washington, DC, pp 206–219

Imhof MO, Rusconi S, Lezzi M (1993) Cloning of a *Chironomus tentans* cDNA encoding a protein (cEcRH) homologous to the *Drosophila melanogaster* ecdysteroid receptor (dEcR). Insect Biochem Mol Biol 23:115–124

Jindra M, Huang JY, Malone F, Asahina M, Riddiford LM (1997) Identification and mRNA developmental profiles of two ultraspiracle isoforms in the epidermis and wings of *Manduca sexta*. Insect Mol Biol 6:41–53

Jones G, Sharp PA (1997) Ultraspiracle: an invertebrate nuclear receptor for juvenile hormones. Proc Natl Acad Sci USA 94:13499–13503

Kamimura M, Tomita S, Fujiwara H (1996) Molecular cloning of an ecdysone receptor (B1 isoform) homologue from the silkworm, *Bombyx mori*, and its mRNA expression during wing disc development. Comp Biochem Physiol 113B:341–347

Kapitskaya M, Wang S, Cress DE, Dhadialla TS, Raikhel AS (1996) The mosquito *ultraspiracle* homologue, a partner of ecdysteroid receptor heterodimer: cloning and characterization of isoforms expressed during vitellogenesis. Mol Cell Endocrinol 121:119–132

Karlson P (1967) The chemistry of insect hormones and insect pheromones. Pure Appl Chem 14:75–87

Kataoka H, Toschi A, Li JP, Carney RL, Schooley DA, Kramer SJ (1989) Identification of an allatotropin from adult *Mandula sexta*. Science 243:1481–1483

Koelle MR, Talbot WS, Segraves WA, Bender MT, Cherbas P, Hogness DS (1991) The *Drosophila EcR* gene encodes an ecdysone receptor, a new member of the steroid receptor superfamily. Cell 67:59–77

Koelle MR, Segraves WA, Hogness DS (1992) DHR3: a *Drosophila* steroid receptor homolog. Proc Natl Acad Sci USA 13:6167–6171

Kothapalli R, Palli SR, Ladd TR, Sohi SS, Cress D, Dhadialla TS, Tzertzinis G, Retnakaran A (1995) Cloning and developmental expression of the ecdysone receptor gene from the spruce budworm, *Choristoneura fumiferana*. Dev Genet 17:319–330

Lafont R (1997) Ecdysteroids and related molecules in animals and plants. Arch Insect Biochem Physiol 35:3–20

Leid M, Kastner P, Chambon P (1992) Multiplicity generates diversity in the retinoic acid signalling pathways. Trends Biochem Sci 17:427–433

Locke M (1998) Epidermis. In: Harrison FW, Locke M Microscopic anatomy of invertebrates, vol 11A: Insecta. Wiley-Liss, New York, pp 75–138

Mangelsdorf DJ, Ong ES, Dyck JA, Evans RM (1990) Nuclear receptor that identifies a novel retinoic acid response pathway. Nature 345:224–229

Marchiondo AA, Riner JL, Sonenstrive DE, Rowe KF, Slussen JH (1990) Ovicidal and larvicidal modes of action of fenoxycarb against the cat flea (Siphonaptera: Pulicidae). J Med Entomol 27:913–921

Martinez A, Scanlon D, Gross B, Perara SC, Palli SR, Greenland AJ, Windass J, Pongs O, Broad P, Jepson I (1999a) Transcriptional activation of the cloned *Heliothis virescens* (Lepidoptera) ecdysone receptor (HvEcR) by muristeroneA. Insect Biochem Mol Biol 29: 915–30

Martinez A, Spark C, Hart CA, Thompson J, Jepson I (1999b) Ecdysone agonist inducible transcription in transgenic tobacco plants. Plant J 19:97–106

Mouillet JF, Delbecque JP, Quennedey B, Delachambre J (1997) Cloning of two putative ecdysteroid receptor isoforms from *Tenebrio molitor* and their developmental expression in the epidermis during metamorphosis. Eur J Biochem 248:856–863

Nijhout HF (1994) Insect hormones. Princeton University Press, Princeton

No D, Yao TP, Evans RM (1996) Ecdysone-inducible gene expression in mammalian cells and transgenic mice. Proc Natl Acad Sci USA 93:3346–3351

O'Connor JD (1985) Ecdysteroid action at the molecular level. In: Kerkut GA, Gilbert LI (eds) Comprehensive insect physiology, biochemistry and pharmacology, vol 8. Pergamon Press, Oxford, pp 85–98

Oro AE, McKeown M, Evans RM (1990) Relationship between the product of the *Drosophila ultraspiracle* locus and the vertebrate retinoid X receptor. Nature 347:298–301

Palli SR, Hiruma K, Riddiford LM (1991) Juvenile hormone and "retinoic acid" receptors in *Manduca* epidermis. Insect Biochem 21:7–15

Palli SR, Hiruma K, Riddiford LM (1992) An ecdysteroid-inducible *Manduca* gene similar to the *Drosophila* DHR3 gene, a member of the steroid hormone receptor superfamily. Dev Biol 150:306–318

Palli SR, Touhara K, Charles J-P, Bonning BC, Atkinson JK, Trowell SC, Hiruma K, Goodman WG, Kyriakides T, Prestwich GD, Hammock BD, Riddiford LM (1994) A nuclear juvenile hormone binding protein from larvae of *Manduca sexta*: a putative receptor for the metamorphic action of juvenile hormone. Proc Natl Acad Sci USA 91:6191–6195

Palli SR, Primavera M, Tomkins W, Lambert D, Retnakaran A (1995) Age-specific effects of a non-steroidal ecdysteroid agonist, RH-5992, on the spruce budworm, *Choristoneura fumiferana* (Lepidoptera: Tortricidae). Eur J Entomol 92:325–332

Palli SR, Ladd TR, Sohi SS, Cook BJ, Retnakaran A (1996) Cloning and developmental expression of *Choristoneura* hormone receptor 3, an ecdysone inducible gene and a member of the steroid hormone receptor superfamily. Insect Biochem Mol Biol 26:485–499

Palli SR, Ladd TR, Ricci AR, Sohi SS, Retnakaran A (1997) Cloning and developmental expression of *Choristoneura* hormone receptor 75: a homologue of the *Droscophila* E75 A gene. Dev Genet 20:36–46

Palli SR, Retnakaran A (1999) Molecular and biochemical aspects of chitin synthesis inhibition. In: Jollès P, Muzzarelli RAA (eds) Chitin and chitinases. Birkhäuser, Basel, pp 85–98

Palma KG, Mesla SM, Mesla RW (1993) Mode of action of pyriproxifen and methoprene on eggs of *Ctenocephalides felis* (Siphonaptera: Pulicidae). J Med Entomol 30:421–426

Perera SC, Palli SR, Ladd TR, Krell P, Retnakaran A (1998) The *ultraspiracle* gene of the spruce budworm, *Choristoneura fumiferana*: cloning of cDNA and developmental expression of mRNA. Dev Gen 22:169–179

Perera SC, Ladd TR, Dhadialla TS, Krell PJ, Sohi SS, Retnakaran A, Palli SR (1999) Studies on two ecdysone receptor isoforms of the spruce budworm, *Choristoneura fumiferana*. Mol Cell Endocrinol 152:73–84

Restifo LL, Wilson TG (1998) A juvenile hormone agonist reveals distinct developmental pathways mediated by ecdysone-inducible broad complex transcription factors. Dev Gen 22:141–159

Retnakaran A (1970) Blocking of embryonic development in the spruce budworm, *Choristoneura fumiferana* (Lepidoptera: Tortricidae), by some compounds with juvenile hormone activity. Can Entomol 102:1592–1596

Retnakaran A (1973a) Ovicidal effect in the white pine weevil, *Pissodes strobi* (Coleoptera: Curculionidae), of a synthetic analogue of juvenile hormone. Can Entomol 105:591–594

Retnakaran A (1973b) Hormonal induction of supernumerary instars in the spruce budworm, *Choristoneura fumiferana* (Lepidoptera: Tortricidae). Can Entomol 105:459–461

Retnakaran A (1974) Induction of sexual maturity in the white pine weevil, *Pissodes strobi* (Coleoptera: Curculionidae), by some analogues of juvenile hormone. Can Entomol 106:831–834

Retnakaran A (1975) Hormone-mimetic and pharmacological effects of some juvenile hormone analogues on the embryonic respiration of the spruce budworm, *Choristoneura fumiferana* (Clemens). Comp Biochem Physiol 50C:81–87

Retnakaran A, Granett J, Ennis T (1985) Insect growth regulators. In: Kerkut GA, Gilbert LI (eds) Comprehensive insect physiology, biochemistry and pharmacology, vol 12. Pergamon Press, Oxford, pp 529–601

Retnakaran A, Hiruma K, Palli SR, Riddiford LM (1995) Molecular analysis of the mode of action of RH-5992, a lepidopteran-specific, non-steroidal ecdysteroid agonist. J Inset Physiol 25:109–117

Retnakaran A, MacDonald A, Tomkins W, Davis C, Brownwright AJ, Palli SR (1997a) Ultrastructural effects of a non-steroidal ecdysone against RH-5992, on the sixth instar larva of the spruce budworm, *Choristoneura fumiferana*. J Insect Physiol 43:55–68

Retnakaran A, Smith LFR, Tomkins WL, Primavera MJ, Palli SR, Payne N, Jobin L (1997b) Effect of RH-5992, a non-steroidal ecdysone agonist, on the spruce budworm, *Choristoneura fumiferana* (Lepidoptera: Tortricidae): laboratory, greenhouse and ground spray trials. Can Entomol 129:871–885

Riddiford LM (1976) Hormonal control of insect epidermal cell commitment in vitro. Nature 259:115–117

Riddiford LM (1991) Hormonal control of sequential gene expression in insect epidermis. In: Binnington K, Retnakaran A (eds) Physiology of the insect epidermis. CSIRO, Melbourne, pp 46–54

Riddiford LM (1996) Juvenile hormone: the status of its "status quo" action. Arch Insect Biochem Physiol 32:271–286

Riddiford LM, Truman JW (1993) Hormone receptors and the regulation of insect metamorphosis. Am Zool 33:340–347

Saleh DS, Zhang J, Wyatt GR, Walker VK (1998) Cloning and characterization of an ecdysone receptor cDNA from *Locusta migratoria*. Mol Cell Endocrinol 143:91–99

Sevala V, Davey KG (1989) Action of juvenile hormone on the follicle cells of *Rhodnius prolixus*: evidence for a novel regulatory mechanism involving protein kinase C. Experientia 45:355–356

Sevala V, Davey KG, Prestwich GD (1995) Photoaffinity labeling and characterization of a juvenile hormone binding protein in the membrane of follicle cells of *Locusta migratoria*. Insect Biochem Mol Biol 25:267–273

Shea MJ, King DL, Conboy MJ, Mariani BD, Kafatos FC (1990) Proteins that bind to *Drosophila* chorion cis-regulatory elements: a new C2H2 zinc finger protein and a C2C2 steroid receptor-like component. Genes Dev 4:1128–1140

Shemshedini L, Lanoue M, Wilson TG (1990) Evidence for a juvenile hormone receptor involved in protein synthesis in *Drosophila melanogaster*. J Biol Chem 265:1913–1918

Smagghe G, Degheele D (1994) Action of the nonsteroidal ecdysteroid mimic RH-5849 on larval development and adult reproduction of insects of different orders. Invert Reprod Dev 25:227–236

Socha R, Marec F (1989) Genotoxicity of the anti-juvenile hormone agent precocene II as revealed by the *Drosophila* wing spot test. Mutagenesis 4:216–220

Steel CGH, Davey KG (1985) Integration in the insect endocrine system. In: Kerkut GA, Gilbert LI (eds) Comprehensive insect physiology, biochemistry and pharmacology, vol 8. Pergamon Press, Oxford, pp 1–35

Sugumaran M (1988) Molecular mechanisms of cuticular sclerotization. Adv Insect Physiol 21:179–231

Suhr ST, Gil EB, Senut MC, Gage FH (1998) High level transactivation by a modified *Bombyx* ecdysone receptor in mammalian cells without exogenous retinoid X receptor. Proc Natl Acad Sci USA 95:7999–8004

Sundaram MS, Palli SR, Ishaaya I, Krell PJ, Retnakaran A (1998a) Toxicity of ecdysone agonists correlates with the induction of CHR3 mRNA in the spruce budworm. Pesticide Biochem Physiol 62:201–208

Sundaram M, Palli SR, Sohi SS, Dhadialla TS, Retnakaran A (1998b) Basis for selective action of a synthetic molting hormone agonist, RH-5992, on lepidopteran insects. Insect Biochem Mol Biol 28:693–703

Svoboda JA, Thompson MJ, Robbins WE (1972) Azasteroids, potent inhibitors of insect molting and metamorphosis. Lipids 7:553–556

Swevers L, Drevet JR, Lunke MD, Iatrou K (1995) The silkmoth homolog of the *Drosophila* ecdysone receptor (B1 isoform): cloning and analysis of expression during follicular cell differentiation. Insect Biochem Mol Biol 25:57–866

Thomas HE, Stunnenberg HG, Stewart AF (1993) Heterodimerization of the *Drosophila* ecdysone receptor with retinoid X receptor and *ultraspiracle*. Nature 362:471–475

Thummel CS (1995) From embryogenesis to metamorphosis: the regulation and function of *Drosophila* nuclear receptor superfamily members. Cell 6:871–877

Thummel CS (1997) Dueling orphans – interacting nuclear receptors coordinate *Drosophila* metamorphosis. Bioessays 8:669–72

Tobe SS, Stay B (1985) Structure and regulation of the corpus allatum. Adv Insect Physiol 18:305–432

Truman JW (1990) Neuroendocrine control of ecdysis. In: Oshinishi E, Ishizaki H (eds) Molting and metamorphosis. Springer, Berlin Heidelberg New York, pp 67–82

Truman JW (1992) The eclosion hormone system of insects. Prog Brain Res 92:361–374

Tsai C-C, Kao HY, McKeown M, Evans RM (1999) SMRTER, a Drosophila nuclear receptor coregulator, reveals that EcR-mediated repression is critical for development. Mol Cell 4:175–178

Tzertzinis G, Malecki A, Kafatos FC (1994) BmCF1, a *Bombyx mori* RXR-type receptor related to the *Drosophila* ultraspiracle. J Mol Biol 238:479–486

Verras M, Mavroidis M, Kokolakis G, Gourzi P, Zacharopoulou A, Mintzas AC (1999) Cloning and characterization of CcEcR. An ecdysone receptor homolog from the Mediterranean fruit fly *Ceratitis capitata*. Eur J Biochem 265:798–808

Vogtli M, Imhof MO, Brown NE, Rauch P, Spindler-Barth M, Lezzi M, Henrich VC (1999) Functional characterization of two ultraspiracle forms (CtUSP-1 and CtUSP-2) from *Chironomus tentans*. Insect Biochem Mol Biol 10:931–42

West TF, Campbell GA (1950) DDT and newer persistent insecticides, 2nd edn. Chapman and Hall, London

Williams CM (1967) Third generation pesticides. Sci Am 217:13–17

Willy PJ, Umesono K, Ong ES, Evans RM, Heyman RA, Mangelsdorf DJ (1995) LXR, a nuclear receptor that defines a distinct retinoid response pathway. Genes Dev 9:1033–1045

Wilson TG, Ashok M (1998) Insecticide resistance resulting from an absence of target-site gene product. Proc Natl Acad Sci USA 95:14040–14044

Wilson TG, Fabian J (1986) A *Drosophila melanogaster* mutant resistant to a chemical analog of juvenile hormone. Dev Biol 118:190–201

Wing KD (1988) RH-5849, a nonsteroidal ecdysone agonist: effects on a *Drosophila* cell line. Science 241:467–469

Wing KD, Slawecki RA, Carlson GR (1988) RH-5849, a nonsteroidal ecdysone agonist: effects on larval lepidoptera. Science 241:470–472

Wyatt GR, Davey KG (1996) Cellular and molecular actions of juvenile hormone. II. Roles of juvenile hormone in adult insects. Adv Insect Physiol 26:1–155

Yao T-P, Segraves WA, Oro AE, McKeown M, Evans RM (1992) *Drosophila ultraspiracle* modulates ecdysone receptor function via heterodimer formation. Cell 71:63–72

Yao T-P, Forman BM, Jlang Z, Cherbas L, Chen J-D, McKeown M, Cherbas P, Evans RM (1993) Functional ecdysone receptor is the product of *EcR* and *ultraspiracle* genes. Nature 366:476–479

Imaginal Discs and Tissue Cultures as Targets for Insecticide Action

HERBERT OBERLANDER[1] and GUY SMAGGHE[2]

1
Introduction

It is a significant challenge to selectively control insect pests that are of agricultural, medical or veterinary importance, while at the same time protecting the environment. In recent decades, insect resistance, concerns for the environmental, and regulatory actions have reduced the availability and/or the effectiveness of classical insecticides. The discovery of insect growth regulators (IGRs), which act on the physiology and development of insects, has provided some new opportunities to develop selective, environmentally acceptable pesticides. Research on IGRs requires bioassays that reflect the selectivity and specific mode of action desired to control pest insects while, at the same time, minimizing any effects on non-target organisms. Thus, it is important to evaluate novel bioassays that show promise for determining the selectivity and mode of action of candidate insecticides. In this chapter, we focus primarily on the use of imaginal discs, in vivo and in vitro, for investigating the action of IGRs.

2
Imaginal Discs as Targets of Insect Hormones in Vivo and in Vitro

Holometabolous insects have devised an impressive strategy for dealing with the striking consequences of complete metamorphosis by setting aside cells destined for adult structures and assembling these progenitor cells into discrete epithelial pouches, which we call imaginal discs. Imaginal discs were first described more than two centuries ago by Lyonet (1762) in his anatomical work on Lepidoptera. However, the developmental role of imaginal discs was not recognized until work by Weismann (1864) in his research on Diptera. The imaginal discs, then, are maintained as a diploid population of proliferating cells that are ultimately committed to adult differentiation, but only in response to the appropriate hormonal signals that are present at metamorphosis (Oberlander 1985).

[1] Center for Medical, Agricultural and Veterinary Entomology, Agricultural Research Service, US Department of Agriculture, P.O. Box 14565, Gainesville, FL 32604, USA
[2] Laboratory of Agrozoology, Ghent University, Coupure Links 563, B-9000 Ghent, Belgium

Isaac Ishaaya (Ed.): Biochemical Sites of
Insecticide Action and Resistance
© Springer-Verlag Berlin, Heidelberg 2001

Much of the recent work on imaginal discs as targets of insecticides depends on the specific responses that these discs have to developmental hormones. Thus, a variety of IGRs act through either mimicking or inhibiting developmental processes that are ordinarily triggered by natural hormones. Hence, it is necessary to begin with an overview of research on the interactions of insect hormones and imaginal discs. Until recently, only a few publications on imaginal discs and hormonal regulation were available. However, this early work was important. In a series of elegant experiments based on transplantation in vivo of imaginal discs of *Drosophila melanogaster* and *Drosophila virilis*, Bodenstein (1943) demonstrated that imaginal discs that are transplanted into adult male hosts cease to develop unless ring glands, a source of molting hormone (ecdysone), are simultaneously transplanted.

During the 1950s and 1960s, imaginal discs became a favorite model system for developmental studies, particularly in Hadorn's laboratory in Switzerland (e.g., Nöthiger and Oberlander 1967; Ursprung and Nöthiger 1972; Oberlander 1972). There were also endocrine studies that focused on transplanted imaginal discs in vivo and the use of transplanted glands as sources of hormones. This work primarily addressed the role of ecdysteroids in stimulating differentiation of transplanted imaginal discs for the purposes of investigating developmental phenomena. Nevertheless, this research demonstrated the potential utility of imaginal discs for bioassays of insect hormones.

The dependence of differentiation of the imaginal discs on molting hormone was confirmed when ecdysone became available for investigational purposes simultaneously with advances in insect tissue culture which made it possible to examine hormonal effects in vitro. These tissue culture experiments were carried out with *Galleria mellonella* (Oberlander and Fulco 1967), *Chilo supressalis* (Agui et al. 1969) and *D. melanogaster* (Mandaron 1970; Postlethwait and Schneiderman 1970; Fristrom 1972). This combination of a highly sensitive target tissue, the imaginal discs, with known quantities of insect hormones in tissue culture proved to be a highly effective means of exploring the actions of molting hormones. Thus, the use of imaginal discs cultured in vitro became the basis for subsequent studies on IGRs and other developmental mimics/inhibitors with insecticidal potential.

One problem with these early investigations is that, while the first available IGRs were related to juvenile hormone (JH), the basic research on insect hormones and imaginal discs focused on ecdysteroids or glandular sources of the molting hormone. However, Vogt (1943) demonstrated that removal of larval ring glands, a source of both JH and molting hormone in Drosophila, prevented metamorphosis when transplanted simultaneously with imaginal discs into adults, and, conversely, removal of ring glands from larvae accelerated metamorphosis. Thus, the hypothesis was that the larval ring glands were producing juvenile hormone, and that this prevented metamorphosis. Analogous studies were conducted by Sehnal (1968) with corpora allata from *G. mellonella* larvae. A comprehensive series of investigations was then conducted on the ability of imaginal wing discs of the Indian meal moth, *Plodia interpunctella*, to respond to both ecdysteroids and JH in vivo and in vitro. In this work it was clear that the imaginal discs were highly sensitive to both 20 E and to JH-I, but that demonstration of ecdysteroid effects was more easily obtainable in tissue culture, while JH-I worked well in vivo. Therefore, experiments were carried out in which the imaginal discs were exposed in vivo to JH followed by testing of wing disc responsiveness to 20 E in vitro. These experiments showed that JH-I reversibly inhibited wing disc responses to 20 E only during the early part of the last larval instar (Oberlander and Tomblin 1972; Silhacek and Oberlander 1975; Oberlander and Silhacek 1976).

The early successes noted above, that utilized organ cultures of explanted imaginal discs to test ecdysteroids and even juvenile hormone, led to an extensive series of publications on the action of ecdysteroids on the differentiation of wing imaginal discs of *P. interpunctella* (e.g., Oberlander 1976; Dutkowski et al. 1977). These imaginal discs responded in tissue culture to 20 E with evagination and the production of pupal cuticle including the synthesis of chitin (Oberlander et al. 1978; Ferkovich et al. 1981). These findings were important, as the range of IGR actions was expanded subsequently to include inhibitors of chitin synthesis.

The use of imaginal discs in organ cultures made it possible to study hormone-tissue interactions in isolation from the intact organism and under defined conditions. This would become relevant subsequently in the understanding of the mode of action of several classes of IGRs. However, the organ culture approach is limited by its dependence on freshly dissected tissues for each experiment, which, as a practical matter, reduces the amount of tissue available. In addition, replications are only as repeatable as the standardized conditions of rearing of the source of the biological material. By contrast, medical research has made extensive use of continuous cell lines to investigate the action of putative therapeutic agents as part of pharmacological investigations. Thus, by analogy, the advent of specific cell lines, derived from imaginal discs, or other discrete tissues, that maintained a responsiveness to insect hormones, would be highly desirable. The advantages of cell lines are that a large amount of biological material is available at any given time for experiments, and replications are more accessible. However, there is one caveat that over hundreds of cell divisions the parent cell line may change and hormone responsiveness over many cell passages must be monitored.

Numerous insect cell lines have been established from ovaries or embryos, and some of these cell lines have proved to be hormonally responsive, including those derived from *D. melanogaster* (e.g., see reviews by Best-Belpomme et al. 1980; Cherbas et al. 1980; Dinan et al. 1990), *Manduca sexta* (Marks and Holman 1979; Oberlander et al. 1981; Marks et al. 1984), *Trichoplusia ni* and *Spodoptera littoralis* (Kislev et al. 1984), among others. However, the specific cell type that was the source of these cell lines was not known. In principle, it seemed that imaginal discs would be an ideal source of cells for establishing cell lines that were hormonally responsive. By their nature, imaginal discs are essentially "in culture" in the hemolymph of larval stage holometabolous insects, and are a diploid dividing population of cells that are destined to differentiate into adult tissues upon exposure to specific hormonal cues. The major question would then be whether such cell lines would retain their hormonal responsiveness. An early effort to develop cell lines specifically from imaginal discs of Lepidoptera indicated that imaginal wing discs in organ culture formed extensive monolayers due to cell migration and mitosis, and the cells isolated from these explants survived for 5 weeks (Kurtti and Brooks 1970). Moreover, imaginal discs in the second half of the last larval instar had the ability to respond to 20 E, and this responsiveness increased as the instar progressed (Oberlander and Silhacek 1976). Thus, cell lines would be useful for research on hormonal action if they were established from imaginal discs at a stage in which hormonal responsiveness to ecdysteroids had already been achieved. However, for studies of JH it should be remembered that full competence to respond to ecdysteroids occurred as responsiveness to JH-I declined. This makes it problematic to have a single cell line that would be useful for investigating the action of both 20 E and JH.

There have been intensive efforts to establish cell lines from Lepidoptera to study the action of hormones and IGRs. To do this, imaginal wing discs from the mid-last larval instar of *P. interpunctella*, *Spodoptera frugiperda* and *Trichoplusia ni* were explanted and over time developed into continuous cell lines (IAL-TND1, IAL-PID2, IAL-SFD1), all of which responded to ecdysteroids (Lynn et al. 1982; Lynn and Oberlander 1983). The nature of the response of these lepidopteran cell lines to ecdysteroid hormones has been well documented and has been reviewed by Oberlander and Miller (1987) and Oberlander and Ferkovich (1994). In addition, a *D. melanogaster* cell line derived from wing imaginal discs displayed reduced proliferation when exposed to 20 E, an effect ameliorated by JH-III (Cottam and Milner 1998).

Manifestly, by the 1990s, both organ cultures and cell lines derived from imaginal discs of Lepidoptera had been well studied with respect to ecdysteroid responsiveness, with evidence that imaginal discs synthesize chitin in culture. However, there were few indications that these tissue culture systems would be advantageous for studying reactions to JH and related IGRs.

3
Insecticide Action In Vitro: Juvenile Hormone Mimics

Mimics of JH were the first IGRs investigated for practical control of insect pests, based entirely on activity studies in vivo. Experiments on JH in vitro were difficult because of problems with solubility and stability, which had to be resolved before experiments in tissue culture could be fruitful. Thus, the testing of JH mimics in tissue culture, particularly with imaginal discs or hormonally responsive cell lines, has been quite limited. In *D. melanogaster* cell lines, the JH mimics, ethyl dichlorofarnesoate and methoprene, prevent the arrest of proliferation induced by 20 E (Wyss 1976; Cherbas et al. 1989). Also, methoprene inhibited vesicle formation in *D. melanogaster* cell lines treated with ecdysone (Milner and Dübendorfer 1982). These results are consistent with in vivo tests with farnesyl methyl ether, a mimic of JH, that was tested by Dewes (1975) on genital and wing imaginal discs of *Ephestia kuehniella*. Exposure of last larval instar to the JH mimic delayed growth and differentiation of the imaginal discs, an effect that was reversed when the larvae were placed on an IGR-free diet.

Recently, the action of JH and JH mimics on an imaginal-disc-derived cell line, IAL-PID2, has been studied (Oberlander et al. 2000). Methoprene, fenoxycarb, and farnesol inhibited cell proliferation after 3 days of exposure of the cells to each of the compounds, while linoleic acid had no effect. The concentrations at which the JH compounds were effective (50 µg/ml and greater) were in the same range of concentrations that were effective in inhibiting ecdysteroid-induced development in intact *P. interpunctella* imaginal discs in vitro (Oberlander and Tomblin 1972). The suitability of this cell line as a routine JH mimic assay was constrained by the lack of correspondence between the degree of inhibition of cell proliferation in vitro and the relative activity of the JH compounds in vivo. Thus, although farnesol was 48× less active than JH-I and more than 100× less active than methoprene in a *Rhodnius prolixus* assay (Patterson and Schwarz 1977), the compounds were equally effective in the in vitro experiments. However, in a cell strain derived from the parent line of *Plodia* cells, Willis (1997) reported that as little as 50 nM of JH-I prevented ecdysteroid-induced detachment of PID cells from the culture dish surface.

4
Insecticide Action in Vitro: Chitin Synthesis Inhibitors

4.1
Organ Cultures

The successful stimulation of chitin synthesis by ecdysteroids in tissue culture provided favorable prospects for investigating the action of CSIs in vitro. Organ cultures of imaginal discs of *P. interpunctella* have been used as a model system to study ecdysteroid-induced chitin synthesis and its inhibition by agents which block RNA and protein synthesis or interfere with microtubule integrity (Oberlander et al. 1978, 1980, 1983). Similarly, cultured integument of *P. interpunctella*, *Galleria mellonella* (Ferkovich et al. 1981) and *Tenebrio molitor* (Soltani et al. 1987) synthesized chitin in tissue culture. Based on research on both imaginal discs and epidermis in vitro, Ferkovich et al. (1981) proposed that 20 E enhanced the uptake of chitin precursors, induced the production of chitin synthetase, and stimulated the synthesis of cuticular proteins. Clearly, these cultured tissues that synthesize chitin could be used to investigate the mode of action of chitin synthesis inhibitors (CSIs). For example, chitin synthesis in cultured integument from *Periplaneta americana* was inhibited by diflubenzuron (Nakagawa and Matsumura 1994; Nakagawa et al. 1993). For an overview, see Oberlander and Silhacek (1998).

A comprehensive examination of the effects of benzoylphenyl urea compounds on chitin synthesis was carried out with cultured wing imaginal discs of *Spodoptera frugiperda* (Mikolajczyk et al. 1994). In these experiments wing discs from last-instar larvae were treated with chlorfluazuron, diflubenzuron or teflubenzuron. All three CSIs inhibited 20 E-dependent GlcNAc incorporation into chitin in the cultured wing discs. Interestingly, inhibition occurred with as little as a 4-h exposure to the CSI before, during, or at the conclusion of treatment with 20 E. The treatment of freshly dissected wing discs with chlorfluazuron for only 15 min resulted in a 90% inhibition of chitin synthesis when measured 3 days later (Mikolajczyk et al. 1994). These findings are in contrast to the expectation that the effects of benzoylphenyl ureas on chitin synthesis would be reversible (Binnington and Retnakaran 1991).

The hypothesis that inhibition of chitin synthesis by benzoylphenyl ureas might be caused by an inhibition of uptake of GlcNAc from the culture medium was tested in imaginal disc organ culture. In cultured wing discs of *P. interpunctella* diflubenzuron and teflubenzuron inhibited incorporation of radiolabeled GlcNAc into chitin. In these experiments there was no correlation with inhibition of uptake of GlcNAc or of the transported but non-metabolized sugars, 3-*O*-methyl-D-glucose and 2-deoxy-D-glucose (Oberlander et al. 1991). There also was no correspondence between inhibition of uptake of GlcNAc and inhibition of chitin synthesis in *S. frugiperda* wing discs treated with diflubenzuron, teflubenzuron or chlorfluazuron (Mikolajczyk et al. 1994). Thus, these tissue culture experiments failed to support the hypothesis that the inhibition of uptake of chitin precursors is the basis for the inhibition of chitin synthesis by benzoylphenyl ureas.

Mitsui et al. (1984, 1985) evaluated a possible role for membrane transport in the action of diflubenzuron. They reported that diflubenzuron inhibited chitin synthesis in cultures of midgut from last-instar larvae of *Mamestra brassicae*. Diflubenzuron did not inhibit chitin synthesis in cultured midguts that were turned inside out and ligated at both ends and incubated with GlcNAc and UDP-GlcNAc. Thus, if the chitin-producing cells were exposed to a direct chitin precursor, diflubenzuron was not inhibitory. Thus, Mitsui concluded that diflubenzuron inhibited the transport of UDP-GlcNAc across the midgut epithelium and thereby prevented chitin synthesis. It is not clear, however, if this hypothesis is applicable to 20E-dependent chitin synthesizing systems (Oberlander and Silhacek 1998).

4.2
Cell Lines

The feasibility of using insect cell lines to study the mode of action of inhibitors of chitin synthesis was suggested by Londershausen et al. (1987) and Oberlander (1989). Also, Ward et al. (1988) showed that the cockroach cell line, UMBGE, synthesized a chitin-like material which was stimulated by 20 E and inhibited by diflubenzuron. In another example, chitin synthesis in an epithelial cell line from *Chironomus tentans* was inhibited by 20 E and SIR 8514, a benzoylphenyl urea derivative (Spindler-Barth et al. 1989; Londershausen et al. 1993; Spindler-Barth 1993).

CSIs were tested in the IAL-PID2 cell line derived from *P. interpunctella* wing imaginal discs to determine if they would inhibit uptake of chitin precursors in these target cells (Porcheron et al. 1988). The PID2 cells responded to treatment with 20 E with increased uptake of GlcNAc, *N*-acetylgalactosamine and D-glucosamine, precursors of chitin, but not D-glucose or D-mannose (Porcheron et al. 1988). Teflubenzuron, a CSI, did not reduce GlcNAc uptake by the PID2 cells, while diflubenzuron had a small inhibitory effect (Oberlander et al. 1991). These results are consistent with those obtained with intact imaginal discs (Oberlander and Silhacek 1998).

5
Insecticide Action In Vitro: Ecdysteroid Agonists

5.1
Organ Cultures

The advent of non-steroidal agonists of ecdysteroids provided an excellent opportunity for taking advantage of existing tissue culture systems that had been developed for the study of 20 E (see reviews by Oberlander et al. 1995; Dhadialla et al. 1998). For example, the non-steroidal ecdysteroid agonist, RH-5849, stimulated incorporation of GlcNAc into chitin in wing imaginal discs from last-instar larvae of *P. interpunctella*. RH-5849 was 10–100× less potent than 20 E for stimulation of chitin synthesis and evagination in vitro (Oberlander et al. 1990; Silhacek et al. 1990). Similar results were obtained with integumentary fragments in culture (Oikawa et al. 1993). Concentrations of 0.1 to 1.0 µM were stimulatory; however, higher concentrations were less effective. More recently, Smagghe et al. (1999a) reported that RH-5992, at doses lower than required for 20 E, induced the premature synthesis of chitin as measured by the incorporation of ^{14}C-GlcNAc in cultured claspers of the European corn borer, *Ostrinia nubilalis.* There is ultrastructural evidence for new cuticle formation in imaginal wing discs of the beet armyworm, cultured in the presence of RH-5992. However, the development of treated caterpillars is abnormal and the newly formed cuticle is imperfect and unable to survive a molt (Smagghe et al. 1996b). Thus, RH-5849 and RH-5992 mimicked 20 E by stimulating chitin synthesis in cultured tissues, although the cuticle formed was not normal. From these results we conclude that the major insecticidal mode of action of these compounds, particularly for lepidopteran pests, resides in their effects on cuticle formation, where they mimic the action of 20 E, even though they are non-steroidal. During metamorphosis in holometabolous insects, cellular disintegration and/or reorganization is mediated by lysosomal activity triggered by 20 E. In this context, Ashok and Dutta-Gupta (1991) showed that RH-5849 stimulated the production of acid phosphatase in cultures of fat body of the rice moth, *Corycra cephalonica,* suggesting that these dibenzoylhydrazines may have additional effects in insects.

5.2
Cell Lines

RH-5849, RH-5992 and 20 E act similarly in several cell lines. Wing (1988) demonstrated the ecdysteroid-like action of RH-5849 at the cellular level in experiments with *D. melanogaster* Kc cells (Echalier and Ohannessian 1970). RH-5849 caused a cessation of proliferation, clumping, and the formation of extended cellular processes, effects that were indistinguishable from the effects of 20 E. As with the organ culture results discussed above, 20 E was 100 times as effective as RH-5849. Likewise, in the imaginal disc cell line from *P. interpunctella,* IAL-PID2, 20 E was 100 times as effective as RH-5849 in inhibiting cell proliferation (Silhacek et al. 1990). Similarly, Clément et al. (1993) showed that in a dipteran tumorous blood cell line (Gateff et al. 1980), RH-5849 reduced proliferation, but at substantially higher concentrations than 20 E. On the other hand, RH-5992 inhibited proliferation in the PID2 cells at the same concentration as 20 E (2×10^{-7} M) (Oberlander et al. 1995).

The hormone 20 E causes changes in morphology in cell lines that grow as multicellular vesicles. For example, in a *Chironomous tentans* cell line (Wyss 1982), the vesicles shrank in response to either 20 E or RH-5849 (Spindler-Barth et al. 1991). The situation was more complex for the *Trichoplusia ni* imaginal disc derived cell line, IAL-TNDI (Lynn et al. 1982). After the 30th passage of this continuously maintained cell line, the fluid-filled multicellular vesicles spontaneously changed to solid clusters of cells. The vesicle form could be restored by the addition of a vesicle promoting factor (VPF), which was isolated from the hemolymph as a 16.9-kDa peptide (Ferkovich et al. 1987, 1995). Interestingly, the action of VPF was blocked by either 20 E (Oberlander et al. 1987) or RH-5849 (Oberlander et al. 1995). In addition RH-5849 could substitute for 20 E in stimulating incorporation of *N*-acetyl-D-glucosamine into chitin, a process inhibited by diflubenzuron or teflubenzuron (Oberlander et al. 1991; Fig. 1). Thus, a series of quite specific effects of 20 E on cell lines were duplicated by RH-5849, though higher concentrations were usually required.

Biochemical changes in tissue cultures were also reported. For example, both 20 E and RH-5849 included increased acetylcholinesterase activity in dipteran cell lines (Wing 1988; Spindler-Barth et al. 1991). Also, 20 E, as well as RH-5849 and RH-5992, inhibited the spontaneous synthesis of chitin by the *C. tentans* cell line (Spindler-Barth et al. 1991; Spindler et al. 1994). More typically, the IAL-PID2 imaginal disc cell line is stimulated to synthesize chitin by 20 E and by RH-5849 (Silhacek et al. 1990).

If RH-5849, RH-5992 and related compounds mimic 20 E through the same mode of action as the natural hormone, then they should have the same molecular mode of action. In insects the molecular target for ecdysteroids consists of at least two proteins, the ecdysteroid receptor (EcR) and the product of another gene, ultraspiracle (USP) (Yao et al. 1993). Both EcR and USP are members of the steroid hormone receptor superfamily (Hangelsdorf et al. 1995). Ecdysteroids functionally bind to EcR only when EcR and USP exist as a heterodimeric complex (Yao et al. 1993). However, additional factors may also

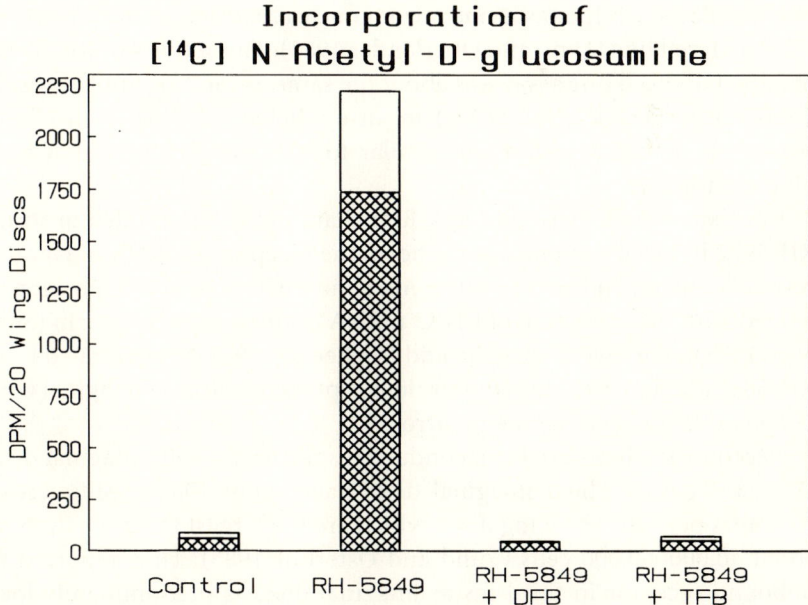

Fig. 1. Incorporation of [^{14}C]-*N*-acetyl-D-glucosamine into wing discs incubated in vitro with 100 µM RH-5849 and either 10 µM diflubenzuron or 10 µM teflubenzuron. The results are expressed as dpm/20 wing discs, ±SD. (Oberlander et al. 1991)

be involved for ligand binding and ecdysteroid-dependent gene regulation. The expression of specific EcR isoforms has been correlated with particular responses, suggesting that different EcR isoforms may govern distinct steroid-induced responses (Schubiger et al. 1998). These observations support the hypothesis that different EcR isoforms control cell-type specific responses during remodeling of the nervous system at metamorphosis.

RH-5849 binds to ecdysteroid-specific receptors that are members of the steroid hormone receptor superfamily (Koelle et al. 1991; Sobek et al. 1993; Dhadialla et al. 1998). RH-5849 was 100 times less effective than 20 E in displacing radiolabeled ponasterone A from receptors in D. melanogaster Kc cells (Wing 1988). Spindler-Barth et al. (1991) made similar observations in the C. tentans cell line, where RH-5849 was four times less effective than 20 E. However, RH-5992 was more effective than 20 E in the same binding assay (Carlson et al. 1994; Spindler et al. 1994). There was also a good correlation between the extent of enzyme induction and the affinity of the corresponding compounds to the ecdysteroid receptor.

Compelling evidence for the interaction of RH-5849 and RH-5992 with ecdysteroid receptors also comes from experiments on cell lines from several species (Dhadialla et al. 1998). When Drosophila Kc cells were incubated for 4 weeks with either 20 E or RH-5849 they became resistant to ecdysteroids (Wing 1988). Both of these resistant populations of Kc cells had reduced capacity to bind [³H]ponasterone A. Wing and Ramsay (1989) also demonstrated that RH-5849 displaced [³H]ponasterone A from nuclear extracts of the P. interpunctella PID2-IAL cell line (Lynn and Oberlander 1983). The activity of RH-5849 in this imaginal disc cell line assay was about the same as for 20 E. Similar results were obtained by Quack et al. (1995) in an epithelial cell line from Chironomus tentans, in which RH-5849 was similar to 20 E, but RH-5992 showed greater binding affinity.

Consistent with these findings, Retnakaran et al. (1995) demonstrated that RH-5992 induced Manduca sexta hormone receptor 3 mRNA in larval epidermis of M. sexta, cultured in vitro. Also, the toxicity of ecdysone agonists correlated with the induction of CHR3 mRNA in the spruce budworm (Sundaram et al. 1998). Moreover, Smagghe and Degheele (1995) demonstrated binding of RH-5849 and RH-5992 to ecdysteroid receptors in cultured imaginal wing discs of Spodoptera exigua larvae in vitro.

Receptor studies have been conducted specifically with imaginal discs from D. melanogaster. These imaginal discs have about 1000 specific ecdysteroid receptors per cell. The wing discs, which are the largest imaginal discs, contain about 40,000–60,000 cells (Yund and Osterbur 1985). They reported that the unbound receptor in third-instar imaginal discs is predominantly located in the nucleus as are vertebrate steroid receptors. Mass-isolated imaginal discs of D. melanogaster larvae were used in these early studies, which investigated in detail ecdysteroid action in dipteran imaginal discs. By contrast, there are relatively few papers on ecdysteroid receptors in imaginal discs isolated by dissection, particularly from non-drosophilids. In recent years there have been several studies using hand-dissected imaginal discs for ecdysteroid receptors in non-drosophilid species; these were performed with leg discs of *Calliphora vicina* (Terentiou et al. 1993) and wing discs of *Spodoptera exigua* (Smagghe and Degheele 1995). Typically, the level of 20E causing 50% competition binding in imaginal discs correlated well with the ecdysteroid peak in the hemolymph of the respective last-instar larvae (Terentiou et al. 1993; Smagghe et al. 1995). The parallel determination of biological activity and receptor binding competition with individual imaginal discs made it possible to evaluate the activity of ecdysteroids and agonists. Additionally, it was of great interest to extend these data to other species of insects.

Recently, Smagghe et al. (1996a) reported on the actions of ecdysteroids and non-steroidal agonists in cultured imaginal wing discs of the Colorado potato beetle *Leptinotarsa decemlineata*, and the greater wax moth *G. mellonella*. Further, Smagghe et al. (1999b) determined the effectiveness of RH-5849, RH-5992, RH-2485 (methoxyfenozide) and RH-0345 (halofenozide) on evagination and binding affinity in imaginal discs of laboratory and multi-resistant field strains of *S. littoralis*. They found that the order of effectiveness of these ecdysone agonists with respect to evagination and receptor binding was the same as their larval toxicity. The imaginal disc assay made possible both the measurement of the potency of ecdysteroid agonists as well as their binding affinity to the ecdysteroid receptor. The specific action on insect selectivity of the dibenzoylhydrazine ecdysteroid agonists, especially the high toxicity of RH-5992 for Lepidoptera and the non-susceptibility of Coleoptera, appears to be related to selective binding to the ecdysteroid receptors.

Smagghe et al. (1999c) investigated the molecular mechanisms involved in the action of these agonists with a quantitative structure-activity relationship (QSAR) study of a series of 45 dibenzoylhydrazines with different ring substituents, including RH-2485. This latter compound is the most Lepidoptera selective and toxic ecdysteroid agonist from the RH-5992 family (Ishaaya et al. 1995; Le et al. 1996; Smagghe et al. 1999d). From these studies, the molting activity expressed by pEC50 (M) (log of the reciprocal of EC50) was compared with the competitive receptor binding activity by pIC50 (M) and followed a linear regression: pEC50 = 0.997 (±0.206) pIC50 + 0.421 (±1.523) with r = 0.994. The concentrations required to elicit disc evagination were comparable to the doses that induced larval toxicity (Table 1). In addition, the molecular structure of ligands from the RH-5992 family for receptor binding in the beet armyworm and the rice stem borer, *Chilo suppresalis* (Oikawa et al. 1994a,b), revealed a high similarity.

This parallelism agrees with the phylogenetic tree based on quantitation of the sequence distances of the ecdysteroid receptor in different species of varying insect orders (Dhadialla and Tzertzinis 1997). Results from this work

Table 1. Activity of 20 E, ponasterone A, RH-5849, RH-5992 and RH-2485 in vivo against last-instar larvae, and in vitro on evagination, and receptor binding in imaginal discs of *Spodoptera exigua*. (Smagghe et al. 1999c)

Compound	pLD_{50} (mmol/insect)[a]	EC_{50} (nM)[b]	IC_{50} (nM)[c]
RH-2485 (methoxyfenozide)	8.27 ± 0.21	1.7 ± 1.0	3.4 ± 1.2
RH-5992 (tebufenozide)	7.06 ± 0.29	10.5 ± 1.0	32.1 ± 11.6
RH-5849	4.91 ± 0.23	777 ± 77	1397 ± 335
20 E	ND	67.1 ± 12.1	238 ± 102
Ponasterone A	ND	2.5 ± 0.5	6.0 ± 1.5

ND=not determined

[a] Larvicidal toxicity is expressed in terms of the log value (mean ± SD) of the reciprocal of LD_{50}s (mmol/insect) representing the doses that cause mortality in 50% of test insect larvae and calculated by probit analysis

[b] Biological activity is expressed as the ability to induce wing disc evagination. Data are means ± SE with EC_{50}s representing the concentration that induces evagination in 50% of the cultured wing discs

[c] Receptor binding affinity was measured as the ability to compete with 6 nM ^{3}H-PonA. Data are means ± SE with IC_{50}s representing the concentration that induces 50% competition of the amount of specifically bound ^{3}H-PonA

showed that *Bombyx mori*, *Manduca sexta*, and *Choristoneura fumiferana* are very closely related, and that these species were more closely related to Diptera than to Coleoptera. On the other hand, the receptor amino acid fragments of crayfish were fully separated. The similarity of effects on Lepidoptera may result from an analogy in the ecdysteroid receptor sequence/structure. Hence, this may explain why RH-5992 and RH-2485 are more specific to Lepidoptera in their toxicity and relatively ineffective with Coleoptera (Aller and Ramsay 1988; Hsu 1991; Darvas et al. 1992; Smagghe and Degheele 1994a,b; Oberlander et al. 1995; Le et al. 1996; Smagghe et al. 1996a; Dhadialla et al. 1998).

Bidmon and Sliter (1990) and Jindra (1994) have proposed that the structure and biochemical binding properties of ecdysteroid receptors vary among insect species, especially between orders. In this context, a distinct ligand molecular structure is important in provoking a differential receptor binding affinity that may explain differences in potency. One would expect that, since dibenzoylhydrazines act as mimics of the ecdysteroid hormones, the interaction of such ecdysteroid agonists with the intracellular ecdysteroid receptor would occur in a lock-and-key manner (Yund and Osterbur 1985). This suggests that the selective larval activity of the tested non-steroidal ecdysteroid agonists is related to a differential ecdysteroid receptor that might be specific to different families or orders. The results of the current assays with cultured insect tissues and cells, and with whole insect larvae, suggest that the molecular mechanism of ecdysteroid agonists is similar among lepidopteran species.

However, further fundamental research in performing comparative assays with a greater number of species, especially with non-Lepidoptera, is necessary before a general conclusion can be drawn.

Moreover, it is remarkable that RH-5992 only induced precocious molting in larvae that had ingested the compound during the early part of the stadium, before the appearance of the ecdysteroid peak. One possible reason for this age-dependent effect may be that ecdysteroids regulate molting and metamorphosis by controlling the expression of genes by different mechanisms. There are three different isoforms of EcR (EcR-A, EcR-B1 and EcR-B2) present before and after the commitment peak of ecdysteroids in *Drosophila melanogaster* (Talbot et al. 1993). Similarly, *Manduca sexta* has EcR-A and EcR-B1 (Jindra et al. 1996; Lan et al. 1999), and these are similar to the *Drosophila* EcR isoforms. Both EcR-A and EcR-B1 appeared in the larval wing discs after pupal commitment and persisted throughout pupal differentiation. Both isoforms of EcR appeared in every molt, with EcR-B1 present during the commitment and pre-differentiative phases and EcR-B1 present at the onset of cuticle synthesis. Thus, the interactions of non-steroidal ecdysteroid agonists and the various EcR isoforms can be expected to be complex with respect to species, stage of development and tissue type.

Acknowledgements. Dr. Guy Smagghe acknowledges the National Fund of Scientific Research, Brussels, for a post-doctoral fellowship. The authors thank S. Dyby, C. E. Leach, and S. Valles for reviewing the manuscript.

References

Agui N, Yagi S, Fugaya M (1969) Effects of ecdysterone on the in vitro development of wing discs of rice stem borer, *Chilo supressalis.* Appl Entomol Zool 4:158–159

Aller HE, Ramsay JR (1988) RH-5849 – a novel insect growth regulator with a new mode of action. Proc BCPC-Pests Diseases 5:511–518

Ashok A, Dutta-Gupta A (1991) In vitro effects of non-steroidal ecdysone agonist RH-5849 on fat body acid phosphatase activity in rice moth, *Corcyra cephalonica* (Insecta). Biochem Int 24:69–75

Best-Belpomme M, Courgeon AM, Echalier G (1980) Development of a model for the study of ecdysteroid action: *Drosophila melanogaster* cells established in vitro. In: Hoffmann JA (ed) Progress in ecdysone research. Elsevier/North Holland Biomedical Press, Amsterdam, pp 379–392

Bidmon H-J, Sliter TJ (1990) The ecdysteroid receptor. Invert Reprod Dev 18:13–27

Binnington K, Retnakaran A (1991) Epidermis – a biologically active target for metabolic inhibitors. In: Binnington K, Retnakaran A (eds) Physiology of the insect epidermis. CSIRO, Melbourne, pp 307–326

Bodenstein D (1943) Hormones and tissue competence in the development of *Drosophila*. Biol Bull 84:34–58

Carlson GR, Dhadialla TS, Thompson C, Ramsay T, Thirugnanam M, James W, Slawecki R (1994) Insect toxicity, metabolism and receptor binding characteristics of the non-steroidal ecdysone agonist, RH-5992. Proc 11th Ecdysone Workshop, p 43

Cherbas L, Cherbas P, Savakis C, Demetri G, Manteuffel-Cymborowska M, Yonge CD, Williams CM (1980) Studies of ecdysteroid action on a *Drosophila* cell line. In: Kurstak E, Maramorosch K,

Dübendorfer A (eds) Invertebrate systems in vitro. Elsevier/North Holland Biomedical Press, Amsterdam, pp 217–228

Cherbas L, Koehler MD, Cherbas P (1989) Effects of juvenile hormone on the ecdysone response of *Drosophila* Kc Cells. Dev Genet 10:177–188

Clément CY, Bradbrook DA, Lafont R, Dinan L (1993) Assessment of a microplate-based bioassay for the detection of ecdysteroid-like or antiecdysteroid activities. Insect Biochem Mol Biol 23:187–193

Cottam DM, Milner MJ (1998) The effect of juvenile hormone on the response of the Drosophila imaginal disc cell line C1 8+ to moulting hormone. J Insect Physiol 44:1137–1144

Darvas B, Polgar L, Tag El-din MH, Eröss K, Wing KD (1992) Developmental disturbances in different insect orders caused by an ecdysteroid agonist, RH-5849. J Econ Entomol 85:2107–2112

Dewes E (1975) Entwicklungsleistungen von Imaginalscheiben unter dem Einfluss von Farnesylmethlyäther bei *Ephestia kuehniella*. Z Verh Dtsch Zool Ges 1974:75–80

Dhadialla TS, Tzertzinis G (1997) Characterization and partial cloning of ecdysteroid receptor from a cotton boll weevil embryonic cell line. Arch Insect Biochem Physiol 35:45–57

Dhadialla TS, Carlson GR, Le DP (1998) New insecticides with ecdysteroidal and juvenile hormone activity. Annu Rev Entomol 43:545–569

Dinan L, Spindler-Barth M, Spindler K-D (1990) Insect cell lines as tools for studying ecdysteroid action. Invertebr Reprod Dev 18:43–53

Dutkowski AB, Oberlander H, Leach CE (1977) Ultrastructure of cuticle deposited in *Plodia interpunctella* wing discs after various β-ecdysone treatments in vitro. Wilhelm Roux' Arch 183:155–164

Echalier G, Ohanessian A (1970) In vitro culture of *Drosophila melanogaster* embryonic cells. In Vitro 6:162–172

Ferkovich SM, Oberlander H, Leach CE (1981) Chitin synthesis in larval and pupal epidermis of the Indian meal moth, *Plodia interpunctella* (Hübner), and the greater wax moth, *Galleria mellonella* (L.). J Insect Physiol 27:509–514

Ferkovich SM, Oberlander H, Dillard C, Leach CE (1987) Purification and properties of a factor from insect hemolymph that promotes multicellular vesicle formation in vitro. Arch Insect Biochem Physiol 67:73–83

Ferkovich SM, Miller SG, Oberlander H (1995) Multicellular-vesicle-promoting polypeptide from *Trichoplusia ni*: tissue distribution and *N*-terminal sequence. Arch Insect Biochem Physiol 29:381–390

Fristrom JW (1972) The biochemistry of imaginal disk development. In: Ursprung H, Nöthiger R (eds) Results and problems in cell differentiation. Springer, Berlin Heidelberg New York, pp 109–154

Gateff E, Gissmann L, Shrestha R, Plus N, Pfister H, Schröder J, Zur Hausen H (1980) Characterization of two tumorous blood cell lines of *Drosophila melanogaster* and the viruses they contain. In: Kurstak E, Maramorosch, K, Dübendorfer A (eds) Invertebrate systems. In vitro. Elsevier/North-Holland Biomedical Press, Amsterdam, pp 517–534

Hanglesdorf DJ, Thummel C, Beato M, Herrlich P, Schuetz G, Umesono K, Blumberg B, Kastner P, Mark M, Chambon P, Evans RM (1995) The nuclear receptor superfamily: the second decade. Cell 83:835–839

Hsu AC-T (1991) 1,2-Diacyl-1-alkylhydrazines: a new class of insect growth regulators. In: Baker DR, Fenyes JG, Molberg WK (eds) ACS symposium series, vol 443: synthesis and chemistry of agrochemicals II. American Chemical Society, Washington, DC, pp 478–4904

Ishaaya I, Yablonski S, Horowitz AR (1995) Comparative toxicity of two ecdysteroid agonists, RH-2485 and RH-5992, on susceptible and pyrethroid-resistant strains of the Egyptian cotton leafworm, *Spodoptera littoralis*. Phytoparasitica 23:139–145

Jindra J (1994) Gene regulation by steroid hormones: vertebrates and insects. Eur J Entomol 91:163–187

Jindra M, Malone F, Hiruma K, Riddiford LM (1996) Developmental profiles and ecdysteroid regulation of the RNAs for two ecdysone receptor isoforms in the epidermis and wings of the tobacco hornworm, *Manduca sexta*. Dev Biol 10:258–272

Kislev N, Segal I, Edelman M (1984) Ecdysteroids induce morphological changes in continuous cell lines of Lepidoptera. Wilhelm Roux' Arch 193:252–256

Koelle MR, Talbot WS, Segraves WS, Bender WA, Cherbas P (1991) The *Drosophila* EcR gene encodes an ecdysone receptor, a new member of the steroid receptor super family. Cell 67:59–77

Kurtti TJ, Brooks MA (1970) Growth of lepidopteran epithelial cells and hemocytes in primary cultures. J Invert Pathol 15:341–350

Lan Q, Hiruma K, Hu XA, Jindra M, Riddiford LM (1999) Activation of a delayed early gene encoding MHR3 by the ecysone receptor heterodimer EcR-B1-USP-1, but not by EcR-B1-USP-2. Mol Cell Biol 19:4897–4906

Le DP, Thirugnanam M, Lidert Z, Carlson GR, Ryan JB (1996) RH-2485: a new selective insecticide for caterpillar control. BCPC-Pests Dis 2:481–485

Londershausen M, Spindler-Barth M, Kammann V, Spindler KD, Thomas H (1987) Insect cell lines as models for the study of insecticides interfering with cuticle formation. Zentralbl Bakteriol Mikro Hyg [A] 267:302

Londershausen M, Turberg A, Buss U, Spindler-Barth M, Spindler KD (1993) Comparison of chitin synthesis from an insect cell line and embryonic tick tissues. In: Mussarelli RAA (ed) Chitin enzymology. European Chitin Society, Ancona, pp 101–108

Lynn DE, Oberlander H (1983) The establishment of cell lines from imaginal wing discs of *Spodoptera frugiperda* and *Plodia interpunctella*. J Insect Physiol 29:591–596

Lynn DE, Miller SG, Oberlander H (1982) Establishment of a cell line from lepidopteran wing imaginal discs: induction of newly synthesized proteins by 20-hydroxyecdysone. Proc Natl Acad Sci USA 79:2589–2593

Lyonet P (1762) Traité anatomique de la chenille qui ronge le bois de saule. La Haye

Mandaron P (1970) Development in vitro de disques imaginaux de la drosophile, aspects morpholgiques et histologiques. Dev Biol 22:298–320

Marks EP, Holman GM (1979) Ecdysone action on insect cell lines. In Vitro 15:300–307

Marks EP, Balke J, Klosterman H (1984) Evidence for chitin synthesis in an insect cell line. Arch Insect Biochem Physiol 1:225–230

Mikolaczyk P, Oberlander H, Silhacek DL, Ishaaya I, Shaaya E (1994) Chitin synthesis in *Spodoptera frugiperda* wing imaginal discs. I. Chlorfluazuron, diflubenzuron, and teflubenzuron inhibit incorporation but not uptake of [^{14}C]N-acetyl-D-glucosamine. Arch Insect Physiol Biochem 25:245–258

Milner MJ, Dübendorfer (1982) Tissue specific effects of the juvenile hormone analogue ZR-515 during metamorphosis in *Drosophila* cell cultures. J Insect Physiol 28:661–666

Mitsui T, Nobusawa C, Fukami J (1984) Mode of inhibition of chitin synthesis by diflubenzuron in the cabbage armyworm, *Mamestra brassicae* L. J Pestic Sci 9:19–26

Mitsui T, Hitturu T, Nobusawa C, Yamaguchi I (1985) Inhibition of UPD-N-acetylglucosamine transport by diflubenzuron across biomembranes of the midgut epithelial cells in the cabbage armyworm, *Mamestra brassicae* L. J Pestic Sci 10:55–60

Nakagawa Y, Matsumura F (1994) Diflubenzuron affects gamma-thioGTP stimulated Ca^{2+} transport in vitro in intracellular vesicles from the integument of the newly molted American cockroach, *Periplaneta americana* L. Insect Biochem Mol Biol 24:1009–1015

Nakagawa Y, Matsumura F, Hashino Y (1993) Effect of diflubenzuron on incorporation of [^{3}H]-N-acetylglucosamine ([^{3}H-NAGA]) into chitin in the intact integument from the newly molted American cockroach, *Periplaneta americana*. Comp Biochem Physiol C106:711–715

Nakagawa Y, Shimizu B-I, Oikawa N, Akamatsu M, Nishimura K, Kurihara N, Ueno T, Fujita T (1995) Three-dimensional quantitative structure-activity analysis of steroidal and dibenzoylhydrazine-type ecdysone agonists. In: Hansch C, Fujita T (eds) Classical and three-dimensional QSAR in agrochemistry. ACS symposium series, vol 606. American Chemical Society, Washington, DC, pp 288–301

Nöthiger R, Oberlander H (1967) Differentiation of pulsating regions in genital imaginal discs after culture in vivo (*Drosophila melanogaster*). J Exp Zool 164:61–68

Oberlander H (1972) The hormonal control of development of imaginal disks. In: Ursprung H, Nöthiger R (eds) Results and problems in cell differentiation, vol 5. Springer, Berlin Heidelberg New York, pp 156–172

Oberlander H (1976) Hormonal control of growth and differentiation of insect tissues cultured in vitro. In Vitro 12:225–235

Oberlander H (1983) Imaginal disks. In: Downs RGH, Laufer H (eds) Endocrinology of insects. Liss, New York, pp 503–507

Oberlander H (1985) The imaginal discs. In: Kerkut GA, Gilbert LI (eds) Comprehensive insect physiology, biochemistry and pharmacology, vol 2. Pergamon Press, Oxford, pp 151–182

Oberlander H (1989) Insect cell lines as models for the study of biorational pesticides. In Vitro Cell Dev Biol 25:18A

Oberlander H, Ferkovich SM (1994) Physiological and developmental capacities of insect cell lines. In: Maramorosch K, McIntosh AH (eds) Insect cell biotechnology. CRC Press, Boca Raton, pp 129–140

Oberlander H, Fulco L (1967) Growth and partial metamorphosis of imaginal disks of the greater wax moth, *Galleria mellonella*, in vitro. Nature 216:1140–1141

Oberlander H, Miller SG (1987) Lepidopteran cell lines: tools for research in physiology, development and genetics. In: Maramorosch K (ed) Advances in cell culture, vol 5. Academic Press, New York, pp 187–207

Oberlander H, Silhacek DL (1976) Action of juvenile hormone on imaginal discs of the Indian meal moth. In: Gilbert LI (ed) The juvenile hormones. Plenum Press, New York, pp 220–233

Oberlander H, Silhacek DL (1998) New perspectives on the mode of action of benzoylphenylurea insecticides. In: Ishaaya I, Degheele D (eds) Applied agriculture: insecticides with novel modes of action. Springer, Berlin Heidelberg New York, pp 92–105

Oberlander H, Tomblin C (1972) Cuticle deposition in imaginal disks: effects of juvenile hormone and fat body in vitro. Science 177:441–442

Oberlander H, Ferkovich SM, Van Essen F, Leach CE (1978) Chitin biosynthesis in imaginal discs cultured in vitro. Wilhelm Roux' Arch 185:95–98

Oberlander H, Ferkovich S, Leach E, Van Essen F (1980) Inhibition of chitin biosynthesis in cultured imaginal discs; effects of alpha-amanitin, actinomycin-D, cycloheximide, and puromycin. Wilhelm Roux' Arch 188:81–86

Oberlander H, Leach CE, Lynn DE (1981) Effects of cycloheximide on cellular elongation in a *Manduca sexta* cell line. Wilhelm Roux' Arch 190:60–61

Oberlander H, Lynn DE, Leach CE (1983) Inhibition of cuticle production in imaginal discs of *Plodia interpunctella* (cultured in vitro): effects of colcemid and vinblastine. J Insect Physiol 29:47–53

Oberlander H, Leach CE, Lanka S, Willis JH (1987) Ecdysteroid action on moth epithelial tissues and cell lines. Arch Insect Biochem Physiol 5:81–89

Oberlander H, Silhacek DL, Porcheron P (1990) Action of a non-steroidal ecdysteroid mimic, RH-5849, on the Indian meal moth in vivo and in vitro. Invertebr Reprod Dev 18:124

Oberlander H, Silhacek DL, Leach E, Ishaaya I, Shaaya E (1991) Benzoylphenylureas inhibit chitin synthesis without interfering with amino sugar uptake in imaginal wing discs of *Plodia interpunctella* . Arch Insect Biochem Physiol 18:219–227

Oberlander H, Silhacek DL, Porcheron P (1995) Non-steroidal agonists: tools for the study of hormonal action. Arch Insect Biochem Physiol 28:209–223

Oberlander H, Silhacek DL, Leach CE (1998) Interactions of ecdysteroid and juvenoid agonists in *Plodia interpunctella* (Hübner). Arch Insect Biochem Physiol 38:91–99

Oberlander H, Leach CE, Shaaya E (2000) Juvenile hormone and juvenile hormone mimics inhibit proliferation in a lepidopteran imaginal disc cell line. J Insect Physiol 46:259–265

Oikawa N, Nakagawa Y, Soya Y, Nishimura K, Kurihara N, Ueno T, Fujita T (1993) Enhancement of *N*-acetylglucosamine incorporation into the cultured integument of *Chilo suppressalis* by molting hormone and dibenzoylhydrazine insecticides. Pest Biochem Physiol 47:165–170

Oikawa N, Nakagawa Y, Nishimura K, Ueno T, Fujita T (1994a) Quantitative structure-activity analysis of larvacidal 1-(substituted benzoyl)-2-benzoyl-1-*tert*-butylhydrazines against *Chilo suppressalis*. Pestic Sci 41:139–148

Oikawa N, Nakagawa Y, Nishimura K, Ueno T, Fujita T (1994b) Quantitative structure-activity studies of insect growth regulators. X. Substituent effects on larvacidal activity of 1-*tert*-butyl-

1-(2-chlorobenzoyl)-2-(substituted benzoyl)hydrazines against *Chilo suppressalis* and design synthesis of potent derivatives. Pestic Biochem Physiol 48:135–144

Patterson JW, Schwarz M (1977) Chemical structure, juvenile hormone activity and persistence within the insect of juvenile hormone mimics for *Rhodnius prolixus*. J Insect Physiol 23:121–129

Porcheron P, Oberlander H, Leach CE (1988) Ecdysteroid regulation of amino sugar uptake in a lepidopteran cell line derived from imaginal discs. Arch Insect Biochem Physiol 7:145–155

Postlethwait JH, Schneiderman HA (1970) Induction of metamorphosis by ecdysone analogues; *Drosophila* imaginal discs cultured in vivo. Biol Bull 138:46–55

Quack S, Fretz A, Spindler-Barth M, Spindler K-D (1995) Receptor affinities and biological responses of non-steroidal ecdysteroid agonists on the epithelial cell line from *Chironomus tentans*. Eur J Entomol 92:341–347

Retnakaran A, Hiruma K, Palli SR, Riddiford LM (1995) Molecular analysis of the mode of action of HR-5992, a lepidopteran-specific, non-steroidal ecdysteroid agonist. Insect Biochem Mol Biol 25:109–117

Schubiger M, Wade AA, Carney GE, Truman JW, Bender M (1998) *Drosophila* EcR-b ecdysone receptor isoforms are required for larval molting and for neuron remodeling during metamorphosis. Development 125:2053–2062

Sehnal F (1968) Influence of the corpus allatum on the development of internal organs in *Galleria mellonella* L. J Insect Physiol 14:73–85

Silhacek DL, Oberlander H (1975) Time-dosage studies of juvenile hormone action on the development of *Plodia interpunctella*. J Insect Physiol 21:233–243

Silhacek DL, Oberlander H, Porcheron P (1990) Action of RH-5849, a non-steroidal ecdysteroid mimic, on *Plodia interpunctella* (Hübner) in vivo and in vitro. Arch Insect Biochem Physiol 15:201–212

Smagghe G, Degheele D (1994a) Action of a novel non-steroidal ecdysteroid mimic, tebufenozide (RH-5992), on insects of different orders. Pestic Sci 42:85–92

Smagghe G, Degheele D (1994b) Action of the non-steroidal ecdysteroid mimic RH-5849 on larval development and adult reproduction of insects of different orders. Invert Reprod Dev 25:227–236

Smagghe G, Degheele D (1994c) Effects of the ecdysteroid agonists RH-5849 and RH-5992, alone and in combination with a juvenile hormone analogue, pyriproxyfen, on larvae of *Spodoptera exigua*. Entomol Exp Appl 72:115–123

Smagghe G, Degheele D (1995) Biological activity and receptor-binding of ecdysteroids and the ecdysteroid agonists RH-5849 and RH-5992 in imaginal wing discs of *Spodoptera exigua* (Lepidoptera: Noctuidae). Eur J Entomol 92:333–340

Smagghe G, Böhm G-A, Richter K, Degheele D (1995) Effect of non-steroidal ecdysteroid agonists on ecdysteroid titer in *Spodoptera exigua* and *Leptinotarsa decemlineata*. J Insect Physiol 41:971–974

Smagghe G, Eelen H, Verschelde E, Richter K, Degheele D (1996a) Differential effects of non-steroidal ecdysteroid agonists in Coleoptera and Lepidoptera: analysis of evagination and receptor binding in imaginal discs. Insect Biochem Mol Biol 26:687–695

Smagghe G, Viñuela E, Budia F, Degheele D (1996b) In vivo and in vitro effects of the non-steroidal ecdysteroid agonist tebufenozide on cuticle formation in *Spodoptera exigua*: an ultrastructural approach. Arch Insect Biochem Physiol 33:121–134

Smagghe G, Gelman D, Tirry L (1999a) In vivo and in vitro effects of tebufenozide and 20-hydroxyecdysone on chitin synthesis. Arch Insect Biochem Physiol 41:33–41

Smagghe G, Carton G, Heirman A, Tirry L (1999b) Toxicity and impact of kinetics, metabolism and binding of nonsteroidal ecdysone agonists in a susceptible and resistant strain of the cotton leafworm. Proc 14th Int Plant Protection Congr, Jerusalem, Israel, p 152

Smagghe G, Nakagawa Y, Carton B, Mourad AK, Tirry L (1999c) Comparative ecdysteroid action of ring-substituted dibenzoylhydrazines in *Spodoptera exigua*. Arch Insect Biochem Physiol 41:42–53

Smagghe G, Carton B, Wesemael W, Ishaaya I, Tirry L (1999d) Ecdysone agonists – mechanism and application on *Spodoptera* species. Pestic Sci 55:386–389

Sobek L, Böhm G-A, Penzlin H (1993) Ecdysteroid receptors in last-instar larvae of the wax moth *Galleria mellonella* L. Insect Biochem Mol Biol 23:125–129

Soltani N, Quennedey A, Delbecque JP, Delachambre J (1987) Diflubenzuron-induced alterations during in vitro development of *Tenebrio molitor* pupal integument. Arch Insect Biochem Physiol 5:201–209

Spindler K-D, Quack S, Fretz A, Spindler-Barth M (1994) Correlation of ecdysteroid receptor affinity and biological effects of non-steroidal ecdysteroid agonists on the epithelial cell line from *Chironomus tentans*, 11th Ecdysone Workshop, Proceedings

Spindler-Barth M (1993) Hormonal regulation of chitin metabolism in insect cell lines. In: Muzzarellli RAA (ed) Chitin enzymology. European Chitin Society, Acona, pp 75–82

Spindler-Barth M, Spindler KD, Londerhausen M, Thomas H (1989) Inhibition of chitin synthesis in an insect cell line. Pestic Sci 25:115–121

Spindler-Barth M, Turberg A, Spindler K-D (1991) On the action of RH-5849, a non-steroidal ecdysteroid agonist, on a cell line from *Chironomus tentans*. Arch Insect Biochem Physiol 16:11–18

Sundaram M, Palli SR, Ishaaya I, Krell PJ, Retnakaran A (1998) Toxicity of ecdysone agonists correlates with the induction of CHR3 mRNA in the spruce budworm. Pestic Biochem Physiol 62:201–208

Talbot WS, Swyryd EA, Hogness DS (1993) Drosophila tissues with different metamorphic responses to ecdysone express different ecdysone receptor isoforms. Cell 73:1323–1337

Terentiou P, Blattmann M, Bradbrook D, Käuser G, Koolman J (1993) Biological activity and receptor binding of ecdysteroids in imaginal discs of *Callliphora vicina*: a comparison. Insect Biochem Mol Biol 23:131–136

Ursprung H, Nöthiger R (1972) The biology of imaginal disks. Results and problems in cell differentiation, vol 5. Springer, Berlin Heidelberg New York

Vogt M (1943) Aus Produktion und Bedeutung metamorphosefördernder Hormone während der Larvenentwicklung von *Drosophila*. Biol Zentralbl 63:395–446

Ward GB, Newman SM, Klosterman HJ, Marks EP (1988) Effect of 20-hydroxyecdysone and diflubenzuron on chitin production by a cockroach cell line. In Vitro Cell Dev Biol 24:326–332

Weismann A (1864) Die nach embryonale Entwicklung der Musciden nach Beobachtungen an *Musca vomitoria* und *Sarcophaga carnaria*. Z Wiss Zool 14:187–336

Willis JH (1997) An imaginal disc cell line that responds to ecdysteroids and juvenoids. In Vitro Cell Dev Biol 33:86A

Wing KD (1988) RH-5849, a non-steroidal ecdysone agonist: effects on a *Drosophila* cell line. Science 241:467–469

Wing KD, Aller HE (1990) Ecdysteroid agonists as novel insect growth regulators. In: Casida JE (ed) Pesticides and alternatives. Elsevier, Amsterdam, pp 251–257

Wing K, Ramsay JR (1989) Other hormonal agents: ecdysone agonists. In: Progress and prospects in insect control. BCPC Monogr 43:107–118

Wing KD, Slawecki RA, Carlson GR (1988) RH-5849, a non-steroidal ecdysone agonist: effects on larval Lepidoptera. Science 241:470–472

Wyss C (1976) Juvenile hormone analogue counteracts growth stimulation and inhibition by ecdysones in clonal *Drosophila* cell line. Experientia 32:1272–1274

Wyss C (1982) *Chironomus tentans* epithelial cell lines sensitive to ecdysteroids, juvenile hormone, insulin and heat shock. Exp Cell Res 139:309–319

Yao TP, Forman BM, Jiang Z, Cherbas L, Chen JD, McKeown M, Cherbas P, Evans RM (1993) Functional ecdysone receptor is the product of EcR and ultraspiracle genes. Nature 366:476–479

Yund MA, Osterbur DL (1985) Ecdysteroid receptors and binding proteins. In: Kerkut GA, Gilbert LI (eds) Comprehensive insect physiology, biochemistry and pharmacology, vol 7. Pergamon Press, Oxford, pp 473–490

Insect Neuropeptide Antagonists: a Novel Approach for Insect Control

MIRIAM ALTSTEIN[1] and CHAIM GILON[2]

1
Introduction

The success of modern agriculture in achieving and maintaining high-yield crops strongly depends on controlling insect pests via the intensive utilization of insecticides. To date, organo-synthetic chemical insecticides are the main means for the protection of crops against the damage caused by insects. Uncontrolled application of chemical insecticides in recent decades has led to acquired resistance in insects, has contaminated the environment with toxic residues that endanger humans and other life forms, and has disrupted the ecological balance in crop fields. The growing concern regarding the toxic effects of insecticides has led to the implementation of strict regulations in the Western World which are being adopted also in Third World countries. These regulations limit the application of the existing organo-chemical insecticides and ban further application of more toxic ones.

The strategic approach which is directing worldwide R&D efforts is aimed at the identification and development of novel families of insect-specific compounds which are safe and compatible with integrated pest management (IPM) programs, and which will eventually replace the toxic organo-synthetic chemicals as the mainstream pest control compounds.

The initial products that have emerged from this effort during the past two decodes are based on a variety of chemical and biological technologies. These bio-insecticidal, bio-rational products include: bacterial toxins (BT), microorganisms, mating disruption pheromones, insect growth regulators (IGRs), genetically engineered pest-resisting crops, natural enemies, and natural products extracted from plants. The main drawback of these technologies and products, which constitute approximately 2–3% of the insecticides market, is that they cannot replace the organo-chemicals as the mainstream insecticides, because of their high production cost and limited applicability.

In the quest for a novel group of selective insecticides which could eventually replace the toxic organo-chemical compounds and overcome the limita-

---

[1] Department of Entomology, The Volcani Center, ARO, Bet Dagan 50250, Israel
[2] Institute of Organic Chemistry, The Hebrew University, 91804 Jerusalem, Israel

Isaac Ishaaya (Ed.): Biochemical Sites of
Insecticide Action and Resistance
© Springer-Verlag Berlin, Heidelberg 2001

tions introduced by the existing bio-insecticides, entomological studies concentrate on the search for targets and compounds which will serve as a basis for the development of highly effective and environmentally friendly insect-control agents and will emerge as a new mainstream of insecticide technology.

Insect neuropeptides are a prime target in the development of novel insect-control agents, since they regulate most of the key functions in insects such as: embryonic and post-embryonic development, homeostasis, osmoregulation, migration, oviposition, mating, etc. (for review see Gäde 1997). The multiple functional capacities of insect neuropeptides provide opportunities for new insect-control strategies based on interference with the steps associated with the activities of neuropeptides, namely, their biosynthesis (transcription, processing, post-translational modification – PTM), release, transport, binding, activation of the receptor on the target cell, signal transduction and degradation. Although, theoretically, all of these steps could serve as targets for exploitation, the one which is most effective, insect-specific and susceptible to manipulation is the binding of the neuropeptide to its target receptor. Inhibition of this stage may be elicited by competitive and/or allosteric antagonists.

Antagonists, which are selective inhibitors of the neuropeptide receptors, may disrupt and interfere with the normal growth, development, homeostasis and behavior of the insect by blocking the receptor of the neuropeptide; therefore, they can form receptor-selective, insect-specific insecticides. Such antagonists are derived from and resemble the natural peptides but are peptidomimetic in nature. A similar innovative approach has recently been applied to human neuropeptides in the pharmaceutical industry.

Although neuropeptide-based antagonists exhibit a high potential for insect management, their application in pest control has not been implemented so far because of two major limitations. The first is the linear peptidic nature of neuropeptides, which renders them non-selective, highly susceptible to proteolytic degradation, and impenetrable. The second is the lack of an approach to antagonist design, because of lack of information regarding the three-dimensional (3-D) structure of the receptor-agonist complex and the mechanism of receptor activation.

Recently, we have developed a novel approach, termed the backbone cyclic neuropeptide-based antagonist (BBC-NBA) approach, for the discovery of neuropeptide antagonists which overcomes the limitations mentioned above. This approach has been applied to the insect pheromone biosynthesis activating neuropeptide (PBAN) and resulted in the discovery of highly potent, stable and selective antagonists (Gilon et al. 1997; Altstein et al. 1999a; Zeltser et al. 2000).

The basic concepts behind the BBC-NBA technology (backbone cyclization and cycloscan) are explained, and the advantages that are introduced by these methods, as well as their application to the insect PBAN for the discovery of highly potent antagonists, are described.

2
Backbone Cyclic Neuropeptide-Based Antagonist (BBC-NBA) Approach

The BBC-NBA approach is based on backbone cyclization of peptides and the cycloscan concept (see below) and comprises the following steps:

1. Identification of the neuropeptide that controls the required function.
2. Elucidation of the shortest sequence of the neuropeptide that constitutes the active site.
3. Discovery of a peptidic linear lead antagonist on the basis of the sequence found in (2).
4. Discovery of a potent BBC antagonist devoid of agonistic activity, based on (3).
5. Determination of the structural requirements for antagonistic activities, on the basis of (4).
6. Design and synthesis of an insecticide prototype (a small, metabolically stable, selective bioavailable, cost-effective peptidomimetic compound), based on the information obtained in (5), ready for formulation, toxicology and field trials.

The BBC-NBA approach depends on the availability of simple and quantitative in vitro and in vivo bioassays for monitoring bioactivity and on the availability of advanced chemistry to design, synthesize and determine the structure of compounds of interest. The approach is general and can be applied to any insect neuropeptide.

The rationale behind the steps and the basic concepts (backbone cyclization and cycloscan) that make up the BBC-NBA technology are described below.

2.1
Determination of the Active Sequence in the Neuropeptide

There are two types of antagonists: a competitive antagonist, that binds to the same site as the agonist but does not elicit signal transduction; and a type that inhibits binding and/or signal transduction by means of allosteric effects. Antagonists of the first type are usually derived from agonists but exhibit different structural and conformational features; those of the second type are usually non-peptides and are identified by screening of natural product libraries or combinatorial libraries. From this perspective it is clear that an initial requirement for the development of a competitive antagonist is the identification of the smallest active sequence in the neuropeptide that binds and activates the receptor; this sequence is the basis on which modifications can be made, leading to the development of a lead antagonist.

Elucidation of the active site of a neuropeptide is achieved by means of structure-activity relationship (SAR) studies. In the case of large neuropeptides (having more than 15 amino acid residues), peptide mapping is performed. This involves determination of the activities of peptide fragments containing up to 10 amino acid residues, that span the entire sequence of the neuropeptide. In the case of smaller neuropeptides, or when active peptides are discovered by peptide mapping, des-amino acid scanning is performed, i.e., SAR studies of a library of peptides that lack one to five amino acids from either the C- or the N-terminus. It should be mentioned that in a neuropeptide family in which there is sequence homology among the various members of the family, the homologous sequence is usually the active region. Elucidation of the shortest sequence of the neuropeptide that constitutes the active site enables progression toward the second step: discovery of a lead antagonist.

2.2
Development of a Competitive Lead Antagonist

Lead antagonists are usually partial antagonists that bind to the receptor site but can only partially activate the transduction system. Improvement in antagonistic activity (high potency, selectivity and lack of agonistic activity) can be achieved by SAR studies or by imposing conformational constraints (see below).

Most of the lead competitive antagonists discovered hitherto are based on vertebrate neuropeptide agonists. The following empirical practices for the conversion of agonists to antagonists have emerged from these studies: (1) systematic replacement of the naturally occurring L-amino acids by their non-natural D-isomers or replacement of amino acid residues with D-hydrophobic amino acid residues, such as D-Phe or D-Trp (Vale et al. 1972; Rees et al. 1974; Piercey et al. 1981; Sawyer et al. 1981; Rosell et al. 1983; Folkers et al. 1984; Vevrek and Stewart 1985; Heinz-Erian et al. 1987; Hruby et al. 1990; Rhaleb et al. 1991; Hruby 1992; Cody et al. 1995; Collins et al. 1996; Maretto et al. 1998); (2) omission of amino acid residues from agonistic sequences, or omission or replacement of functional side chains (e.g., [Sar^1]-angiotensin II (1–7)amide in which Asp^1 in angiotensin was replaced with Sar and Phe^8 was replaced with an amide group, or [D-Phe^6,Des-met^{14}]-bombesin (6–14)ethyl amide in which the six N-terminal amino acids and Met^{14} were omitted) (Coy et al. 1989); (3) replacement of a C-terminal amide with a free acid (as in bombesin and gastrin) (Rodriguez et al. 1987; Llinares et al. 1999); (4) reduction of peptide bonds, as in the case of bombesin (Coy et al. 1988); (5) conformational and/or topographical alteration (for reviews see Hruby 1981a,b, 1992; Goodman and Ro 1995; Collins et al. 1996; Becker et al. 1999). Implementation of the above empirical practices for given agonists necessitates detailed knowledge of the SAR, and any available information regarding the bioactive conformation.

To the best of our knowledge, these approaches have been applied to a very limited number of insect neuropeptides for the discovery of antagonists.

The D-Phe approach was applied to the insect neuropeptide proctolim and resulted in the discovery of a few peptides with antagonistic activity (Kuczer et al. 1999), and recently, we discovered a lead antagonist for the insect PBAN by applying these empirical rules (see below and Gilon et al. 1997; Zeltser et al. 2000).

2.3
Improvement of the Antagonistic Activity by Conformational Constraint

One major problem that impedes the use of neuropeptides (e.g., the linear lead antagonists discussed above) as insect-control agents is related to the fact that they are susceptible to proteolytic degradation, have low bioavailability and are non-selective because of conformational flexibility. An approach to overcoming these limitations involves the introduction of conformational constraint into peptides, which leads to a slower equilibrium rate, thus reducing the flexibility of the peptide.

Conformational constraint can be imposed by various methods (for reviews see Hruby 1981a,b; 1992; Goodman and Ro 1995). Cyclization of peptides is one of the commonest and most attractive methods to introduce conformational constraint into peptides and thus to restrict their conformational space (Kessler 1982). The conformational constraint confers the following attributes on the peptides: (1) selectivity: the cyclic structure may restrict the conformational space to a conformation which mediates one function of the peptide and excludes those which mediate other functions; (2) enhanced metabolic stability: the cyclic structure may exclude the conformation which is recognized by degrading enzymes from the conformational space, thus preventing enzymatic degradation; (3) increased biological activity: the rigidified structure will be more potent than the linear one, since it spends more time in the bioactive conformation because of the much slower equilibrium between the conformations. However, this is only true when the conformational space of the cyclic peptide overlaps the bioactive conformation; in most cases, cyclization will yield an inactive peptide because of mismatching; (4) improved bioavailability because of reduction in polarity.

Nature too has chosen the cyclization route for restricting the conformation of peptides, and many natural cyclic peptides are known today, some of which are in therapeutic use (e.g., insulin, oxytocin and cyclosporin).

There are four modes of cyclization in peptides: (1) covalent bonding between two side chains (e.g., an amide bond between aspartic acid and lysine, or a disulphide bond between two cysteine residues); (2) covalent linking between the amino and carboxy termini of the peptide; (3) bonding between an Asp or Glu side chain and the amino terminus; and (4) bonding between a lysine or ornithine side chain and the carboxy terminus (for review see Gilon et al. 1991).

The four natural modes of cyclization cannot easily be applied to most peptides for two reasons: (i) not every peptide contains functional amino acid residues which can be covalently interconnected, and (ii) more importantly, even when there are such residues, in most cases they are crucial for the biological activity of the peptide and using them for ring closure causes a loss of or marked reduction in the bioactivity. The same holds true for the amino and carboxy termini, which might be important for the activity of naturally occurring peptides. In addition, the natural modes of cyclization result in only a small number of conformational combinations, which are too few to effectively screen the conformational space available for a linear peptide with a given sequence.

2.4
Backbone Cyclization: a Tool for Imposing Conformational Constraint on Peptides

In order to overcome the above-mentioned problems, the concept of backbone cyclization was developed (Gilon et al. 1991), in which the cyclization takes place by means of a covalent interconnection of the peptide backbone atoms (N^{α} and/or C^{α}) to each other, to side chains, or to amino or carboxyl termini. By means of backbone cyclization, the functionality and the activity of the side chains can be retained during performance of the cyclization. BBC peptides have the following advantages over linear peptides: enhanced stability against proteolytic digestion (Byk et al. 1996; Gilon et al. 1998a), high selectivity (Byk et al. 1996; Gilon et al. 1998a), more potent biological activity (Altstein et al. 1999a), and improved bioavailability. The advantages introduced by the cyclic peptides make them excellent leads for development of insecticide prototypes. In addition, BBC peptides have a constraint conformation which facilitates easy determination of their bioactive conformation by nuclear magnetic resonance (NMR) (Saulitis et al. 1992; Golic-Gradadolnik et al. 1994; Gilon et al. 1998a) and X-ray (Kasher et al. 1999) – provided they are active as the endogenous (parent) peptide. This information is most important for the further design of non-peptide small molecules.

2.5
Cycloscan: Conformationally Constrained BBC Peptide Libraries

In order to obtain the optimal BBC peptides based on a given sequence, namely the one which best matches the bioactive conformation (and thus exhibits the highest antagonistic activity), it is necessary to synthesize a large number of BBC peptides in order to screen the conformational space of the peptide systematically. Cycloscan (Gilon et al. 1998b) is the methodology that has been developed for this purpose. Cycloscan is defined as a selection method based on conformationally constrained BBC peptide libraries intended for efficient screening of the conformational space and thus for fast identification of a BBC peptide lead compound that overlaps the bioactive conformation.

Cycloscan is performed by designing and synthesizing libraries of BBC peptides and screening them with the appropriate bioassay. All the peptides in each library bear the same sequence, and differ from each other in distinct parameters which affect their conformation and hence their bioactivity. This is achieved by the gradual introduction of discrete modifications so as to ensure efficient screening of the conformational space of the parent peptide. The majority of the peptides in such a library should be inactive, because they do not overlap the bioactive conformation, but those which do fit into the bioactive conformation should be very potent and should have all the advantages mentioned above.

The main difference between a cycloscan library and the conventional combinatorial peptide libraries (either chemical or phage display) is that in the latter every peptide in the library has a different sequence whereas in the former all the peptides in the library have the same sequence, and they differ from each other only in their conformation. It is, therefore, possible to generate a large BBC peptide library for each biologically active lead peptide discovered by peptide mapping or via combinatorial libraries.

Cycloscan can be performed in two general ways: (1) sequence-biased cycloscan, in which all the peptides have the same primary sequence but differ in their bridge size, chemistry and location; and (2) combinatorial cycloscan, in which further diversification is achieved by sequential replacement of amino acids. In cases where sequence-biased cycloscan is applied, the amino acid sequence of the linear lead antagonist is used as a basis for library construction.

The diversity of sequence-biased cycloscan is not sequential but conformational, and includes the following parameters: (i) the modes of backbone cyclization; (ii) the position of the backbone bridge along the peptide sequence; (iii) the size of the bridge; and (iv) the chemistry of the bridge. Each of these diversity parameters have been shown to affect the conformation and hence the biological activity (Bitan et al. 1996, 1997; Byk et al. 1996; Altstein et al. 1999a).

The concepts of backbone cyclization and cycloscan were used initially in our laboratory to obtain conformationally constrained analogs of naturally occurring vertebrate neuropeptides. The model peptide chosen for the development of these techniques was substance P (SP) (Gilon et al. 1991). A variety of building units were prepared and incorporated into sequence-biased libraries of BBC SP analogs. The cycloscan parameters, such as ring size and ring chemistry, had substantial effects on the activity of the various analogs (Bitan et al. 1996, 1997; Byk et al. 1996). Overall, the extensive research on BBC SP analogs has clearly proved the feasibility and effectiveness of the concepts of backbone cyclization and cycloscan, and enabled these techniques to be employed on somatostatin (Gilon et al 1998a) and on the family of pyrokinin/PBAN insect neuropeptides (Gilon et al. 1997; Altstein et al. 1999a, see below).

3
Pheromone Biosynthesis Activating Neuropeptide

Pheromone biosynthesis activating neuropeptide (PBAN) is an important neuropeptide that mediates some of the key functions in insects. It was first reported by Raina and Klun (1984) as the neuropeptide that regulates sex pheromone production in female moths, and its amino acid sequence was determined in 1989 in *Helicoverpa zea* (Hez-PBAN; PBAN 1–33NH$_2$; Table 1; Raina et al. 1989). Since then, six other PBAN molecules have been isolated from five additional moth species (*Bombyx mori, Lymantria dispar, Mamestra brassicae, Agrotis ipsilon* and *H. assulta*). Their entire primary structures have been determined (Kitamura et al. 1989, 1990; Masler et al. 1994; Choi et al. 1998; Duportets et al. 1998; Jacquin-Joly et al. 1998; Table 1), and the c-DNA and genes of PBAN have been cloned (Davis et al. 1992; Kawano et al. 1992, 1997; Ma et al. 1994; Choi et al. 1998; Jacquin-Joly et al. 1998; Duportets et al. 1998, 1999). PBAN molecules were found to be C-terminally amidated neuropeptides consisting of 33–34 amino acid residues, and comparison of their primary struc-

Table 1. Amino acid sequence of PBAN and peptides of the pyrokinin/PBAN family. Bold letters indicate conserved amino acid sequences

Code name	Species	Amino acid sequence
Hez-PBAN	*Helicoverpa zea*	LSDDMPATPADQEMYRQDPEQIDSRTKY**FSPRL-NH**$_2$
Bom-PBAN-I	*Bombyx mori*	LSEDMPATPADQEMYQPDPEEMESRTRY**FSPRL-NH**$_2$
Bom-PBAN-II	*Bombyx mori*	RLSEDMPATPADQEMYQPDPEEMESRTRY**FSPRL-NH**$_2$
Lyd-PBAN	*Lamantria dispar*	LADDMPATMADQEVYRPEPEQIDSRNKY**FSPRL-NH**$_2$
Haz-PBAN	*Helicoverpa assulta*	LSDDMPATPADQEMYRQDPEQIDSRTKY**FSPRL-NH**$_2$
Agi-PBAN	*Agrotis ipsilon*	LADDTPATPADQEMYRPDPEQIDSRTKY**FSPRL-NH**$_2$
Mab-PBAN	*Mamestra brassicae*	LADDMPATPADQEMYRPDPEQIDSRTKY**FSPRL-NH**$_2$
Bom-DH	*Bombyx mori*	TDMKDESDRGAHSERGALCF**GPRL-NH**$_2$
Hez-DH	*Helicoverpa zea*	NDVKDGAASGAHSDRLGLWF**GPRL-NH**$_2$
Has-DH	*Helicoverpa assulta*	NDVKDGAASGAHSDRLGLWF**GPRL-NH**$_2$
Pss-PT	*Pseudaletia separata*	KLSYDDKVFENVEF**TPRL-NH**$_2$
Lom-PK-I	*Locusta migratoria*	pQDSGDGWPQQPF**VPRL-NH**$_2$
Lom-PK-II	*Locusta migratoria*	pQSVPTF**TPRL-NH**$_2$
Lem-PK	*Leucophaea maderae*	pQTSF**TPRL-NH**$_2$
Lom-MT-I	*Locusta migratoria*	GAVPAAQF**SPRL-NH**$_2$
Lom-MT-II	*Locusta migratoria*	EGDF**TPRL-NH**$_2$
Lom-MT-III	*Locusta migratoria*	RQQPF**VPRL-NH**$_2$
Lom-MT-VI	*Locusta migratoria*	RLHQNGMPF**SPRL-NH**$_2$

tures revealed that they share a high degree of homology and an identical pentapeptide C-terminal sequence (Phe-Ser-Pro-Arg-Leu-NH$_2$, Table 1), which also constitutes the active core required for their biological activity (Raina and Kempe 1990, 1992; Altstein et al. 1995, 1996a, 1997; Nagasawa et al. 1994; Kochansky et al. 1997). Since 1984, the presence of PBAN-like activity has been demonstrated in a variety of moths and in other non-Lepidopteran species, and its mode of action has been studied extensively (for review see Raina 1993; Gäde 1997).

Further studies on the regulation of sex pheromone biosynthesis in moths have revealed that this biosynthesis can be elicited by additional neuropeptides, isolated from various insects, all of which share the common C-terminal pentapeptide of PBAN (Phe-Xxx-Pro-Arg-Leu-NH2; Xxx = Ser, Gly, Thr, Val, Table 1) (Fónagy et al. 1992; Kuniyoshi et al. 1992a,b; Abernathy et al. 1995; Teal et al. 1996). Among these peptides are the pyrokinins (Lem-PK, Lom-PK-I and Lom-PK-II) and the locustamyotropins (Lom-MT-I to IV) (myotropic peptides isolated from the Madeira cockroach *Leucophaea maderae* (Fabricius) and the migratory locust, *Locusta migratoria* (Nachman et al. 1993; Schoofs et al. 1993); pheromonotropin (Pss-PT), an 18 amino acid peptide isolated from *Pseudaletia* (*Mythimna*) *separata* (Walker) (Matsumoto et al. 1992), and diapause hormone (DH), isolated from the silkworm, *B. mori* (L.) (Imai et al. 1991), *H. zea* (Davis et al. 1992) and *H. assulta* (Choi et al. 1998). These peptides have recently been designated to the pyrokinin/PBAN family (Table 1). In addition to their ability to stimulate sex pheromone biosynthesis in moths, members of this family have been found to control a variety of other physiological and behavioral functions, such as melanization and reddish coloration in moth larvae (Matsumoto et al. 1990; Altstein et al. 1996b), contraction of the locust oviduct (Schoofs et al. 1991), myotropic activity of the cockroach and locust hindgut (Nachman et al. 1986; Schoofs et al. 1991), egg diapause in the silkworm (Imai et al. 1991), and acceleration of pupariation in fleshfly *Sarcophaga bullata* (Parker) larvae (Nachman et al. 1997).

Three reasons urged us to apply the BBC-NBA approach to the generation of receptor-selective antagonists to the pyrokinin/PBAN family: (1) the major role of the pyrokinin/PBAN family in the physiology of moths and other insects; (2) knowledge of the amino acid sequence and the large body of information derived from SAR studies; and (3) the availability of biological assays which enable antagonistic activities to be determined quantitatively.

4
Implementation of the BBC-NBA Strategy to the Pyrokinin/PBAN Family

The first stage in the implementation of the BBC-NBA strategy for the discovery of receptor-selective antagonists for the insect pyrokinin/PBAN family involved optimization of two in vivo biological assays for evaluation of agonistic and/or antagonistic activities of linear and BBC peptides (Gazit et al. 1990; Altstein et al. 1993, 1996b). With the aid of these assays, and by using the BBC-NBA approach, we achieved the following:

1. Identification (by SAR studies) of the minimal active sequence of PBAN (MINI-PBAN) that constitutes the active core of the pyrokinin/PBAN molecule (Altstein et al. 1995, 1996a,b, 1997).
2. Design and synthesis of a biased library of linear peptides based on the information obtained in step 1 for the identification of a linear lead antagonist (Gilon et al. 1997; Zeltser et al. 2000).
3. Discovery (by SAR studies) of a peptidic linear lead antagonist from the biased linear library obtained in step 2 (Gilon et al. 1997; Zeltser et al. 2000).
4. Design and synthesis of BBC peptide libraries based on the sequence of the lead antagonist found in step 3 for the discovery of a cyclic antagonist (Gilon et al. 1997; Altstein et al. 1999a).
5. Discovery (by SAR studies) of several selective and highly potent BBC anti-PBAN antagonists, devoid of agonistic activity, which inhibit (at 1–3 nmol) PBAN/pyrokinin mediated functions (sex pheromone biosynthesis and melanization) from the cyclic libraries in step 4 (Gilon et al. 1997; Altstein et al. 1999a).

The results obtained provide a solid proof that neuropeptide antagonists can inhibit biological activities elicited by endogenous neuroendocrine mechanisms and provide valuable information on the structural requirements of pyrokinin/PBAN antagonists (Altstein et al. 1999a).

5
Conversion of Neuropeptide Antagonists into Insecticide Prototypes

The design of neuropeptide antagonists for insecticidal applications requires, in addition to the antagonistic properties, the development of materials that will conform with the common practices of the insecticide industry, i.e., non-peptide compounds of low molecular weight, which exhibit high penetrability through the insect cuticle and gut, and are environmentally stable and cost effective in production.

Design and synthesis of a non-peptide insecticide prototype by means of the chemical combinatorial library approach requires: (1) development of technologies for the conversion of backbone cyclic, conformationally constrained peptides to small non-peptide organic molecules; (2) synthesis of small non-peptide combinatorial libraries; and (3) development of a high-throughput screening (HTS) assay for fast screening of the libraries and selection of biologically active compounds.

Structural data derived from the active and non-active conformationally constrained BBC antagonists (such as those described above) can form the basis for the design and synthesis of a non-peptide insecticide prototype. The approach for optimization of BBC antagonists comprises the conversion of such antagonists into small molecule non-peptide leads using a modified version of the small molecule library described by Souers et al. (1999) (which is based on a triazacyclooctathiazolidindione scaffold). In the case of the pyrokinin/PBAN family, the biased library will consist of a collection of scaffolds in which the various substituents derived from the amino acid side chains of the BBC PBAN antagonistic lead are introduced in a combinatorial fashion.

Screening of these libraries for biologically active compounds can be performed by an HTS assay. Radio-receptor assays (RRAs) (such as the one that has recently been developed in our laboratory for the pyrokinin/PBAN family of peptides) (Altstein et al. 1999b) can form the basis for the development of HTS assays which can be further processed toward an "ELISA-based" automated HTS assay (Devlin 1997).

The development of a technology for the conversion of cyclic peptides into small non-peptide organic molecules and the availability of a reliable HTS assay should complete the development of the overall "neuropeptide antagonist-based insecticides" technology by bringing it to a stage which is in line with the common practice of the insecticide industry: namely, generation of compounds of a non-peptidic nature which can be produced on the multi-ton scale at a relatively low cost. This part of the strategy is now under examination in our laboratory.

6
Concluding Remarks

In summary, this review presents a novel general approach that combines rational design and a selection method for the generation of antagonistic cyclic peptides based on the sequence of an insect neuropeptide. This approach, applied to PBAN, has led to the discovery of several antagonists which effectively inhibit sex pheromone biosynthesis in female *H. peltigera* moths and formation of cuticular melanin in *Spodoptera littoralis* larvae.

Beyond the immediate benefits introduced by the cyclic peptides as selective antagonists, the information on the bioactive conformations of the antagonists may serve as a basis for the design of improved (small, cost effective and having enhanced metabolic stability and bioavailability), non-peptide, mimetic antagonists. Such compounds are potential candidates for agrochemical applications, and could serve, after formulation and preliminary field experiments, as prototypes for the development of a novel group of highly effective, insect-specific and environmentally friendly insecticides.

The BBC-NBA approach is one of several possible approaches to neuropeptide-based insect control, which is based on interference with the neuropeptide activity by blocking its receptors with antagonists. Other potential approaches to neuropeptide interference may involve inhibition or induction of other steps such as translation, processing, PTM, release, signal transduction and degradation. These approaches, however, are not very promising and will most likely fail to lead to insect-specific compounds, since the mechanisms that regulate these processes are, to the best of our very limited knowledge, common to vertebrates and invertebrates. If further studies were to reveal insect-specific mechanisms, then they would emerge as additional, novel highly potent targets for application to insect management.

Additional neuropeptide-based approaches to interference involve exposure of insects to genetically engineered superagonist/agonists or antagonists, generated by transgenic plants or insect-specific vectors (such as viruses, nematodes, fungi or bacteria) which express neuropeptide genes or which encode for neuropeptide agonists and antagonists. Such approaches involve exposure of insects to the agonists or antagonists via ingestion (transgenic plants) or exposure to the insect-specific vectors. In the latter case, insect control may be enhanced by a combination of two approaches: biological control contributed by the microorganisms and neurochemical control contributed by the neuropeptide agonists or antagonists.

The approaches described above are still theoretical and, to the best of our knowledge, there are no commercial insect-control agents based on neuropeptide agonists and antagonists. In order to exploit the potential hidden in these approaches, there is a need to guide biological and biochemical research, to develop novel chemical and molecular technologies, to develop meaningful in vivo, in vitro and HTS assays, and to improve recombinant DNA and expression technologies. We can anticipate that integration of the above knowledge will lead: (1) to a better understanding of the modes of action of neuropeptides; (2) will make possible the design and synthesis of biologically active, highly potent, insect-specific, peptidomimetic compounds; (3) will improve our understanding and effectiveness of the transgenic route; and (4) will enable us to overcome some of the major obstacles in the way of the implementation of insect neuropeptides as control agents.

Acknowledgements. This research was supported by the Israel Ministry of Science and Technology and by the Israel Science Foundation, administered by the Israel Academy of Sciences and Humanities. This contribution is from the Institute of Plant Protection, Agricultural Research Organization, The Volcani Center, Bet Dagan, Israel. No 517/00 series.

References

Abernathy RL, Nachman RJ, Teal PEA, Yamashita O, Tumlinson JH (1995) Pheromonotropic activity of naturally occurring pyrokinin insect neuropeptides (FXPRLamide) in *Helicoverpa zea*. Peptides 16:215–219

Altstein M, Gazit Y, Dunkelblum E (1993) Neuroendocrine control of sex pheromone biosynthesis in *Heliothis peltigera*. Arch Insect Biochem Physiol 22:153–168

Altstein M, Dunkelblum E, Gabay T, Ben-Aziz O, Schafler I, Gazit Y (1995) PBAN-induced sex pheromone biosynthesis in *Heliothis peltigera*: structure, dose and time-dependent analysis. Arch Insect Biochem Physiol 30:309–317

Altstein M, Ben-Aziz O, Gabay T, Gazit Y, Dunkelblum E (1996a) Structure-function relationship of PBAN/MRCH. In: Carde RT, Minks AK (eds) Insect pheromone research: new directions. Chapman and Hall, London, pp 56–63

Altstein M, Gazit Y, Ben Aziz O, Gabay T, Marcus R, Vogel Z, Barg J (1996b) Induction of cuticular melanization in *Spodoptera littoralis* larvae by PBAN/MRCH: development of a quantitative bioassay and structure function analysis. Arch Insect Biochem Physiol 31:355–370

Altstein M, Dunkelblum E, Gazit Y, Ben Aziz O, Gabay T, Vogel Z, Barg J (1997) Structure-function analysis of PBAN/MRCH:A basis for antagonist design. In: Rosen D, Tel-Or E, Hadar Y, Chen Y (eds) Modern agriculture and the environment. Kluwer, London, pp 109–116

Altstein M, Ben-Aziz O, Daniel S, Schefler I, Zeltser I, Gilon C (1999a) Backbone cyclic peptide antagonists, derived from the insect pheromone biosynthesis activating neuropeptide (PBAN), inhibit sex pheromone biosynthesis in moths. J Biol Chem 274:17573–17579

Altstein M, Gabay T, Ben Aziz O, Daniel S, Zeltser I, Gilon C (1999b) Characterization of the pheromone biosynthesis activation neuropeptide (PBAN) receptor from the pheromone gland of *Heliothis peltigera*. Invertebr Neurosci 4:33–40

Becker JAJ, Wallace A, Garzon A, Ingallinella P, Bianchi E, Cortese R, Simonin F, Kieffer BL, Pessi A (1999) Ligands for κ-opioid and ORL1 receptors identified from a conformationally constrained peptide combinatorial library. J Biol Chem 274:27513–27522

Bitan G, Zeltser I, Byk G, Halle D, Mashriki Y, Gluhov EV, Sukhotinsky I, Hanani M, Selinger Z, Gilon C (1996) Backbone cyclization of the C-terminal part of substance P: the important role of the sulphur in position 11. J Peptide Sci 2:261–269

Bitan G, Sukhotinsky I, Mashriki Y, Hanani M, Selinger Z, Gilon C (1997) Synthesis and biological activity of novel backbone-bicyclic substance P analogs containing lactam and disulfide bridges. J Peptide Res 49:421–426

Byk G, Halle D, Zeltser I, Bitan G, Selinger Z, Gilon C (1996) Synthesis and biological activity of NK-1 selective, N-backbone cyclic analogs of the C-terminal hexapeptide of substance P. J Med Chem 39:3174–3178

Choi MY, Tanaka M, Kataoka H, Boo KS, Tatsuki S (1998) Isolation and identification of the cDNA encoding the pheromone biosynthesis activating neuropeptide and additional neuropeptides in the oriental tobacco budworm, *Helicoverpa assulta* (Lepidoptera: Noctuidae). Insect Biochem Mol Biol 28:759–766

Cody WL, He JX, DePue PL, Waite LA, Leonard DM, Sefler AM, Kaltenbronn JS, Haleen SJ, Walker DM, Flynn MA, Welch KM, Reynolds EE, Doherty AM (1995) Structure-activity relationships of the potent combined endothelin-A/endothelin-B receptor antagonist Ac-DDip[16]-Leu-Asp-Ile-Ile-Trp[21]: development of endothelin-B receptor selective antagonists. J Med Chem 21:2809–2819

Collins N, Flippen-Anderson JL, Haaseth RC, Deschamps JR, George C, Kövér K, Hruby VJ (1996) Conformational determinants of agonist versus antagonist properties of [D-Pen[2], D-Pen[5]]enkephalin (DPDPE) analogs at opioid receptors. Comparison of X-ray crystallographic structure, solution ^{1}H NMR data and molecular dynamic simulations of [L-Ala[3]]DPDPE and [D-Ala[3]]DPDPE. J Am Chem Soc 118:2143–2152

Coy DH, Heinz-Erian P, Jiang N-Y, Sasaki J, Taylor J, Moreau J-P, Wolfey JD, Gardner JD, Jensen RT (1988) A novel bombesin antagonist with reduced peptide bond. J Biol Chem 263:5055–5060

Coy DH, Taylor J, Jiang N-Y, Kim SH, Wang L-H, Huang SC, Moreau J-P, Gardner JD, Jensen RT (1989) Short chain bombesin receptor antagonists with $IC_{50}S$ for cellular secretion and growth approaching the picomolar region. In: Rivier JE, Marshall GR (eds) Peptides 11. Escom, Leiden, pp 65–67

Davis M-TB, Vakharia VN, Henry J, Kempe TG, Raina AK (1992) Molecular cloning of the pheromone biosynthesis-activating neuropeptide in *Helicoverpa zea*. Proc Natl Acad Sci USA 89:142–146

Devlin JP (1997) High throughput screening: the discovery of bioactive substances. Marcel Dekker, New York

Duportets L, Gadenne C, Dufour MC, Couillaud F (1998) The pheromone biosynthesis activating neuropeptide (PBAN) of the black cutworm moth, *Agrotis ipsilon*: immunohistochemistry, molecular characterization and bioassay of its peptide sequence. Insect Biochem Mol Biol 28:591–599

Duportets L, Gadenne C, Dufour MC, Couillaud F (1999) A cDNA, from *Agrotis ipsilon*, that encodes the pheromone biosynthesis activating neuropeptide (PBAN) and other FXPRL peptides. Peptides 20:899–905

Folkers K, Jakanson R, Horig J, Xu JC, Leander S (1984) Biological evaluation of substance P antagonists. Br J Pharmacol 83:449–456

Fónagy A, Schoofs L, Matsumotor S, De Loof A, Mitsui TM (1992) Functional cross-reactivity of some locustamyotropins and *Bombyx* pheromone biosynthesis activating neuropeptide. J Insect Physiol 38:651–657

Gäde G (1997) The explosion of structural information on insect neuropeptides. Prog Chem Org Nat Products 71:1–128

Gazit Y, Dunkelblum E, Benichis M, Altstein M (1990) Effect of synthetic PBAN and derived peptides on sex pheromone biosynthesis in *Heliothis peltigera* (Lepidoptera: Noctuidae). Insect Biochem 20:853–858

Gilon C, Halle D, Chorev M, Selinger Z, Byk G (1991) Backbone cyclization: a new method for conferring conformational constraint on peptides. Biopolymers 31:745–750

Gilon C, Zeltser I, Daniel S, Ben-Aziz O, Schefler I, Altstein M (1997) Rationally designed neuropeptide antagonists: a novel approach for generation of environmentally friendly insecticides. Invertebr Neurosci 3:245–250

Gilon C, Huonges M, Matha B, Gellerman G, Hornik V, Rosenfeld R, Afargan M, Amitay O, Ziv O, Feller E, Gamliel A, Shohat D, Wanger M, Arad O, Kessler H (1998a) A backbone-cyclic, receptor 5-selective somatostatin ananlogue: synthesis, bioactivity, and nuclear magnetic resonance conformational analysis. J Med Chem 41:919–929

Gilon C, Muller D, Bitan G, Salitra Y, Goldwasser I, Hornik V (1998b) Cycloscan: conformational libraries of backbone cyclic peptides. In: Ramage R, Epton R (eds) Peptides: chemistry, structure and biology. Mayflower Scientific, England, pp 423–424

Golic-Gradadolnik S, Mierke DF, Byk G, Zeltser I, Gilon C, Kessleer H (1994) Comparison of the conformations of active and nonactive backbone cyclic analogs of substance P as a tool to elucidate features of the bioactive conformation: NMR and molecular dynamics in water and dimethyl sulfoxide. J Med Chem 37:2145–2152

Goodman M, Ro S (1995) Peptidomimetics for drug design. In: Wolff E (ed) Medicinal chemistry and drug discovery, vol 1, 5th edn. . Wiley, New York, pp 803–861

Heinz-Erian P, Coy DH, Tamura M, Jones SW, Gardener JD, Jensen RT (1987) [D-Phe^{12}]bombesin analogues: a new class of bombesin receptor antagonists. Am J Physiol 252:G439–G442

Hruby VJ (1981a) Structural and conformation related to the activity of peptide hormones. In: Eberle A, Geiger R, Weiland T (eds) Perspectives in peptide chemistry. Karger, Basel, pp 207–220

Hruby VJ (1981b) Relation of conformation to biological activity in oxytocin, vasopressin and their analogues. In: Burgen SV, Roberts GCK (eds) Topics in molecular pharmacology, vol 1. Elsevier, Amsterdam, pp 99–126

Hruby VJ (1992) Strategies in the development of peptide antagonists. Prog Brain Res 92:215–224

Hruby VJ, Al-Obeidi F, Kazmierski W (1990) Emerging approaches in the molecular design of receptor-selective peptide ligands: conformational, topographical and dynamic considerations. Biochem J 268:249–262

Imai K, Konno T, Nakazawa Y, Komiya T, Isobe M, Koga K, Goto T, Yaginuma T, Sakakibara K, Hasegawa K, Yamashita O (1991) Isolation and structure of diapause hormone of the silkworm, *Bombyx mori*. Proc Jpn Acad 67B:98–101

Jacquin-Joly E, Burnet M, Franxois MC, Ammar D, Nagnan-Le Meillour P, Descoins C (1998) cDNA cloning and sequence determination of the pheromone biosynthesis activating neuropeptide of *Mamestra brassicae*: a new member of the PBAN family. Insect Biochem Mol Biol 28:251–258

Kasher R, Oren DS, Barda Y, Gilon C (1999) Protein miniaturization: the backbone cyclic proteinomimetic approach. J Mol Biol 292:421–429

Kawano T, Kataoka H, Nagasawa H, Isogai A, Suzuki A (1992) c-DNA cloning and sequence determination of the pheromone biosynthesis activating neuropeptide of the silkworm, *Bombyx mori*. Biochem Biophys Res Commun 189:221–226

Kawano T, Kataoka H, Nagasawa H, Isogai A, Suzuki A (1997) Molecular cloning of a new type of c-DNA for pheromone biosynthesis activating neuropeptide in the silkworm, *Bombyx mori*. Biosci Biotech Biochem 61:1745–1747

Kessler H (1982) Conformation and biological activity of cyclic peptides. Angew Chem Int Ed Engl 21:512–523

Kitamura A, Nagasawa H, Kataoka H, Inoue T, Matsumoto S, Ando T, Suzuki A (1989) Amino acid sequence of pheromone-biosynthesis-activating neuropeptide (PBAN) of the silkworm *Bombyx mori*. Biochem Biophys Res Commun 163:520–526

Kitamura A, Nagasawa H, Kataoka H, Ando T, Suzuki A (1990) Amino acid sequence of pheromone biosynthesis activating neuropeptide-II (PBAN-II) of the silkmoth *Bombyx mori*. Agric Biol Chem 54:2495–2497

Kochansky JP, Raina AK, Kempe TG (1997) Structure-activity relationship in C-terminal fragment analogs of pheromone biosynthesis activating neuropeptide in *Helicoverpa zea*. Arch Insect Biochem Physiol 35:315–332

Kuczer M, Rosinski G, Issberner J, Osborne R, Konopinska D (1999) Further proctolin analogues modified in the position 2 of the peptide chain and their myotropic effects in insects tenebrio molitor and schistocerca gregaria. Pol J Pharmacol 51:79–85

Kuniyoshi H, Nagasawa H, Ando T, Suzuki A (1992a) N-terminal modified analogs of C-terminal fragments of PBAN with pheromonotropic activity. Insect Biochem Mol Biol 22:399–403

Kuniyoshi H, Nagasawa H, Ando T, Suzuki A, Nachman RJ, Holman MG (1992b) Cross-activity between pheromone biosynthesis activating neuropeptide (PBAN) and myotropic pyrokinin insect peptides. Biosci Biotech Biochem 56:167–168

Llinares M, Devin C, Chloin O, Azay J, Noel-Artis A-M, Bernad N, Fehrentz J-A, Martinez J (1999) Synthesis and biological activities of potent bombesin receptor antagonists. J Peptide Res 53:275–283

Ma PWK, Knipple DC, Roelofs WL (1994) Structural organization of the *Helicoverpa zea* gene encoding the precursor protein for pheromone biosynthesis-activating neuropeptide and other neuropeptides. Proc Natl Acad Sci USA 91:6506–6510

Maretto S, Schievano E, Mammi S, Bisello A, Nakamoto C, Rosenblatt M, Chorev M, Peggion E (1998) Conformational studies of a potent Leu[11], D-Trp[12]-containing lactam-bridged parathyroid hormone-related protein-derived antagonist. J Peptide Res 52:241–248

Masler EP, Raina AK, Wagner RM, Kochansky JP (1994) Isolation and identification of a pheromonotropic neuropeptide from the brain-suboesophageal ganglion complex of *Lymantria dispar*: a new member of the PBAN family. J Insect Biochem Mol Biol 24:829–836

Matsumoto S, Kitamura A, Nagasawa H, Kataoka H, Orikasa C, Mitsui T, Suzuki A (1990) Functional diversity of a neurohormone produced by the suboesophageal ganglion: molecular identity of melanization and reddish colouration hormone and pheromone biosynthesis activating neuropeptide. J Insect Physiol 36:427–432

Matsumoto S, Fonagy A, Kurihara M, Uchiumi K, Nagamine T, Chijimatsu M, Mitsui T (1992) Isolation and primary structure of a novel pheromonotropic neuropeptide structurally related to leucopyrokinin from the armyworm larvae, *Pseudaletia separata*. Biochem Biophys Res Commun 182:534–539

Nachman RJ, Holman MG, Cook BJ (1986) Active fragments and analogs of the insect neuropeptide leucopyrokinin: structure-function studies. Biochem Biophys Res Commun 137:936–942

Nachman RJ, Holman MG, Haddon WF (1993) Leads for insect neuropeptide mimetic development. Arch Insect Biochem Physiol 22:181–197

Nachman RJ, Zdareej J, Holman MG, Hayes TK (1997) Pupariation acceleration in fleshfly (*Sarcophaga bullata*) larvae by the pyrokinin/PBAN neuropeptide family. Ann NY Acad Sci 814:73–79

Nagasawa H, Kuniyoshi H, Arima R, Kawano T, Ando T, Suzuki A (1994) Structure and activity of *Bombyx* PBAN. Arch Insect Biochem Physiol 25:261–270

Piercey MF, Schroeder LA, Einspahr FJ (1981) Behavioral evidence that substance P may be a spinal cord nociceptor neurotransmitter. In: Rich DH, Gross E (eds) Peptides: synthesis-structure-function. Pierce Chemical, Rockford, IL, pp 589–592

Raina AK (1993) Neuroendocrine control of sex pheromone biosynthesis in Lepidoptera. Annu Rev Entomol 3:329–349

Raina AK, Kempe TG (1990) A pentapeptide of the C-terminal sequence of PBAN with pheromonotropic activity. Insect Biochem 20:849–851

Raina AK, Kempe TG (1992) Structure activity studies of PBAN of *Helicoverpa zea* (Lepidoptera: Noctuidae). Insect Biochem Mol Biol 22:221–225

Raina AK, Klun JA (1984) Brain factor control of sex pheromone production in the female corn earworm moth. Science 225:531–533

Raina AK, Jaffe H, Kempe TG, Keim P, Blacher RW, Fales HM, Riley CT, Klun JA, Ridgway RL, Hayes TK (1989) Identification of a neuropeptide hormone that regulates sex pheromone production in female moths. Science 244:796–798

Rees RWA, Foell TJ, Chai S-Y, Grant N (1974) Synthesis and biological activities of analogues of the luteinizing hormone-releasing hormone (LH-RH) modified in position 2. J Med Chem 17:1016–1019

Rhaleb N-E, Télémaque S, Roussi N, Dion S, Jukic D, Drapeau G, Regoli D (1991) Structure-activity studies of bradykinin and related peptides. Hypertension 17:107–115

Rodriguez M, Dubreuil P, Laur J, Bali JP, Martinez J (1987) Synthesis and biological activity of partially modified retro-inverso pseudopeptide derivatives of the C-terminal tetrapeptide of gastrin. J Med Chem 30:758–763

Rosell S, Björkroth U, Xu JC, Folkers K (1983) The pharmacological profile of a substance P (SP) antagonist. Evidence for the existence of subpopulations of SP receptors. Acta Physiol Scand 117:445–449

Saulitis J, Mierke DF, Byk G, Gilon C, Kessler H (1992) Conformation of cyclic analogues of substance P: NMR and molecular dynamics in dimethyl sulfoxide. J Am Chem Soc 114:4818–4823

Sawyer WH, Pang PKT, Seto J, McEnroe M (1981) Vasopressin analogs that antagonize antidiuretic responses by rats to the antidiuretic hormone. Science 212:4951–4953

Schoofs L, Holman MG, Nachman RJ, Hayes TK, DeLoof A (1991) Isolation, primary structure, and synthesis of locusta pyrokinin: a myotropic peptide of *Locusta migratoria*. Gen Comp Endocrinol 81:97–104

Schoofs L, Vanden JB, De Loof A (1993) The myotropic peptides of *Locusta migratoria*: structures, distribution, functions and receptors. Insect Biochem Mol Biol 23:859–881

Souers AJ, Virgilio AA, Rosenquist A, Fenuek W, Ellman JA (1999) Identification of a potent heterocyclic ligand to somatostatin receptor subtype 5 by the synthesis and screening of β-turn mimetic libraries. J Am Chem Soc 121:1817–1825

Teal PEA, Abernathy RL, Nachman RJ, Fang N, Meredith JA, Tumlinson JH (1996) Pheromone biosynthesis activating neuropeptides: functions and chemistry. Peptides 17:337–344

Vale W, Grant G, Rivier JE, Monahan M, Amoss M, Blackwell R, Borgos R, Guillemin R (1972) Synthetic polypeptide antagonists of the hypothalamic luteinizing hormone releasing hormone. Science 176:933–934

Vevrek RJ, Stewart JM (1985) Competitive antagonists of bradykinin. Peptides 6:161–164

Zeltser I, Gilon C, Ben-Aziz O, Schefler I, Altstein M (2000) Discovery of a linear lead antagonist to the insect pheromone biosynthesis activating neuropeptide (PBAN). Peptides (in press)

Ion Balance in the Lepidopteran Midgut and Insecticidal Action of *Bacillus thuringiensis*

J. L. GRINGORTEN[1]

1
Introduction

The use of *Bacillus thuringiensis* (Bt) for insect control ranks as one of the more notable achievements in the effort to develop safe and environmentally benign pesticides as alternatives to synthetic chemicals. Numerous strains of the organism have been investigated for their potential as biological insecticides in agriculture and forestry against insect defoliators and in vector control against mosquitoes and blackflies (Dulmage and cooperators 1981; Feitelson et al. 1992; Becker and Margalit 1993; Keller and Langenbruch 1993; Navon 1993; van Frankenhuyzen 1993, 2000). To date, Bt-based products are the most widely used biopesticides, although their share of the total commercial insecticide market (1–2%) is still very small (Harris 1997; Schnepf et al. 1998).

Bacillus thuringiensis is a soil- and phylloplane-dwelling, Gram-positive bacterium comprising many subspecies and strains that is found worldwide (Martin and Travers 1989; Smith and Couche 1991; Bernhard et al. 1997; Chaufaux et al. 1997). As a species it is closely related to *B. cereus* and, indeed, opinion is divided as to whether the two should be considered a single species (Schnepf et al. 1998). In addition to the numerous insecticidal strains, other strains have been isolated that are active against mites, nematodes, flatworms and protozoa (Feitelson et al. 1992).

The insecticidal strains of Bt synthesize δ-endotoxin proteins during sporulation that act as stomach poisons when ingested by a susceptible host. The proteins coalesce as a parasporal crystalline inclusion which is released along with the spore at the time of vegetative cell lysis (reviewed by Norris 1971; Fast 1981). The crystals vary in morphology and composition, depending on the Bt strain, and may contain from one to several different types of δ-endotoxins. Approximately 150 δ-endotoxin genes have been sequenced to date and are grouped into two broad taxons, "*cry*" and "*cyt*", based on amino acid sequence homology of the expressed proteins ("Cry" and "Cyt" proteins, respectively) and toxin mode of action (Crickmore et al. 1998). Cyt proteins are hemolytic, whereas Cry proteins are not. Several investigators have summarized host

---

[1] Canadian Forest Service, Great Lakes Forestry Centre, 1219 Queen Street East, Sault Ste. Marie, Ontario P6A 5M7, Canada

Isaac Ishaaya (Ed.): Biochemical Sites of Insecticide Action and Resistance
© Springer-Verlag Berlin, Heidelberg 2001

specificity data (Keller and Langenbruch 1993; Lereclus et al. 1993; van Frankenhuyzen 1993) and discussed the phylogenetic relationships among the different proteins (Feitelson et al. 1992; Bravo 1997; Crickmore et al. 1998; Schnepf et al. 1998). A unique characteristic of the Cry proteins, one that sets them apart from the broad-spectrum chemical insecticides, is their relatively narrow host specificity and complete absence of toxicity to mammals, features that reduce the risk to non-target organisms and enhance their appeal as bio-rational control agents.

Of the 22 classes of *cry* genes documented in Crickmore et al. (1998), *cry1* is the largest and most diverse, and the Cry1 toxin proteins expressed by this class are, with few exceptions (Bravo 1997), specifically active against Lepidoptera. Even within this class, however, the various proteins differ in host specificity and, despite close structural homologies (up to >90% amino acid sequence identity), their activity spectra against insects and cell lines can be very different (van Frankenhuyzen et al. 1991, 1993; Gringorten et al. 1999). Milne et al. (1990) have discussed the various toxin and host factors that contribute to Bt specificity.

Bacillus thuringiensis δ-endotoxin proteins target the midgut and create lesions that disable vital ion-regulating mechanisms upon which larval survival depend. These same mechanisms are also essential for initiating the pathogenic action of Bt and are the subject of this chapter. Reference will be made mostly to the Cry1 proteins, which have been studied the longest and about which most is known.

2
Pathogenesis

Bacillus thuringiensis was first isolated by Ishiwata (1901, cited by Steinhaus 1961) from the silkworm, *Bombyx mori*, and the early work on its pathogenesis was carried out in Japan, the incentive for which were the periodic outbreaks of Bt infection in silkworm colonies. That a toxin was involved in the insecticidal activity, and not just a septicemia, was also suggested by Ishiwata (see Mitani and Watarai 1916). Its alkali-soluble property was reported by Mitani and Watarai (1916) who concluded that the alkaline nature of the silkworm digestive tract was the factor responsible for solubilizing the toxin, a necessary condition for the toxemia to proceed. That information languished in obscurity for many years and the toxin remained unidentified until Hannay (1953) suggested a possible association with the parasporal protein crystal. Subsequently, Angus (1954, 1956a,b) established unequivocally that the crystal was indeed the alkali-soluble toxin and extractable in larval gut juice. In a series of elegant experiments using ligatures, a barium X-ray technique and histopathological observations, Heimpel and Angus (1959) demonstrated that the site of action was the larval midgut.

The basic unit of the Cry1 δ-endotoxins is a 120–130 kDa molecule (Huber and Lüthy 1981) that itself is not insecticidal, but is actually a protoxin (Angus 1956b; Heimpel and Angus 1959; Lecadet and Martouret 1967). Conversion to active toxin occurs through solubilization of the protoxin crystal in the insect midgut and a stepwise proteolytic digestion of the C-terminal half of the molecule by gut-juice enzymes (Choma et al. 1990), as well as cleavage of the first 28 N-terminal amino acids. What remains is a relatively protease-resistant moiety of M_r approximately 65 kDa that can pass through the peritrophic membrane, diffuse across the ectoperitrophic space and interact with the midgut epithelium. Comprehensive descriptions of the biosynthesis, structure, chemistry and activation of the crystal protein can be found in the reviews of Norris (1971), Cooksey (1971), Huber and Lüthy (1981), and Fast (1981). More recently, a 20-kb pair DNA fragment was identified in the crystal to which protoxin molecules bind (Clairmont et al. 1998). Biochemical analysis of the complex suggested that the DNA was necessary for both crystal formation and the incremental, stepwise digestion of the protoxin.

Most susceptible lepidopteran insects that ingest Bt undergo a rapid gut paralysis and feeding inhibition as a result of the cytolytic action of activated toxin on midgut epithelial cells. In some species, such as *B. mori*, Bt intoxication also causes a general paralysis. Although the toxin itself can cause mortality, most species do not succumb directly to the damage caused to the midgut, but subsequently to starvation or septicemia. Lüthy and Ebersold (1981) have reviewed the histopathological effects of the toxin on the midgut in vivo and on insect cells in vitro.

3
Dependence of Host and Pathogen on Midgut pH

A critical factor in initiating the pathogenesis of the lepidopteran-specific strains of Bt is the high alkalinity of the larval midgut. Mean pH values of 10–11 are common (Dow 1984; Gringorten et al. 1993) and values over pH 12 have been recorded (Dow 1984; Schultz and Lechowicz 1986). These are among the highest known pH values for any biological system and are typical of the different families of the Lepidoptera (Gringorten et al. 1993). The pH profile exhibits both a longitudinal and radial gradient. It rises in the anterior region of the midgut, reaches a peak in the middle region, and then declines in the posterior region (Dow and Harvey 1988; Gringorten et al. 1993). It is also higher in the ectoperitrophic space next to the epithelium than in the centre of the lumen (Skibbe et al. 1996).

Phytophagous insects have evolved a variety of strategies in adapting to a leaf-feeding habit and the extreme midgut alkalinity of the Lepidoptera in the larval stage is thought to be an evolutionary adaptation to a tannin-rich, high-potassium leaf diet. The high pH promotes the dissociation of tannins from leaf proteins and prevents them from forming complexes with gut-juice enzymes, thereby allowing efficient protein digestion (Berenbaum 1980). At the same time, binding of K^+ (potassium ions) to proteins is enhanced, keeping its activity in the midgut and the chemical gradient between lumen and hemolymph relatively low (Dow and Harvey 1988), and permitting a more controlled release during digestion. In their own course of evolution, lepidopteran pathogens such as Bt and occluded viruses have fully exploited the high-pH condition as the starting point for infection following ingestion by a susceptible host. For viruses, the high pH is necessary to dissolve the occlusion protein; for Bt it is essential for solubilizing the crystal protein, possibly aided by the reducing conditions of the larval midgut. The high pH is also a factor in maximizing the cytolytic action of Bt toxin (Gringorten et al. 1992) and may contribute to a germination-enhancing effect of gut juice on Bt spores (Wilson and Benoit 1990).

The high pH is the result of the larva's ability to secrete potassium carbonate (K_2CO_3) into the midgut lumen, a function maintained by the goblet cells of the midgut epithelium and dependent on an energy-driven K^+ pump located in the apical (lumen-side) membrane of these cells. (Aqueous solutions of K_2CO_3 have a pH of approximately 11.6.) The cytolytic properties of the toxin begin with altered midgut permeability and disruption of K^+ regulation. Its action at the molecular level and the effect on midgut K^+ and H^+ gradients (lumen-cell and lumen-hemolymph) are best understood from a consideration of the homeostatic mechanisms that generate these gradients.

4
Midgut K^+ and H^+ Regulation

4.1
The K^+ Pump

Midgut K^+ and H^+ regulation in larval Lepidoptera are intricately linked and directly sustain physiological processes that are essential to larval survival, including digestion, nutrient uptake, nerve and muscle function, and excretion. The high K concentration (free + bound) in the leaf diet is reflected in the relatively high systemic K^+ activity in the larva – higher than any other free cation – and the Lepidoptera appear to have made evolutionary capital of the condition by developing unique mechanisms to utilize the K^+ to their advantage. In addition to its role in generating the exceptionally high pH of the midgut, it is the predominant cation component of symport systems that effect the "uphill" absorption of amino acids from the midgut (Giordana et al. 1982, 1989; Wolfersberger et al. 1987; Hanozet et al. 1989; Sacchi and Wolfersberger 1996).

That carbonate is responsible for the high pH in the midgut of larval Lepidoptera has been known for about 85 years, although it was first misidentified as Na_2CO_3 instead of K_2CO_3 (Mitani and Watarai 1916). Much of our present knowledge of the role of K^+ as the important alkali cation regulating pH has come from experiments on voltage-clamped larval midguts of the Cecropia moth, *Hyalophora cecropia*, and tobacco hornworm, *Manduca sexta*. The early experiments (on *H. cecropia*) had revealed the existence of a very high potential difference (PD) across the midgut (lumen positive) maintained almost exclusively by a strong K^+ current (Harvey and Nedergaard 1964; Harvey et al. 1968). Flux measurements with radiolabelled K^+ on voltage-clamped midgut preparations showed that nearly all (80 to >95%) of the short-circuit current (I_{sc}) could be accounted for by active secretion of K^+ from the basal side to the luminal side of the midgut (Harvey and Nedergaard 1964; Harvey and Zerahn 1972; Cioffi and Harvey 1981). Stepwise microelectrode impalements of the midgut epithelium linked the electrogenic "K^+ pump" to the apical membrane (Wood et al. 1969). Through ultrastructural examination of the larval midgut and the development of membrane fractionation techniques, the pump was precisely pinpointed to numerous 10–12 nm particles (portasomes) lining the cytoplasmic side of the goblet cell apical membrane (GCAM) (Cioffi 1979; Harvey 1980; Wolfersberger et al. 1982; Cioffi and Wolfersberger 1983), and later identified as a cation- (predominantly K^+-)stimulated ATPase (Wieczorek et al. 1986). Further characterization, however, revealed that it was not an alkali metal ion pump, as first thought, but a proton-pumping, vacuolar-type ATPase (V-ATPase) (Schweikl et al. 1989; Klein et al. 1991) coupled with a $1K^+/2H^+$ membrane antiporter (Chao et al. 1991; Wieczorek et al. 1989, 1991; Azuma et al. 1995). The combined pump-antiporter system ensures that K^+, not H^+, ends up being secreted by the goblet cell, leading to alkalization, rather than acidification, of the midgut lumen.

The stoichiometry of the antiporter ($1K^+/2H^+$) was at first based on free-energy considerations and calculations that a 1:1 ratio would, on average, result in acidification rather than alkalization of the lumen (Chao et al. 1991; Wieczorek et al. 1991); the 1:2 ratio was later verified experimentally by Azuma et al. 1995. Although the pump is, strictly speaking, a primary proton pump, the presence of the antiporter in the GCAM produces K^+ at the output end, and the system as a whole is still commonly referred to as the "K^+-pump".

4.2
The 2K⁺/1ATP Model for Midgut Alkalization

The components of the pump-antiport system in this model were described by Wieczorek et al. (1989, 1991) and reviewed by Dow (1992), Lepier et al. (1994) and Wieczorek et al. (1999a,b). According to the model (Fig. 1), the energy derived from the hydrolysis of ATP drives cytoplasmic H^+ toward the cavity side of the GCAM, creating an electrical charge across it (cavity side positive). The transmembrane voltage in turn energizes the antiporter in which protons are recycled to the cytoplasmic side in exchange for K^+. The K^+ ions are secreted

Lumen pH 11.3

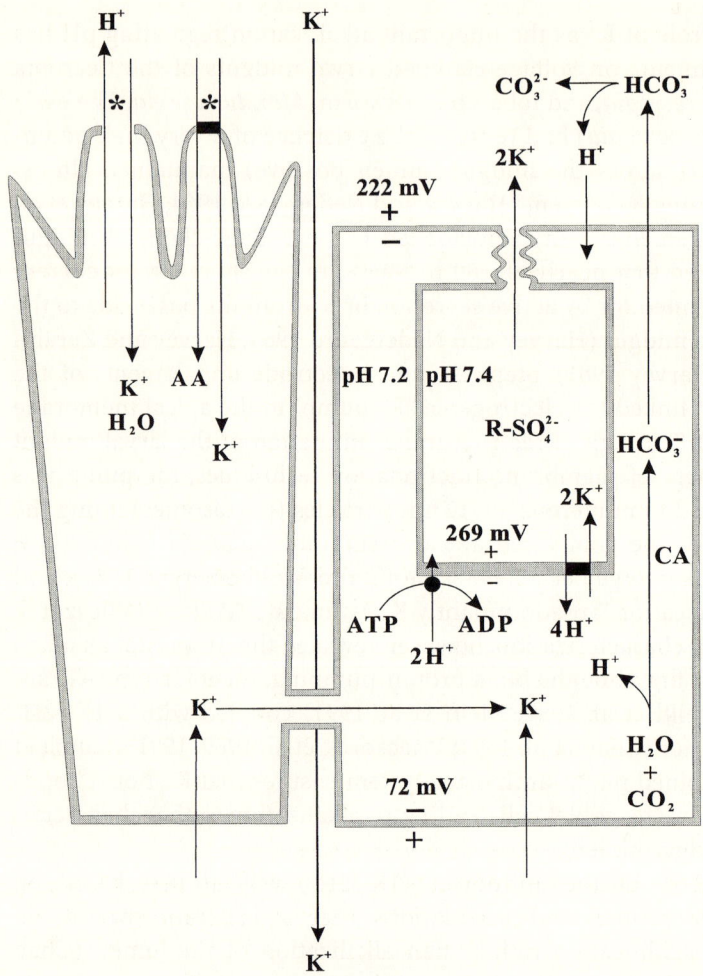

Hemolymph pH 6.8

Fig. 1. $2K^+/1ATP$ model for alkalization of the midgut lumen of larval Lepidoptera and primary target sites (*) of Bt δ-endotoxin. The basic elements of the alkalization process (K^+ secretion and CO_3^{2-} formation) are located in the goblet cell (on the *right*). Illustrated are the primary H^+ pump and K^+/H^+ antiporter in the GCAM surrounding the cavity, along with their respective stoichiometries. A negatively charged sulfated mucopolysaccharide occupies the cavity and balances secreted K^+. The transport route for bicarbonate and the mechanism by which it crosses the apical membrane are uncertain. Gap junctions between cells provide electrochemical coupling between goblet and columnar cells (Moffett and Koch 1988; Dow and Peacock 1989). K^+ can leak to the hemolymph via paracellular shunts. Under steady-state conditions, the >200 mV transapical voltage gradient drives a K^+/amino acid symporter in the apical membrane of columnar cells. The immediate target site of Bt is the CCAM, where the toxin creates pores in the membrane, shuts down the K^+/amino acid symporter, and ultimately destroys cells by alkalization and colloid osmotic lysis. Secondary effects, such as closure of gap junctions between cells and starvation of the K^+ pump, probably also occur. Figures are measurements from *M. sexta*: transmembrane voltages are in vivo values from Dow and Peacock (1989); cell pH figures are open-circuit values from Chao et al. (1991) for the isolated midgut separating asymmetric bathing solutions; lumen and hemolymph pH values are from Dow (1984)

into the goblet cavity where charge balance is provided by a sulfated mucopolysaccharide in the cavity matrix (Dow et al. 1984; Dow 1992). The mechanism by which K^+ leaves the goblet cavity and enters the lumen is not known, but some evidence suggests that it may not be simply by diffusion in free solution through the cavity valve (Moffett et al. 1995).

Overall electroneutrality in the midgut lumen is preserved by concomitant secretion of the bicarbonate ion (HCO_3^-) into the lumen where it is stripped of H^+ in the strong electrical field and converted to carbonate (CO_3^{2-}) (Dow 1984, 1992). The presence of large amounts of carbonic anhydrase in the goblet cells of the anterior and middle sections of the midgut (Ridgway and Moffett 1986) supports this model. The route and mechanism by which HCO_3^- enters the lumen from the goblet cell is uncertain, but it has been suggested that it may be transported through the cytoplasm and across the GCAM without going through the cavity (Dow 1992). The mechanism by which it crosses the GCAM into the lumen may be via HCO_3^-/Cl^- exchange (cf. Chao et al. 1989) or HCO_3^--stimulated ATPase (Deaton 1984; Azuma et al. 1991).

The potential difference across the GCAM between the cavity and cytoplasm in *M. sexta* has been estimated at $\Delta\Psi = 269\,mV$ (cavity side positive) or 26 kJ mol^{-1}, and is virtually equivalent to the total electrochemical gradient for K^+ against which the pump must work (Dow and Peacock 1989). The average free-energy release (ΔG) from ATP hydrolysis in the cell is $52.5\,kJ\,mol^{-1}$ (Dow and Peacock 1989), making possible a K^+/ATP stoichiometry of 2, at maximum pumping capacity and 100% pump-antiport efficiency. The number of protons pumped for each ATP hydrolysed is expected to be 2, based on the relationship between ΔG and $\Delta\Psi$: $\Delta G/nzF = \Delta\Psi$, where n is the H^+/ATP ratio, z is the ion valence (1), and F is the Faraday constant ($0.096487\,kJ\,mV^{-1}\,mol^{-1}$). Solving for n yields an H^+/ATP ratio of 2:1 for the pump. Between the goblet cavity and midgut lumen there is a voltage drop of 47 mV (Dow and Peacock 1989), giving rise to an apical transmembrane PD of 222 mV and a transepithelial potential (TEP, PD between lumen and hemolymph) between lumen and hemolymph of 150 mV (lumen positive; Fig. 1).

Under steady-state conditions, the K^+-pump operates in tandem with a slow K^+ efflux from the lumen back to the hemolymph, mostly through intracellular spaces. The pump-leak steady-state system completes a closed, well-regulated, current-carrying ion circuit. The K^+ conductance of the epithelial apical membrane is very low and the only significant transport across it appears to be via K^+/amino acid symporters.

4.3
The 1K$^+$/1ATP Model for Midgut Alkalization

Klein et al. (1996) argue that the pump-antiporter stoichiometries presented in the above model (Fig. 1) depict an electroneutral system, which could only maintain the voltage gradient between cavity and cytoplasm if the GCAM were completely impermeable to anions, and suggest that this may not be the case. Cl⁻ channel activity has been detected in GCAM vesicles, albeit at high Cl⁻ concentrations (Wieczorek et al. 1989; Wieczorek 1992). They also point out that by drawing two more protons from the goblet cavity matrix than is pumped in each cycle, the antiport would alkalize the cavity. To address these problems, they suggest a model (Fig. 2) in which two H^+ ions are pumped for each K^+ exchanged, rather than 1:1, and the K^+/ATP ratio is 1:1, rather than 2:1 (A. Koch, pers. comm.).

By linking the antiporter to the pump in a 1K⁺/1ATP system, the combination appears to offer the advantage of being electrogenic and, at the same time, allows an anion conductance. Indeed, it would appear that, in this model, an anion efflux, presumably Cl⁻, into the cavity would be essential to prevent K^+ flux from over-charging the membrane capacitance (maximum possible PD = $\Delta G/nzF$ = 544 mV; see Sect. 4.2) and exceeding the dielectric limit for the plasma membrane (probably less than 300 mV; cf. Dow 1992).

The proposed model, however, presents some problems. Theoretical and experimental evidence indicates that the antiporter-pump relationship is more likely to be 2K⁺/1ATP than 1:1 (see Sect. 4.2) since, in vivo, under steady-state conditions, with the amount of K^+ present in the hemolymph, the pump is probably operating at maximum capacity (cf. Harvey et al. 1967). The inclusion of an obligatory Cl⁻ flux across the GCAM could be another problem. Other than the Cl⁻ channels observed by Wieczorek et al. (1989) in GCAM vesicles (at high external Cl⁻ concentration), there is no convincing evidence of their operational presence in vivo (Wieczorek 1992).

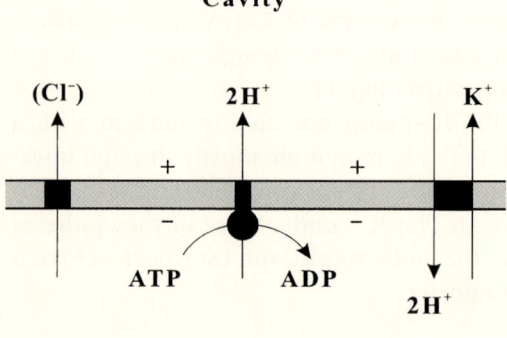

Cavity

Cytosol

Fig. 2. The 1K⁺/1ATP model for midgut alkalization (based on Klein et al. 1996). The V-ATPase pumps 2H⁺ for each K⁺ secreted into the goblet cavity. The pump-antiporter system is potentially more strongly electrogenic than the 2K⁺/1ATP system, and requires a variable anion (Cl⁻) conductance to prevent the membrane capacitance from being over-charged by the K⁺ flux

Are the two models at variance with each other? The V-ATPase pump and antiporter are separate molecular entities (Wieczorek et al. 1991) and there is no evidence of a fixed coupling ratio between them (W.R. Harvey, pers. comm.). It is tempting, therefore, to suggest that their stoichiometric relationship at any given moment may be homeostatically modulated by such factors as trans-membrane PD and ionic balance (K^+, pH and Cl^-). In short, perhaps both models for midgut alkalization are valid and operate at different times, depending on overall physiological conditions. The $2K^+/1ATP$ system is not necessarily electroneutral, although it may appear that way on a macroscale. It is the proton-pumping V-ATPase that first generates the PD that secondarily drives the antiporter, and so a continuously operating pump should, in principle, be able to generate and maintain an electrical charge across the membrane. Nor would its operation necessarily alkalize the goblet cavity. An abundant source of protons exists through CO_3^{2-} formation that could maintain the pH of the cavity matrix, although the mechanism and route of entry of this source of H^+ into the cavity is speculative at present.

4.4
Transmembrane and Transepithelial Ion Gradients

As a direct consequence of the strong transmembrane PD a very steep H^+ gradient is also generated across the midgut with cell cytoplasm/lumen and hemolymph/lumen concentration ratios reaching 10^4 or higher (Table 1, controls). At equilibrium, the pH in the midgut lumen is consistent with a passive distribution of H^+ across the cell membrane, as predicted from the Nernst equation: $\Delta\Psi_{(lumen-cell)} = 59.1 \times (pH_{lumen}-pH_{cell})$ at 25 °C, where $\Delta\Psi_{(lumen-cell)} = 222\,mV$ (Dow and Peacock 1989) and 59.1 is a temperature-dependent, complex number incorporating the gas and Faraday constants. Chao et al. (1991) measured the pH of goblet cell cytoplasm in the isolated posterior midgut of *M. sexta* and recorded a value of 7.16 under open-circuit conditions, with bathing solutions of pH 6.8 on the hemolymph side and pH 8.0 on the luminal side of the preparation. Assuming that the pH is the same for middle

Table 1. Ion activities in the midgut and hemolymph of day-1, fasting, fourth-instar *Bombyx mori* force-fed a lethal dose of Bt activated toxin[a]. (Reproduced with permission from Gringorten 1999 and the Society of Chemical Industry)

| | pH | | K^+ (mM) | | Na^+ (mM)[b] | |
	Midgut	Hemol.	Midgut	Hemol.	Midgut	Hemol.
Control	11.2 ± 0.1	6.6 ± 0.1	92 ± 8	59 ± 4	<0.1	3
Bt treated	8.4 ± 0.2	7.7 ± 0.3	98 ± 13	96 ± 23	0.3	1

[a] Means ±95% CL
[b] Na^+ sample size was too small for confidence limits

midgut goblet cells, rearrangement of the Nernst equation gives an estimate of pH 10.9 for the lumen, in good agreement with observed values in the middle midgut of *M. sexta*: pH 10.1 (Moffett and Cummings 1994) to a peak value of pH 11.3 (Dow 1984). Dow and Peacock (1989) calculated that 21% of the total ATP in *M. sexta* midgut was utilized to pump K^+ into the lumen and alkalize the contents. It is apparent, therefore, that although the apical transmembrane pH gradient in the midgut (cell-lumen) reflects a passive, Nernstian distribution of protons, it comes at a high energy cost.

Compared with the enormous pH gradients across the midgut, K^+ activity ratios are very small, and the force driving K^+ from the lumen to the hemolymph side of the midgut is predominantly the electrical gradient produced by the K^+ pump itself (Dow and Harvey 1988; Dow and Peacock 1989). K^+ levels in leaf-feeding *M. sexta* middle midgut are 25, 62 and 91 mM for hemolymph, lumen and cell cytoplasm, respectively (Dow and Harvey 1988; Moffett and Koch 1988). The lumen-hemolymph chemical gradient is thus $59.1 \times \log (62/25) = 23$ mV at 25 °C, which represents only 13% of the total electrochemical gradient (23 mV + 150 mV TEP) for K^+. The lumen-cell K^+ chemical gradient is negative, indicating that the driving force for K^+ into midgut epithelial cells is entirely electrical. In fasting *B. mori* the lumen/hemolymph K^+ concentration ratio is only 1.6, compared with the proton ratio (hemolymph/lumen) of 4×10^4 (Table 1, controls). The K^+ ratio translates to a chemical gradient of 12 mV and, assuming a TEP value of 150 mV, represents only 7–8% of the total lumen-hemolymph gradient for K^+.

5
Disruption of Midgut Ion Homeostasis by *Bacillus thuringiensis*

5.1
In Vivo Changes

The first action of Bt δ-endotoxins is on the midgut columnar cells where they bind to specific receptors in the columnar cell apical membrane (CCAM) and create cation-selective channels (see Sect. 6). The direct physiological consequence for the larva following ingestion of a lethal dose of Bt is collapse of the electrical gradient between lumen and cell and the disruption of K^+ and pH regulation. Generally, it is found that total K concentration and pH of the hemolymph rise, although wide differences in the extent of these effects exist among different species, with some experiencing very little or no change in pH and only marginal increases in K levels (Nishiitsutsuji-Uwo and Endo 1981a,b). The extent of hemolymph K elevation, in particular, is affected by the nutritional status of the test insect, specifically whether or not there is leaf material present in the midgut.

That the target organ of δ-endotoxin poisoning in the insect is the midgut was demonstrated more than forty years ago by Heimpel and Angus (1959) in *B. mori*. Subsequently, Fast and Angus (1965) presented evidence that the toxin affected midgut permeability. Later, Angus (1968) observed that larvae fed the antibiotic valinomycin, a K^+-specific ionophore, exhibited symptoms that were remarkably similar to those of Bt poisoning (including the increase in hemolymph alkalinity) and made the rather prescient suggestion that Bt acted by altering midgut permeability and seriously disrupting ion regulation. Fast and Morrison (1972) used $^{42}K^+$ to measure K turnover in *B. mori* midgut tissue and found that Bt had no significant effect, leading them to conclude that, although δ-endotoxin and valinomycin affected membrane permeability and produced very similar physiological symptoms, the toxin was not acting as a K^+ ionophore. However, their data suffered from very high variation and their conclusion was considerably weakened by the failure to run a valinomycin control. Crawford and Harvey (1988) attributed the absence of an effect of the toxin on net $^{42}K^+$ accumulation in midgut cells in that study to futile K^+ cycling between lumen and cells as a result of the altered membrane permeability (discussed further in Sect. 5.2). Stronger support for the conclusion that δ-endotoxin was not simply a valinomycin mimic was provided by Nishiitsutsuji-Uwo et al. (1979) who observed distinct differences in their effects on cell ultrastructure, particularly mitochondria, and noted that, unlike valinomycin, δ-endotoxin was just as effective in vitro when K^+ was replaced with Na^+. That their modes of action were quite different became clear when the ion-channel and pore-forming properties of δ-endotoxins were characterized in later investigations (see Sects. 6 and 7) and the relatively narrow host specificity of different toxins became apparent.

Heimpel and Angus (1959) distinguished three main groups of larval Lepidoptera, based on the symptoms of toxemia and the requirement of spores for insecticidal activity. Type I insects, exemplified by *B. mori,* were sensitive to the toxin alone and exhibited both midgut paralysis and a general paralysis. The midgut became more acidic and the hemolymph more alkaline. Type II were less sensitive to toxin and underwent midgut paralysis, with no change in hemolymph pH and usually no general paralysis (midgut pH was not measured). Type III larvae displayed little reaction to the toxin and died only if challenged with both spores and toxin. Only one insect, the Mediterranean flour moth, *Anagasta* (*Ephestia*) *kuehniella*, was classified as Type III at the time. Heimpel and Angus (1959) suggested that the relatively low midgut pH of this larva (≤pH 8.9), compared with Types I and II Lepidoptera, might allow immediate spore germination. It should be noted that there is no sharp division among these three groups with respect to the response to Bt, and even Heimpel and Angus (1959) considered the classification as somewhat arbitrary. The group in which a particular insect is placed can be rather subjective and depend on the dose and type of toxin used (Nishiitsutsuji-Uwo and Endo 1981a,b).

In *B. mori* the electrical gradient across the midgut epithelium, upon which the pH gradient depends, totally collapses when a lethal dose of Bt is ingested. The observed changes in fasting, day-1, fourth instars, as measured with ion-selective electrodes, are significant acidification of the midgut lumen and alkalization of the hemolymph (Table 1), symptoms already noted by Angus and Heimpel (1956) in fourth- and fifth-instar *B. mori* fed Bt-coated foliage. There is also a 60% increase in hemolymph K^+ concentration, but no change in midgut lumen K^+ (Table 1). The toxin-induced changes in hemolymph pH and K^+ concentration are similar to those reported by Nishiitsutsuji-Uwo and Endo (1980) in day-4, fifth-instar *B. mori*. The pH changes are almost identical, but Nishiitsutsuji-Uwo and Endo (1980) measured total K, rather than K^+ activity, and recorded more than a twofold increase. These authors also measured total hemolymph concentrations of Na, Ca and Mg and found that they were unaffected by the toxin.

The hemolymph K^+ data in Table 1 are also qualitatively similar to those reported by Ramakrishnan (1968) and Fast and Morrison (1972). Ramakrishnan (1968) observed more than a fourfold increase in total K concentration in the hemolymph of fifth-instar *B. mori* fed Bt-coated foliage. Fast and Morrison (1972), working with three-day-old, fifth-instar *B. mori*, reported that total hemolymph K concentration almost doubled within two hours in response to the toxin dose. There was no change in total K concentration of midgut tissue or in the midgut tissue concentrations of Na, Ca and Mg. Changes in the luminal contents alone were not measured. In the Ricini moth, *Philosamia ricini* (*Samia cynthia ricini*), total hemolymph K concentration more than doubled in larvae force-fed Bt crystals, while total K concentration of the midgut contents dropped by 24% (Pendleton 1970). Total hemolymph Na concentration, which was also measured, decreased by almost half, while that of the midgut contents, exceedingly low to begin with, remained unchanged. The pH increased marginally from 6.5 to 6.8. Narayanan and Jayaraj (1974) reported more than a fivefold increase in total K concentration in the hemolymph of the citrus leaf caterpillar, *Papilio demoleus*, fed Bt-coated foliage, as well as smaller increases in total Na and (free) Mg^{2+}. The concentration of Ca^{2+}, however, declined by approximately two-thirds. Little or no change in hemolymph pH and only small increases in total K concentration were observed in the greater wax moth, *Galleria mellonella* (Nishiitsutsuji-Uwo and Endo 1981a); imported cabbageworm, *Pieris rapae*; gypsy moth, *Lymantria dispar*; and almond moth, *Ephestia* (*Cadra*) *cautella* (Nishiitsutsuji-Uwo and Endo 1981b).

The finding by Fast and Morrison (1972) that there was no significant accumulation of K^+ in midgut tissue following a lethal dose of Bt toxin led these authors to conclude that the changes in hemolymph ion levels observed in their study and in studies of other authors were secondary effects of a general "breakdown of gut metabolism". In *B. mori*, changes in ion activities do not occur immediately, but correlate with increasing morbidity of the larva. Moreover, time-course measurements in silkworm larvae force-fed Bt reveal that initially there is a significant increase in midgut pH prior to the decline (Fig. 3). At present, it is not known whether this is a direct effect of the toxin, a nonspecific stress response to the force-feeding procedure, or a general response that occurs when fasting larvae are presented with food, regardless of whether or not they actually feed.

5.2
In Vitro Changes

Electrophysiological studies and K^+ flux measurements across the midgut under in vitro conditions have shown that the immediate response to activated toxin applied to the luminal surface is an increase in the CCAM K^+ conductance and leakage of K^+ toward the hemolymph side (Harvey and Wolfersberger 1979; Harvey et al. 1986; Wolfersberger 1992). In the voltage-clamped midgut of *M. sexta*, the I_{sc} drops by 60–70% in less than 40 min in response to partially activated (alkali-dissolved) toxin applied to the luminal side of the preparation (Harvey and Wolfersberger 1979), during which time the K^+ pump may at first be unaffected. Using $^{86}Rb^+$ as a measure of K^+ flux, Harvey and Wolfersberger (1979) estimated that 29% of this initial drop was due to unidirectional K^+ flux from the apical to the basal side of the preparation, which nearly tripled. Most of the decline was attributed to futile K^+ cycling while the K^+ pump was still functional (Crawford and Harvey 1988). In this scenario, K^+ influx from the lumen into the columnar cells through toxin-produced pores

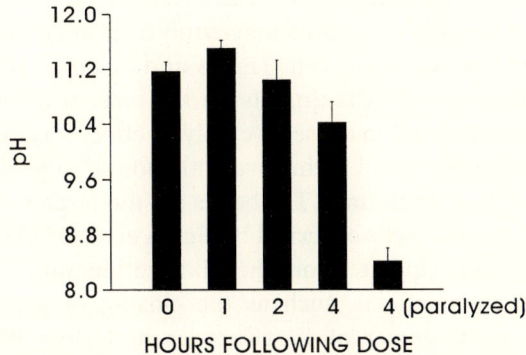

Fig. 3. Time-course changes in midgut pH in *Bombyx mori* larvae force-fed activated toxin. The dose caused 67% mortality in a separate group of insects. Bar = 95% confidence limit. (Reproduced with permission from Gringorten 1999 and the Society of Chemical Industry)

in the CCAM is transported back to the goblet cells through gap junctions between the columnar and goblet cells, and then pumped back out into the lumen. The futile K^+ cycle would also include ions entering the cells from the basal extracellular space (Crawford and Harvey 1988). Such cycling would carry no current. Griego et al. (1979) force-fed purified protoxin crystals to *M. sexta* larvae and examined the effects on the TEP and I_{sc}. Not only were they totally depressed within four hours, but both had reversed polarity and become slightly negative. Active transport of Ca^{2+} or Mg^{2+} toward the hemolymph side of the preparation and Cl^- or HCO_3^- toward the luminal side were eliminated as possible causes of the reversal.

Gupta et al. (1985) used X-ray spectrum analysis to investigate concentration changes of various elements (K, Na, Mg, P, S, Cl, Ca) in the short-circuited midgut after exposure to δ-endotoxin. The major changes were a sharp drop in total K concentration (dry- and wet-weight basis) in the goblet cavity and a large drop in sulfur. The dry-weight fraction decreased at the same time that swelling was observed, indicating water uptake into the cavity, while the cell cytoplasm appeared to shrink. The concentration of Cl increased in the cavity. There were no significant changes in the other elements in the goblet cells. In the columnar cells there was a marked decrease in K concentration and a large amount of water loss. The authors concluded that the immediate action of the toxin was on the goblet cell, at the apical membrane, rather than on the columnar cell, with the result that either the K^+ pump was immediately inhibited or anion conductance was increased, both of which would cause rapid membrane depolarization.

However, the weight of all other evidence is against this conclusion and in support of the CCAM as the primary and immediate target of δ-endotoxin (Percy and Fast 1983; Bravo et al. 1992a,b; Aranda et al. 1996; Pietrantonio and Gill 1996). The receptors for the toxin are in the CCAM, not the GCAM, and there is no evidence that the toxin is internalized in the epithelial cells or diffuses through the goblet cavity valve (Bravo et al. 1992a,b). Histopathological observations indicate that cytolytic changes in the goblet cells lag behind those of the columnar cells (Endo and Nishiitsutsuji-Uwo 1980, 1981; Bravo et al. 1992a,b). That toxin acts on the plasma membrane and does not have to be internalized to cause its cytolytic effect was demonstrated by Fast et al. (1978), using activated toxin covalently bound to Sephadex beads and assayed against an insect cell line. (The beads are too large to be taken up by cells.) The effects on goblet cells observed by Gupta et al. (1985) could just as easily be explained as consequent upon the alkalization and lysis of columnar cells and on indirect effects, such as the closing of gap junctions, which would starve the antiporter of K^+ (Knowles and Dow 1993). Moreover, the presence of transient, futile K^+ cycling indicates that for a brief period following the initial toxin action, the GCAM is intact and the K^+ pump continues to operate. Crawford and Harvey (1988) attributed the decrease in K concentration in the goblet cavity to secondary shutdown of the K^+ pump as a result of increasing cell alkalinity.

5.3
What is the Source of the Elevated Hemolymph K^+?

The 60% increase in hemolymph K^+ concentration observed in toxin-treated, fasting B. mori (Table 1) is consistent with (but not proof of) a model for ion flux in vivo from lumen to hemolymph. If gut-juice K^+ is the source, however, it is surprising that there is no concomitant change in luminal K^+ concentration (Table 1). Its absence stands in marked contrast to the response in P. ricini, in which the increase in total hemolymph K concentration was accompanied by a 24% decrease in midgut K (Pendleton 1970), implying that the toxin had induced significant leakage of K^+ from the midgut to the hemolymph. Both Ramakrishnan (1968) and Narayanan and Jayaraj (1974) assumed that the source of elevated hemolymph K in Bt-treated insects was the midgut luminal contents, but Pendleton (1970) suggested that it was from intracellular stores. Support for this suggestion, at least in part, was provided by measurement of total hemolymph K concentration in P. ricini purged of K in the midgut lumen and force-fed Bt. Although the initial hemolymph K level before the Bt treatment was considerably lower than that of normal larvae feeding on foliage, the concentration still more than doubled in response to the dose (Pendleton 1970), and so must have come from intracellular stores. However, the experiment does not rule out the possibility that the large pool of K in the midgut of a leaf-feeding larva could also contribute to an elevation of hemolymph K.

One explanation for the failure to observe any change in midgut K^+ in Bt-treated, fasting B. mori (Table 1) – one that would jibe with the suggestion that the high midgut pH promotes binding of K^+ to proteins (Dow and Harvey 1988) – is the possibility that K^+ could have been released from midgut proteins, as lumen pH declined, at the same rate that the free ion leaked into the hemolymph. To test this possibility, experiments (unpublished) were carried out to explore the effect of pH on K^+ activity in the gut juice of fasting B. mori. Total K and K^+ activity were measured in gut juice and the ion activity was compared with that following acidification. Total K in gut-juice samples (pH 10.2) was approximately 247 mMolal. Conversion of the control K^+ concentration in Table 1 (92 mM) to mMolal units (from specific gravity and percentage water measurements) revealed that almost two-thirds of the total K in the gut juice of fasting B. mori is bound. Acidification of the samples with CO_2 gas brought the pH down to 7.4–7.8, but had no effect on K^+ activity. The source of the elevated hemolymph K^+ in fasting B. mori, therefore, is still uncertain and requires further investigation. It may well have been intracellular stores, rather than bound gut-juice K^+, which would be consistent with Pendleton's (1970) finding in P. ricini larvae with K-depleted midguts.

5.4
Larval Paralysis and Mortality Factors

Heimpel and Angus (1959) were able to induce general paralysis in different species of type II larvae by injecting CO_3^{2-} into the hemolymph and raising the pH in situ to levels observed in toxemic *B. mori*. They suggested that the elevated hemolymph pH that occurs in Bt-intoxicated Type I larvae may be the factor responsible for the general paralysis, perhaps by its deleterious effect on muscle control (Angus and Heimpel 1956) or inhibition of nerve function (Heimpel and Angus 1959). However, the suggestion appears to have been made without taking into consideration that the "carbonate buffer" used to raise the hemolymph pH also increased the cation concentration. (The type of carbonate, whether Na_2CO_3 or K_2CO_3, was not stated.) Ramakrishnan (1968) drew attention to the importance of K^+ regulation for nerve and muscle function and considered the paralysis in Bt-treated *B. mori* to be due to elevated K^+ levels. Narayanan and Jayaraj (1974) implicated the overall disturbance in ion balance, including abnormal Ca^{2+}, Mg^{2+} and Na^+ levels, as well as elevated hemolymph K^+, as contributing to the general paralysis in Bt-treated insects. Clearly none of these studies distinguish between K^+ (or other metal ions) and pH as putative paralysis factors. Pendleton (1970), however, noted the rather insignificant change in hemolymph pH in paralyzed *P. ricini*, thereby providing somewhat better evidence that elevated K^+ in the hemolymph of Bt-intoxicated insects is by itself an important factor causing paralysis.

Lethal doses of toxin cause irreversible damage to the midgut epithelium through a combination of cell alkalization (Harvey et al. 1986) and colloid osmotic lysis (Knowles and Ellar 1987). Those insects that receive sufficiently high doses of toxin (without spores) to cause general paralysis do not recover. Some lepidopteran species, such as *B. mori*, are very sensitive to Bt δ-endotoxin ($LD_{50} < 10$ ng) and succumb relatively quickly. In these insects, the relationship between midgut breakdown and systemic paralysis, as well as identification of the important mortality-causing factor(s), requires further investigation. In insects that do not undergo general paralysis, midgut damage from toxin alone may be sufficient to cause death by starvation. With most species, however, both the toxin and spores are required for an effective dose, and insects die from the ensuing septicemia. The spores are able to germinate when the midgut pH falls to a level suitable for vegetative growth.

5.5
δ-Endotoxin Effects on K⁺-Dependent Uptake of Amino Acids

Brush border membrane vesicles (BBMVs), prepared from the microvillar apical membrane of midgut columnar cells, have provided investigators with an exceedingly useful in vitro system for studying membrane transport processes and mode of action of Bt δ-endotoxins. Extensive use has been made of such preparations to elucidate the mechanism of amino acid uptake by the midgut and its inhibition by toxin. The method of Wolfersberger et al. (1987), based on precipitation and differential centrifugation of midgut homogenates, is the one generally followed in preparing BBMVs. The vesicles that are obtained are mostly right-side out (Wolfersberger 1989; Schwab and Culver 1990) and can be preloaded with test compounds or polarized through the use of appropriate ionophores to impose a transmembrane PD. In the presence of an inwardly directed K^+ gradient, BBMVs from the cabbage butterfly, *Pieris brassicae*, were shown to transiently accumulate (i.e., overshoot equilibrium levels) a wide range of amino acids (Wolfersberger et al. 1987), and were used by Sacchi et al. (1986) as a model system to study inhibition by δ-endotoxin of K^+/amino acid cotransport (symport) across the midgut apical membrane. Several studies (Hendrickx et al. 1990; Wolfersberger 1991; Parenti et al. 1995) have demonstrated a positive correlation between toxin inhibition of K^+-gradient-driven amino acid uptake by BBMVs from susceptible larvae and in vivo toxicity. The inhibition by BBMVs of amino acid uptake is considered to be a reliable in vitro system for evaluating larvicidal specificity of Cry proteins (Parenti et al. 1995).

There are two principal means by which δ-endotoxins can inhibit amino acid absorption in the larval lepidopteran midgut (prior to cell lysis): dissipation of the transmembrane potential that drives K^+/amino acid symport and direct interference with the symport system itself. These effects were first separated and analyzed in a study of toxin inhibition of histidine uptake by *P. brassicae* BBMVs (Sacchi et al. 1986). In that study it was demonstrated that in the presence of a K^+ gradient, either chemical or electrical, toxin suppressed both the initial rate and level of transient accumulation of histidine by BBMVs, but in the absence of the K^+ gradient (equal K^+ concentration and isopotential inside and outside the BBMVs), the initial rates of uptake by toxin-treated and untreated BBMVs were practically identical. The authors interpreted the results as indicating that the inhibitory effect of toxin on histidine uptake was due principally to the increase in membrane permeability to K^+, which provided a parallel, lower-resistance route for K^+ transport across the membrane that effectively short-circuited, but did not interfere directly with, the K^+/amino acid symport pathway.

Hendrickx et al. (1990), on the other hand, in a study of alanine uptake by *M. sexta* BBMVs, concluded that inhibition of the transient accumulation of the amino acid by toxin was due mostly to the creation of non-selective, rather than cation-selective, pores in the membrane. As evidence, they noted that K^+-channel blockers Ba^{2+} and Ca^{2+} failed to prevent the inhibitory effect of toxin on K^+-gradient-driven alanine accumulation by the vesicles, and toxin-treated, alanine-loaded vesicles leaked the amino acid more quickly than untreated vesicles. From the vantage point of a now better understanding of how δ-endotoxin interacts with the midgut membrane, it is reasonable to postulate that cation-selective channels probably formed first and that these then widened into non-specific pores (see Sects. 6 and 7). Both the origin of the BBMVs and the particular type of toxin used can be expected to influence the kinetics of ion-channel and pore formation. Moreover, Ba^{2+} and Ca^{2+} can themselves inhibit K^+-gradient-dependent uptake of amino acid by *M. sexta* BBMVs (Wolfersberger 1989), which could explain why these cations did not prevent toxin inhibition of alanine uptake.

That δ-endotoxins can also directly inhibit K^+/amino acid symporters in a manner distinct from their effect on membrane cation permeability was demonstrated in both BBMVs and the isolated larval midgut (Reuveni and Dunn 1991; Giordana et al. 1993; Parenti et al. 1995; Leonardi et al. 1997). Direct and selective interference with a K^+/amino acid symport system by activated toxin was described by Reuveni and Dunn (1991) for leucine and aspartic acid uptake in *M. sexta* BBMVs. They determined that the two amino acids were transported by different symporters and that five times as much toxin was required to inhibit uptake of aspartic acid than leucine. Ionophoric inhibition of uptake by valinomycin, however, occurred to the same extent, confirming that a mechanism restricted only to increasing membrane K^+ permeability would have a non-specific and uniform effect on K^+-gradient-dependent processes, and would not interfere directly with those processes. It supported the conclusion that the preferential inhibition of leucine uptake by toxin reflected a certain amount of direct and selective interference with the K^+/leucine symport pathway.

Giordana et al. (1993) examined the effect of activated toxin on histidine and lysine uptake by *B. mori* BBMVs in the presence of a K^+ gradient and reported that, at sufficiently high doses, toxin totally inhibited not only the K^+-gradient-dependent accumulation of these amino acids, but their equilibrative uptake as well. The inhibitory effect of toxin on equilibrative uptake was confirmed in a K^+-gradient-free system (equal K^+ concentration inside and outside the BBMVs) with valinomycin serving as a negative control. Preclusion of the amino acids from attaining a concentration equilibrium with the BBMVs was a clear sign that the toxin had inhibited the symport systems as well as dissipated the K^+ gradient.

Parenti et al. (1995) obtained similar results in a study of toxin inhibition of leucine uptake by *B. mori* BBMVs under different K^+-gradient and pH conditions, but the kinetic parameters of the inhibition suggested that part of the uptake might not be sensitive to toxin. Comparison of the Hill coefficients obtained with media at alkaline and near-neutral pH indicated that at least two toxin molecules were needed per symporter at the lower pH to inhibit amino acid transport. The toxin-K^+/leucine symporter interaction in *B. mori* was further investigated by Leonardi et al. (1997) who measured leucine uptake into the BBMVs and used a voltage-sensitive dye to monitor simultaneous changes in artificially induced transmembrane PDs. They demonstrated the electrophoretic nature of the symport mechanism using BBMVs with an imposed transmembrane PD (inside-negative, outside-positive) in the presence of an inwardly directed cation gradient. Depolarization of the BBMVs was accelerated upon addition of the amino acid and inhibited by toxin. They also extended the study to an isolated midgut preparation (K^+ pump is functional), and showed that, under symmetric K^+ conditions in the perfusate and superfusate, the TEP produced (90 mV) was abolished when toxin was introduced into the lumen, but subsequent uptake of leucine by midgut epithelial cells was not entirely prevented, although the amount of amino acid accumulated by the cells was inversely related to the toxin dose.

5.6
Correlating δ-Endotoxin Effects on the Isolated Midgut with Insecticidal Activity

Midgut preparations exposed to activated δ-endotoxin undergo a sharp decline in the TEP and in the short-circuited midgut there is a rapid inhibition of the I_{sc} (Griego et al. 1979; Harvey and Wolfersberger 1979). In *M. sexta*, Harvey and Wolfersberger (1979) recorded a drop in midgut electrical resistance of more than 40% and a marked increase in K^+ flux from the apical (lumen) side to the basal side of the preparation. The reliability of midgut preparations as indicators of insecticidal activity was tested by several investigators. Wolfersberger and Spaeth (1987) compared the larvicidal activity against *M. sexta* of 22 Bt strains from 20 different subspecies with the ability of their alkali extracts to inhibit the I_{sc} in the voltage-clamped midgut. They found a good qualitative relationship and concluded that the in vitro system could be used as a reliable and rapid indicator of Bt potency.

In a structure-function study, Chen et al. (1993) correlated I_{sc} inhibition in voltage-clamped *B. mori* midguts with insecticidal activity of Cry1Aa mutant toxins from both the slope of the current change with time and the lag time until the onset of the change after applying the toxin. A detailed account of the method was described by Liebig et al. (1995) who showed that there was a direct correlation between toxin concentration and the rate of current change and an inverse correlation between toxin concentration and the lag time. The correlation between the I_{sc} inhibitory response and toxin concentration was linear for toxins with high and low insecticidal activity, but shifted 20- to 30-fold toward higher concentrations for the latter. The "voltage-clamp assay" was also effective in revealing a synergistic effect with a Cry1Aa-Cry1Ac toxin combination against *L. dispar* that correlated with in vivo results (Lee et al. 1996a).

Using microelectrode impalements on isolated midgut preparations and measuring changes in electrical potential across the apical membrane, Peyronnet et al. (1997) examined toxin specificity in vitro in *L. dispar* and *B. mori* and compared the observations with insecticidal activity in vivo. They observed a high degree of correlation for three toxins tested against *B. mori*, but notable differences in three of eight toxins tested against *L. dispar*. In the latter case, three of the toxins caused significant apical-membrane depolarization in vitro, but were at best only very weakly active in bioassays against *L. dispar*. The incomplete correlation raises several important points. First, and most obvious, is that even with freshly mounted midgut preparations, toxin effects in vitro may not be an accurate predictor of insecticidal activity. Second, if the observations for *L. dispar* are any indication, midgut preparations may be monotonically biased as sometime false-positive, rather than false-negative, indicators of Bt insecticidal activity. If this is true, such preparations could be used to rapidly screen Bt toxins and select only those that show activity for further testing in insect assays. Finally, toxins that are active in vitro but not insecticidal indicate that the host insect is somehow able to inactivate them or prevent their diffusion to the midgut epithelium. Identifying the physiological barriers that protect the insect against potentially insecticidal toxins would be a worthwhile research endeavour.

6
Receptor Binding and Ion-Channel Formation

Extensive research has been carried out in the past few years to characterize the membrane interactions of the lepidopteran-specific strains of Bt that lead to ionic disruption and cell lysis, and relate the effects to insecticidal specificity. Much information has come from in vitro investigations of ion-channel activity and permeabilizing effects caused by toxins in planar lipid bilayers (PLBs), liposomes (phospholipid vesicles), BBMVs and insect cell lines, and attempts to extrapolate observations to the midgut in vivo. The overall sequence of events in vivo consists of (1) binding of the activated toxin to specific receptors in the CCAM, (2) induction of a conformational change in the toxin molecule, (3) membrane insertion and formation of cation channels and pores, manifested in the larval midgut mainly as an increase in K^+ conductance, (4) collapse of midgut voltage and pH gradients, and (5) lysis of epithelial cells. The receptor-binding step is reversible, whereas membrane insertion is irreversible (reviewed by Rajamohan et al. 1998 and Schnepf et al. 1998).

6.1
Receptor Binding

Although the presence of saturable, high-affinity binding sites in the apical membrane appears to be a prerequisite for insecticidal activity (Hofmann et al. 1988a,b; Van Rie et al. 1989, 1990a,b; Ferré et al. 1991; Lee et al. 1992; Tabashnik et al. 1994), there is no strict quantitative relationship and their presence does not guarantee toxicity in vivo. On the contrary, in some instances, the correlation between binding and toxicity may be poor or absent (Ferré et al. 1991; Garczynski et al. 1991; MacIntosh et al. 1991; Gould et al. 1992; Ihara et al. 1993; Masson et al. 1995; Yi et al. 1996), or even inverted (Wolfersberger 1990), and insertion of the toxin molecule and pore formation following binding may, in some instances, be the more important determinants of insecticidal activity (Wolfersberger 1990; Ihara et al. 1993). Thus, Wolfersberger (1991) suggested that more effective pore formation would explain the positive correlation between toxin inhibition of amino acid accumulation by *L. dispar* BBMVs and insecticidal activity of Cry1Ab and Cry1Ac, whereas binding affinity to their common receptor and insecticidal activity were inversely related (Wolfersberger 1990). The validity of this suggestion was later borne out in a two-step kinetic analysis of toxin binding to BBMVs carried out by Liang et al. (1995) that took into account both the reversible and irreversible components of binding. These authors demonstrated that the inverse correlation between binding and insecticidal activity observed by Wolfersberger (1990) applied to the reversible step, which is confined to the receptor, but that a direct correlation existed for the irreversible step, which is indicative of insertion. It is apparent, therefore, that binding affinity, binding-site concentration and the ability of toxin molecules to integrate into the plasma membrane and form ion channels and pores all function as determinants of larvicidal activity and contribute to toxin specificity.

In the absence of receptors, Bt δ-endotoxins are capable of partitioning spontaneously into lipid bilayers and forming ion channels and/or pores, if relatively high toxin concentrations are applied (Haider and Ellar 1989; Slatin et al. 1990; Schwartz et al. 1993, 1997a; Lorence et al. 1995; Martin and Wolfersberger 1995; Grochulski et al. 1995). One function of the receptor, therefore, would seem to be to concentrate toxin molecules at the plasma membrane surface, thereby increasing the probability of insertion and pore formation (Haider and Ellar 1989; Schwartz and Laprade 2000). However, the receptor may also play a more active, catalytic-like role in the insertion mechanism (English et al. 1991), possibly through a sensory effect on α-helix 7 of the toxin molecule that helps trigger the conformational change (see Sect. 7) (Gazit and Shai 1995; Gazit et al. 1998). To date, two types of receptors have been identified: aminopeptidase N (Knight et al. 1994, 1995; Sangadala et al. 1994; Garczynski and Adang 1995; Gill et al. 1995; Valaitis et al. 1995; Lee et al. 1996b; Luo et al. 1996, 1997; Lorence et al. 1997; Yaoi et al. 1997) and a cadherin-like glycoprotein (Vadlamudi et al. 1993, 1995; Francis and Bulla 1997; Keeton and Bulla 1997).

Several authors have discussed the possibility that K^+/amino acid symporters might also function as the toxin receptors for insertion and ion-channel formation (Reuveni and Dunn 1991; Giordana et al. 1993; Parenti et al. 1995; Leonardi et al. 1997). It was noted, for example, that aminopeptidase

N is required for Na$^+$/amino acid cotransport in bovine renal BBMVs. However, a dual role does not appear to be the case in *M. sexta*, at least for one symporter. The Cry1Ac receptor in this insect is aminopeptidase N, but the symporter responsible for uptake of neutral amino acids has been identified as a remote member of the γ-aminobutyric acid superfamily of transport proteins (Castagna et al. 1998).

6.2
Ion-Channel Formation in Artificial Membranes and BBMVs

At high pH (9–10), activated toxins (Slatin et al. 1990; Schwartz et al. 1993, 1997a) and N-terminal toxin fragments (Walters et al. 1993) are capable of forming cation-selective channels in receptor-free PLBs. Slatin et al. (1990) observed single-channel conductances in the order of 4 nS with Cry3A, a coleopteran-specific δ-endotoxin, whereas channels formed by Cry1Ac toxin tended to be smaller, most commonly around 600 pS. Variation in the conductance levels of the channels were attributed to either different degrees of toxin aggregation producing pores of various sizes or cooperative gating causing formation of numerous similar-type channels, many of which the recording apparatus was perhaps unable to resolve. The latter was thought more likely since the selectivity of K$^+$ over Cl$^-$ was similar for the different conductance states. No channel activity was observed at neutral pH. However, Schwartz et al. (1993) obtained anion-selective channels in receptor-free PLBs under mildly acidic conditions (pH 6), and Carroll and Ellar (1993) noted a lack of significant ion selectivity in toxin-treated BBMVs at pH 7.5. Carroll and Ellar (1997) postulated that pH-induced changes in ion selectivity could be the result of altered charge state or distribution within the ion channel.

Lorence et al. (1995) showed that specific toxins caused a fast and sustained increase in cation permeability at low alkaline pH (7.5) in BBMVs prepared from susceptible insects. The response was observed only with toxins that were active against the donor insect in vivo, indicating that toxin specificity was preserved under in vitro conditions. The authors fused PLBs with BBMVs and observed that, at pH 9.0 and in the presence of a KCl gradient, toxin in the nM range caused up to a 40-fold increase in conductance. The failure to observe significant conductance increases in these PLBs in the absence of BBMV components, unless toxin in the µM range was used, reinforced the notion that pore formation in midgut cells caused by δ-endotoxins is mediated by receptors in the apical epithelium.

Martin and Wolfersberger (1995) also incorporated BBMVs into PLB preparations, which resulted in very large, irreversible increases in membrane conductance (up to >200 nS at pH 9.6) when exposed to toxin, an indication more likely of the presence of large pores in place of small, discrete ion channels. These remained permanently open in contrast to the smaller channels that formed in PLBs without BBMV proteins and which alternated between open and closed states. Analogous results were obtained by Schwartz et al. (1997a) who investigated toxin-induced ion-channel activity (maximum conductance <450 pS with symmetric bathing solutions, pH 9.0) using a purified receptor complex reconstituted in PLB preparations. The inclusion of the receptor complex reduced the amount of toxin required to produce the channel activity by approximately 250-fold compared with receptor-free membranes. The channels were cation selective and, moreover, exhibited a specificity toward different toxins that correlated with insecticidal activity. Current rectification occurred with the most potent toxin tested.

English et al. (1991), using a $^{86}Rb^+$ assay, demonstrated toxin-induced permeabilization of liposomes to both K^+ and water. K^+ and water fluxes were markedly enhanced in hybrid BBMV-liposomes, and ion permeability was shown to be preferentially cationic. As measured by the extent of vesicle shrinkage (water efflux) in hypertonic medium, the incorporation of BBMV protein into the liposomes reduced effective toxin concentration by 1000-fold. Sangadala et al. (1994) obtained similar results with a purified receptor complex comprising aminopeptidase N and phosphatase that was extracted from BBMVs and reconstituted into liposomes. The receptor complex increased toxin binding by 35% and toxin-induced permeability to $^{86}Rb^+(K^+)$ by 1000-fold. Later, Luo et al. (1997) reconstituted a purified aminopeptidase N receptor from BBMVs into liposomes and demonstrated that enhanced permeability to $^{86}Rb^+(K^+)$ induced by Cry1 toxins correlated with toxin-binding specificity and insecticidal activity against the donor insect species. Escriche et al. (1998) demonstrated that activated toxin significantly increased the permeability of BBMVs to KCl with a specificity that correlated with insecticidal activity.

The conclusion to be drawn from these studies is that receptors present in the midgut apical membrane are intimately and synergistically involved in toxin insertion and in the formation of cation-selective channels. It should be noted too that maximum effects in the in vitro systems were obtained in experiments carried out at high pH (9–10), approximating the midgut condition in vivo and ensuring that both toxin and membrane receptor remained in active conformation. Although characterized as being generally cation selective, in some instances the ion channels formed by toxins may show a higher selectivity for K^+ than Na^+ (Sacchi et al. 1986; Lorence et al. 1995).

It is also significant that the ion-channel results with lipid membrane-BBMVs show a differential toxin specificity that generally correlates well with insecticidal activity, i.e., toxins that insert into such membranes and form ion channels are also insecticidal and those that do not, or elicit only a weak response, are inactive or poorly active against the insect. As with midgut preparations (Peyronnet et al. 1997), however, the correlation is not perfect. Luo et al. (1999) observed toxins that permeabilized BBMVs from species of

Spodoptera but were only weakly active against the insects in vivo. Nevertheless, it is noteworthy that, as with midgut preparations, BBMV assays appear more likely to be false-positive than false-negative indicators of insecticidal activity. In principle, the correlation between negative toxin insertion in artificial lipid-BBMV membranes and negative insecticidal activity would be expected, a priori, to be very good. However, it is conceivable that the method used to activate the toxin for the in vitro preparation could produce an inverse correlation (negative response in vitro, positive in vivo).

6.3
Insect Cell Lines as Proxies for Midgut Cells In Vivo

Microscopic observations and electrophysiological studies involving the patch-clamp technique and fluorescent probes have been used to investigate the action of δ-endotoxins on insect cells derived from different species and tissues. However, there is an inherent risk of over-interpreting results from experiments with insect cells and caution should be used in drawing conclusions from them regarding insecticidal mode of action. Continuous cell lines assume morphological and physiological characteristics that can be very different from those of the progenitor tissue from which the primary cultures were prepared, and their response to Bt toxins often does not correlate with that of the insects from which the cells were derived (Witt et al. 1986; Gringorten et al. 1999). Even established midgut cell lines bear little resemblance to midgut cells in vivo, and their susceptibility to activated toxins correlates poorly with that of the host insect (Gringorten et al. 1999). Generally, they are sensitive to fewer toxins than the host insect. On the other hand, primary midgut cell cultures appear to be susceptible to a broader spectrum of δ-endotoxins than the host insect (Baines et al. 1997), a feature that, as with midgut preparations and BBMVs, creates a bias toward overrating insecticidal activity from in vitro assays.

An obvious drawback to using insect cell lines to study toxin effects in vitro, and attempting to draw conclusions about mode of action in vivo, has been the inability to reproduce the asymmetric environmental conditions that midgut cells are exposed to, namely the steep pH gradient across the epithelium, with the apical surface of the plasma membrane exposed to a highly alkaline medium and the basal surface exposed to a neutral or slightly acidic medium. All experiments that utilize cultured insect cells, including midgut cells, have to be performed at neutral pH to avoid alkaline injury from the solvent alone. Under such conditions, ion-channel activity may be quite different than at alkaline pH.

Although they may be poor indicators of insecticidal activity, cell lines have proven useful for characterizing toxin activity spectra (Gringorten et al. 1999) and investigating membrane permeabilizing effects, particularly in determining pore size (Knowles and Ellar 1987; Villalon et al. 1998). As with columnar cells in vivo, cultured insect cells respond to toxin injury by swelling and lysing (Murphy et al. 1976; Nishiitsutsuji-Uwo et al. 1979; Himeno 1987). Early studies by Nishiitsutsuji-Uwo et al. (1979) and Himeno (1987) on TN-368 cells, a line derived from virgin adult ovaries of the cabbage looper, *Trichoplusia ni*, suggested that ion-channel formation played a role in the cytotoxic response and contributed to the swelling and lysis. In a pivotal study with CF-1 cells, a line derived from neonate spruce budworm, *Choristoneura fumiferana*, Knowles and Ellar (1987) investigated the cytolytic effect of toxin in the presence of neutral solutes with different hydrodynamic radii and developed their model of membrane pore formation and colloid osmotic lysis as the toxin mechanism of action.

The ion-channel activity induced in cell lines appears to be less selective than in midgut epithelial cells in vivo or in artificial membranes fused with BBMV components, perhaps partly a reflection of the different pH conditions. Thus toxin cytolytic to Sf9, a cloned ovarian cell line from the pupal fall armyworm, *Spodoptera frugiperda*, permeabilized the cells to both cations (K^+, Na^+ and H^+) (Vachon et al. 1995) and anions (Cl^-) (Schwartz et al. 1991; Villalon et al. 1998). Possibly, in vivo, anion-channel activity becomes more prevalent in the midgut over the time course of Bt intoxication as the lumen pH begins to fall. In contrast to the effect in Sf9, toxin-treated CHE cells, a line originating from embryonic tissue of *M. sexta*, responded with reduced K^+ uptake and increased cell acidification, attributed to inhibition of a plasma-membrane K^+-dependent ATPase that energized K^+/H^+ exchange with the external medium (English and Cantley 1985).

A number of studies with insect cells involve intracellular molecular events that could have some relevance in vivo at sublethal doses of Bt. Toxin was found to be an effective uncoupler of oxidative phosphorylation in isolated midgut mitochondria (Travers et al. 1976), which is consistent with the observation that it also rapidly stimulates glucose uptake in vivo by midguts of susceptible larvae (Fast and Donaghue 1971), and depletes susceptible cells in culture of ATP (Fast 1981). However, these would have to be regarded as secondary effects of altered membrane permeability and the disruption of ion balance. As previously noted (Sect. 5.2), there is no evidence that toxin becomes internalized in cells. The fact that Sephadex-bound toxin has the same activity against susceptible cells as free toxin indicates that toxin acts at the cell surface (Fast et al. 1978).

Potvin et al. (1998) noted intracellular Ca^{2+} surges in toxin-treated CF-1 cells, resulting from both an influx of extracellular Ca^{2+} and the release of Ca^{2+} from an intracellular source. The release of intracellular Ca^{2+} in response to Bt toxin was also observed in Sf9 cells (Schwartz et al. 1991) and in UCR-SE-1a (Monette et al. 1994), a cell line derived from neonate beet armyworm, *Spodoptera exigua*. Monette et al. (1997) demonstrated a dose-dependent effect of extra-

cellular Ca^{2+} on Bt toxicity to Sf9 that was suppressed by Ca^{2+} channel blockers Co^{2+} and La^{2+}, but enhanced by inhibition of intracellular storage of Ca^{2+}.
The mechanism mediating the intracellular release of Ca^{2+} is not known and
the toxicological significance of it is far from clear. Potvin et al. (1998) suggested that Ca^{2+} may be the cell signal responsible for the toxin-induced
increase in cyclic AMP in CF-1 that had been observed by others (Knowles and
Farndale 1988), especially as Bt toxins also activate insect-cell adenylate cyclase
(Knowles and Farndale 1988). However, Knowles and Farndale (1988) regarded
an intracellular signalling role for Ca^{2+} in toxemic cells as questionable and the
activation of adenylate cyclase and elevation of cyclic AMP as secondary effects
with no direct role in cell lysis.

7
Membrane Insertion and Pore Formation

Several investigators have used published values for the hydrodynamic radii of
ions and neutral solutes that permeate toxin-treated membranes and cause cell
swelling and lysis, as well as the radii of osmotic protectants that inhibit lysis,
to determine the apparent pore diameter of the primary lesion. The various
estimates are 1.0–2.0 nm in CF-1 cells, determined with neutral solutes and
dependent on toxin type (Knowles and Ellar 1987); 1.0–1.2 nm in Sf9, determined with neutral solutes and one toxin (Villalon et al. 1998); 2.4–2.6 nm in
BBMVs, determined with neutral solutes and one toxin, under alkaline conditions and independent of pH between 8.7 and 9.8 (Carroll and Ellar 1997). A
pH effect on pore size was observed by Martin and Wolfersberger (1995) who
measured ion conductance changes in toxin-exposed PLBs fused with *M. sexta*
BBMVs at pH 8.8 and 9.6. The apparent pore diameter was estimated to be 2.2
nm at pH 9.6, similar to the estimate of Carroll and Ellar (1997), but only 0.9
nm at pH 8.8. Carroll and Ellar (1997) note, however, that pH-induced changes
in charge distribution in the pore could affect ion conductance and thereby
give a misleading estimate of its size.

Martin and Wolfersberger (1995) suggested that, considering the high pH
of the larval midgut, the higher size estimate of the pore at pH 9.6 in their study
may be closer to the size of the pore created in the CCAM. It is apparent that
modulation of toxin activity by ambient pH conditions is mediated through
effects on both the toxin molecule (Gringorten et al. 1992) and the membrane
pore. The fact that under pH conditions resembling more closely the pH of the
midgut, two different techniques, both dependent on the presence of BBMV
protein, but one based on ion conductance in artificial membranes (Martin and
Wolfersberger 1995) and the second on osmotic swelling of vesicles (Carrol
and Ellar 1997), yielded similar estimates of pore size, provides a measure of
confidence in the determinations and in their relevance to the midgut in vivo.
The apparent dependence of pore size on toxin type (Knowles and Ellar 1987),

despite very similar three-dimensional structures among the Cry toxins, may reflect differences in the rate of formation of the primary lesion, with the larger pores indicative of more rapid toxin aggregation and pores already in the process of widening. On the other hand, actual differences in the size of the primary lesion caused by different toxins has not been ruled out in any investigation.

Determination of the three-dimensional structure of the Cry family of toxins (Li et al. 1991; Grochulski et al. 1995) and data from mutagenesis studies have led to speculation and debate about the mechanism of toxin insertion (reviewed by Knowles 1994; Dean et al. 1996; Rajamohan et al. 1998; Schwartz and Laprade 2000). The Cry toxins appear to be structurally very similar (Grochulski et al. 1995), indicating that their mechanism of insertion is also likely similar, although some of the channel properties may differ (Grochulski et al. 1995). The activated toxins display three principal structural domains (Li et al. 1991; Grochulski et al. 1995), with domain I, the principal pore-forming region, comprising an α-helix bundle at the N-terminal end of the molecule, followed by two distinct β-sheet domains, with the first of these (domain II) regarded as the principal binding region. Domain III, forming the C-terminal region, is considered to act as a stabilizing structure for the molecule.

Two models have been proposed to explain how the toxin molecule might partition into bilipid membranes and create a pore (reviewed by Knowles 1994). Both propose that binding to a specific membrane receptor triggers a conformational change in the toxin molecule that promotes insertion. There is also agreement that the receptor itself does not become part of the pore. They differ in the particular helical pair of domain I that initiates the primary lesion and the extent to which the remaining helices become associated with the plasma membrane. The "penknife" model, based on Hodgman and Ellar's (1990) computer simulations and the model for Colicin A insertion, proposes that the hairpin loop formed by helices α5 and α6 of domain I of the activated toxin creates the primary pore, with little involvement of the other helices (at least initially).

There is increasing evidence, however, that favors an "umbrella" model (Fig. 4), which proposes that the α4-α5 pair is the pore-forming hairpin, with other helices of domain I remaining associated with the membrane. The loop forming the α4-α5 hairpin is the most hydrophobic region of domain I (Grochulski et al. 1995), making this helical pair a prime candidate for insertion. Implication of α5 and exclusion of α7 are supported by direct experimental evidence (Gazit et al. 1994; Gazit and Shai 1995) and by computer simulations of voltage-dependent membrane interactions of a synthetic α5 peptide (Biggin and Sansom 1996). Biggin and Sansom's (1996) simulations also demonstrated that α7 would not insert, but would remain associated with the membrane surface. That the α4-α5 hairpin is the likeliest pore former was demonstrated experimentally by Schwartz et al. (1997b), using disulfide

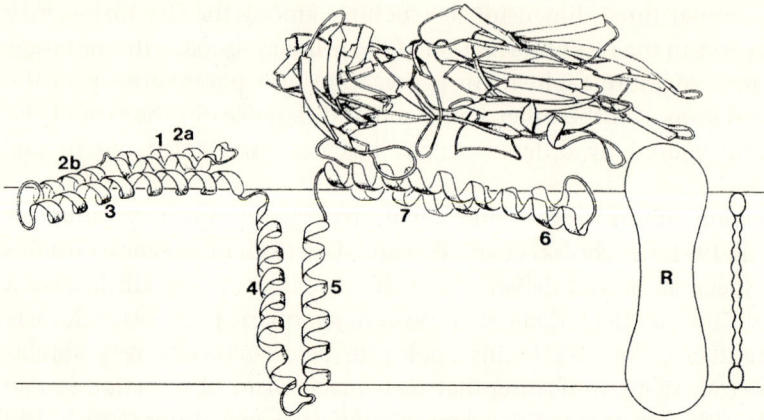

Fig. 4. Umbrella model of insertion of the coleopteran-specific δ-endotoxin, Cry3A, into the plasma membrane. The three-dimensional structures of Cry3A (Li et al. 1991) and the lepidopteran-specific δ-endotoxin, Cry1Aa, (Grochulski et al. 1995) are very similar, especially in their channel-forming regions. This region (domain I) comprises a bundle of seven antiparallel and largely amphipathic α-helices (eight, if α2 is considered to be two separate helices; helices are numbered from the N-terminus). In both toxins, the α4–α5 pair is the likeliest to insert, with the other helices remaining associated with the membrane. The model may represent only the initial insertion step, as there is evidence that more of the molecule, including regions of domain III, may participate in binding and insert into the membrane than previously thought (reviewed by Dean et al. 1996). Recent evidence (Aronson et al. 1999) indicates that the toxin is released from the receptor (*R*) upon partitioning into the membrane. (Reproduced with permission from Knowles 1994 and Academic Press)

bridges to selectively restrict helix and interdomain movement, and by Gazit et al. (1998), using synthetic peptides that corresponded to the α-helices of domain I. The study of Schwartz et al. (1997b) also showed that insertion involved a conformational change in which domain I moved away from the rest of the molecule, a change possibly mediated by α7 serving as a "binding sensor" (Gazit and Shai 1995; Gazit et al. 1998). Further evidence of the involvement of α4 and support for the umbrella model for insertion was provided recently by Masson et al. (1999).

Dean et al. (1996), however, cite evidence that points to a more complex mechanism of insertion. They present data that implicate regions of domain III, in addition to domain II, in receptor binding, and regions of domain III and possibly also domain II, as well as most of domain I (including α7), in membrane insertion and ion-channel formation. Involvement of domain III in binding was also reported by de Maagd et al. (1996, 1999), and further evidence of its participation in membrane insertion was provided by Wolfersberger et al. (1996) and Schwartz et al. (1997c). That there is a far greater degree of integration of the toxin molecule into the plasma membrane than suggested by the umbrella model also appears to be supported by the observation that inserted toxin is resistant to attack by proteases and to binding by monoclonal antibodies (Wolfersberger et al. 1986, cited in Rajamohan et al. 1998). Aronson et al. (1999) reported that, after binding to BBMVs, at least 95% of Cry1Ac appeared to be protected from protease K digestion.

Nevertheless, the umbrella model and the more complex mechanism implied by Dean et al. (1996) are not necessarily mutually exclusive, but may represent different phases of the insertion sequence, with the umbrella model, as depicted in Fig. 4, characterizing the initial membrane interaction. It was suggested by Grochulski et al. (1995), for example, that domain III and possibly domain II might participate in an oligomerization process. They noted that the relatively long distance between domains I and II provided the structural flexibility that would allow such participation. An unpublished observation cited by Rajamohan et al. (1998) that disulfide bridges introduced between $\alpha 2$ and $\alpha 5$ or between $\alpha 5$ and $\alpha 7$ did not alter insertion or toxicity might appear to undermine the umbrella model, even as an initial mechanism, but it is contradicted by Schwartz et al. (1997b) who also constructed an $\alpha 5$-$\alpha 7$ disulfide mutant, as well as an $\alpha 5$-$\alpha 6$ mutant and one with a disulfide bridge linking the interhelical loop of $\alpha 4$-$\alpha 5$ to that of $\alpha 2b$-$\alpha 3$, and found that all lost the ability to form functional ion channels in PLBs.

Aggregation and oligomerization of several toxin molecules in one region of the membrane and insertion of their helical loops could theoretically lead to formation of a barrel-stave pore with an aqueous channel (Knowles 1994). The enormous conductances (>200 nS) and permanently open state of pores observed by Martin and Wolfersberger (1995) in Cry1Ac-treated BBMV-PLBs suggest that oligomerization of the toxin had taken place. The 4-nS conductance states observed by Slatin et al. (1990) in Cry3A-treated PLBs may also have been due, in part, to oligomer formation (as well as to cooperative gating, favored by the authors as an explanation). Evidence that Cry3A is capable of forming a stable dimer in solution was provided by Walters et al. (1994) who used native and denaturing gel systems and ultracentrifugation sedimentation analyses to investigate its stability.

Refinements to the umbrella model suggest a tetramer as the minimal structure, with the $\alpha 4$ helices forming a negatively charged, cation-selective, hydrophilic channel and $\alpha 5$ facing the lipid surface (Schwartz et al. 1997b; Masson et al. 1999). Further toxin aggregation, possibly mediated by $\alpha 5$ (Gazit and Shai 1995; Aronson et al. 1999), as well as secondary partitioning into the plasma membrane of other portions of the toxin molecule, would widen the pore into a non-specific structure. In the midgut, pores created in this manner would allow water and solutes to diffuse into the epithelial cells and also cause a loss of pH homeostasis, all of which would ultimately lead to cell lysis. One of the determinants of toxin activity in the midgut, therefore, would appear to be the extent to which molecular aggregation and oligomerization occur at the insertion point in the CCAM, which would affect pore size and the rate of cytolysis.

An interesting footnote to the discussion of aggregation is a comment made some 20 years ago on the topic by Fast (1981) in an entirely different context. The author drew attention to the "problem" of aggregation of toxin protein in solution and the interference it creates for "almost every analytical or preparative system unless it is prevented". The author pointed out that it cause the loss of toxicity in cell assays, but not necessarily in insect assays, in which toxicity may be restored through molecular disaggregation in the insect midgut (Fast 1981). This observation suggests not only that random aggregation of toxin molecules prior to binding can cause steric hindrance to effective membrane interaction, but also that the aggregation that contributes to pore formation is probably more structurally organized than in free solution and more likely to occur after, rather than before, binding to the midgut receptor.

8
Conclusions and Thoughts

The past two decades have witnessed impressive gains in the understanding of the biology of the larval lepidopteran midgut and its relationship to Bt mode of action. Identification of the unique proton-pumping V-ATPase-K^+/H^+ antiporter system in the GCAM as the most likely mechanism by which the midgut generates the high pH and TEP has provided a physiological framework in which the chemistry of toxin activation and the toxemia itself can be better understood. There remain, of course, many gaps in our knowledge about both midgut function and Bt mode of action. Most of the information on ion function in the intact midgut has been obtained under short-circuit conditions. As Klein et al. (1996) pointed out, however, the short-circuit system can uncouple physiological systems and what is needed are more electrophysiological data on the midgut under open-circuit conditions. Isolation and identification of the antiporter also remains a goal. There is still much to elucidate about the nature of toxin receptors in different insects and the extent to which the toxin molecule integrates further into the midgut membrane after its initial insertion.

Complementing the accomplishments made in the research on the mode of action of Bt δ-endotoxins are the inroads achieved in discerning their structure-function relationships through mutagenesis experiments. The knowledge gained in this area of research makes the prospect of developing an arsenal of bioengineered gene products with improved efficacy and altered specificity a realistic goal. In the context of searching for novel ways of improving Bt efficacy, it would also seem useful to try and gain a better understanding of the larva's physiological defences and recovery mechanisms since, in a susceptible insect, these are the factors that establish the threshold separating a sublethal dose from a lethal one. Targeting defence and recovery mechanisms offers a means of lowering the threshold, thereby increasing the efficacy of a given Bt product.

There are a number of ways in which lepidopteran larvae are known to sequester or inactivate δ-endotoxins in the midgut and remove them from the pool of molecules that interact with the midgut epithelium: (1) binding by the peritrophic membrane (Bravo et al. 1992a,b; Denolf et al. 1993; Yi et al. 1996); (2) proteolytic degradation (Pang and Gringorten 1998); (3) precipitation (Milne et al. 1995, 1998); and (4) binding to a soluble form of the midgut receptor (Lu and Adang 1996). The absence of one or more of these factors under in vitro conditions may account for the broader specificities of toxins observed with midgut preparations, BBMVs and primary midgut cell cultures, and explain why a toxin may be active in vitro but lack insecticidal activity.

One defence mechanism targeted in the past, largely because of its prominence and accessibility, is the peritrophic membrane. Interest has revived in the use of chitinases as a means of potentiating δ-endotoxin efficacy through their degradative effect on the peritrophic membrane (Kramer and Muthukrishnan 1997). Insect recovery mechanisms triggered by sublethal doses of Bt have not been studied to any great extent, but finding ways of interfering with them could also translate into improving Bt efficacy. Sublethal doses that damage the midgut stimulate mitotic activity and division in epithelial stem cells (Nishiitsutsuji-Uwo and Endo 1981a), and an enhanced ability to regenerate epithelial tissue has been implicated as one means by which insects can develop resistance to Bt (Forcada et al. 1999; Martínez-Ramírez et al. 1999). Physiological and anatomical mechanisms that attenuate the effects of δ-endotoxin in the midgut appear to be areas worth exploring.

Acknowledgements. To Drs. Donald Dean, Alan Koch, Raynald Laprade, Jean-Louis Schwartz, Sardar Sohi, Vincent Vachon, Kees van Frankenhuyzen, Helmut Wieczorek and Michael Wolfersberger, I extend sincere thanks for agreeing to review various portions of the work in progress. I am very much in their debt, and am particularly grateful to them and to Dr. William Harvey for their willingness to engage in lengthy and stimulating discussions with me that helped resolve a number of issues. To Dr. Barbara Knowles I express appreciation for granting permission to use Fig. 4. I also thank Mr. Mark Primavera for the graphics in Figs. 1 and 2, and Mr. Michael Gringorten, Ms. Leann Knight and Mr. Carl Nystrom for their computer assistance. Finally, I must thank Dr. Isaac Ishaaya for the invitation to write this chapter and for his sustained interest in the topic.

References

Angus TA (1954) A bacterial toxin paralysing silkworm larvae. Nature 173:545–546
Angus TA (1956a) Association of toxicity with protein-crystalline inclusions of *Bacillus sotto* Ishiwata. Can J Microbiol 2:122–131

Angus TA (1956b) Extraction, purification, and properties of *Bacillus sotto* toxin. Can J Microbiol 2:416–426

Angus TA (1968) Similarity of effect of valinomycin and *Bacillus thuringiensis* parasporal protein in larvae of *Bombyx mori*. J Invertebr Pathol 11:145–146

Angus TA, Heimpel AM (1956) An effect of *Bacillus sotto* on the larvae of *Bombyx mori*. Can Entomol 88:138–139

Aranda E, Sanchez J, Peferoen M, Güereca L, Bravo A (1996) Interactions of *Bacillus thuringiensis* crystal proteins with the midgut epithelial cells of *Spodoptera frugiperda* (Lepidoptera: Noctuidae) J Invertebr Pathol 68:203–212

Aronson AI, Geng C, Wu L (1999) Aggregation of *Bacillus thuringiensis* Cry1A toxins upon binding to target insect larval midgut vesicles. Appl Environ Microbiol 65:2503–2507

Azuma M, Takeda S, Yamamoto H, Endo Y, Eguchi M (1991) Goblet cell alkaline phosphatase in silkworm midgut epithelium: its entity and role as an ATPase. J Exp Zool 258:294–302

Azuma M, Harvey WR, Wieczorek (1995) Stoichiometry of K^+/H^+ antiport helps to explain extracellular pH 11 in a model epithelium. FEBS Lett 361:153–156

Baines D, Schwartz JL, Sohi S, Dedes J, Pang A (1997) Comparison of the response of midgut epithelial cells and cell lines from lepidopteran larvae to Cry1A toxins from *Bacillus thuringiensis*. J Insect Physiol 43:823–831

Becker N, Margalit J (1993) Use of *Bacillus thuringiensis israelensis* against mosquitoes and black flies. In: Entwistle PF, Cory JS, Bailey MJ, Higgs S (eds) *Bacillus thuringiensis*, an environmental biopesticide: theory and practice. Wiley, Chichester, pp 147–170

Berenbaum M (1980) Adaptive significance of midgut pH in larval Lepidoptera. Am Nat 115:138–146

Bernhard K, Jarrett P, Meadows M, Butt J, Ellis DJ, Roberts GM, Pauli S, Rodgers P, Burges HD (1997) Natural isolates of *Bacillus thuringiensis*: worldwide distribution, characterization, and activity against insect pests. J Invertebr Pathol 70:59–68

Biggin PC, Sansom MSP (1996) Simulation of voltage-dependent interactions of α-helical peptides with lipid bilayers. Biophys Chem 60:99–110

Bravo A (1997) Phylogenetic relationships of *Bacillus thuringiensis* δ-endotoxin family proteins and their functional domains. J Bacteriol 179:2793–2801

Bravo A, Jansens S, Peferoen M (1992a) Immunocytochemical localization of *Bacillus thuringiensis* insecticidal crystal proteins in intoxicated insects. J Invertebr Pathol 60:237–246

Bravo A, Hendrickx K, Jansens S, Peferoen M (1992b) Immunocytochemical analysis of specific binding of *Bacillus thuringiensis* insecticidal crystal proteins to lepidopteran and coleopteran midgut membranes. J Invertebr Pathol 60:247–253

Carroll J, Ellar DJ (1993) An analysis of *Bacillus thuringiensis* δ-endotoxin action on insect-midgut-membrane permeability using a light-scattering assay. Eur J Biochem 214:771–778

Carroll J, Ellar DJ (1997) Analysis of the large aqueous pores produced by a *Bacillus thuringiensis* protein insecticide in *Manduca sexta* midgut-brush-border-membrane vesicles. Eur J Biochem 245:797–804

Castagna M, Shayakul C, Trotti D, Sacchi VF, Harvey WR, Hediger MA (1998) Cloning and characterization of a potassium-coupled amino acid transporter. Proc Natl Acad Sci USA 95:5395–5400

Chao AC, Koch AR, Moffett DF (1989) Active chloride transport in isolated posterior midgut of tobacco hornworm (*Manduca sexta*). Am J Physiol 257:R752–R761

Chao AC, Moffett DF, Koch A (1991) Cytoplasmic pH and goblet cavity pH in the posterior midgut of the tobacco hornworm *Manduca sexta*. J Exp Biol 155:403–414

Chaufaux J, Marchal M, Gilois N, Jehanno I, Buisson C (1997) Recherche de souches naturelles du *Bacillus thuringiensis* dans différents biotopes, à travers le monde. Can J Microbiol 43:337–343

Chen XJ, Lee MK, Dean DH (1993) Site-directed mutations in a highly conserved region of *Bacillus thuringiensis* δ-endotoxin affect inhibition of short circuit current across *Bombyx mori* midguts. Proc Natl Acad Sci USA 90:9041–9045

Choma CT, Surewicz WK, Carey PR, Pozsgay M, Raynor T, Kaplan H (1990) Unusual proteolysis of the protoxin and toxin from *Bacillus thuringiensis*: structural implications. Eur J Biochem 189:523–527

Cioffi M (1979) The morphology and fine structure of the larval midgut of a moth (*Manduca sexta*) in relation to active ion transport. Tissue Cell 11:467–479

Cioffi M, Harvey WR (1981) Comparison of potassium transport in three structurally distinct regions of the insect midgut. J Exp Biol 91:103–116

Cioffi M, Wolfersberger MG (1983) Isolation of separate apical, lateral and basal plasma membrane from cells of an insect epithelium. A procedure based on tissue organization and ultrastructure. Tissue Cell 15:781–803

Clairmont FR, Milne RE, Pham VT, Carrière MB, Kaplan H (1998) Role of DNA in the activation of the Cry1A insecticidal crystal protein from *Bacillus thuringiensis*. J Biol Chem 273:9292–9296

Cooksey KE (1971) The protein crystal toxin of *Bacillus thuringiensis*: biochemistry and mode of action. In: Burges HD, Hussey NW (eds) Microbial control of insects and mites. Academic Press, London, pp 247–274

Crawford DN, Harvey WR (1988) Barium and calcium block *Bacillus thuringiensis* subspecies *kurstaki* δ-endotoxin inhibition of potassium current across isolated midgut of larval *Manduca sexta*. J Exp Biol 137:277–286

Crickmore N, Ziegler DR, Feitelson J, Schnepf E, Van Rie J, Lereclus D, Baum J, Dean DH (1998) Revision of the nomenclature for the *Bacillus thuringiensis* pesticidal crystal proteins. Microbiol Mol Biol Rev 62:807–813

Dean DH, Rajamohan F, Lee MK, Wu SJ, Chen XJ, Alcantara E, Hussain SR (1996) Probing the mechanism of action of *Bacillus thuringiensis* insecticidal proteins by site-directed mutagenesis B, a minireview. Gene 179:111–117

Deaton LE (1984) Tissue K^+-stimulated ATPase and HCO_3^--stimulated ATPase in the tobacco hornworm, *Manduca sexta*. Insect Biochem 14:109–114

de Maagd RA, Kwa MSG, van der Klei H, Yamamoto T, Schipper B, Vlak JM, Stiekema WJ, Bosch D (1996) Domain III substitution in *Bacillus thuringiensis* delta-endotoxin CryIA(b) results in superior toxicity for *Spodoptera exigua* and altered membrane protein recognition. Appl Environ Microbiol 62:1537–1543

de Maagd RA, Bakker PL, Masson L, Adang MJ, Sangadala S, Stiekema W, Bosch D (1999) Domain III of the *Bacillus thuringiensis* delta-endotoxin Cry1Ac is involved in binding to *Manduca sexta* brush border membranes and to its purified aminopeptidase N. Mol Microbiol 31:463–471

Denolf P, Jansens S, Peferoen M, Degheele D, Van Rie J (1993) Two different *Bacillus thuringiensis* delta-endotoxin receptors in the midgut brush border membrane of the European corn borer, *Ostrinia nubilalis* (Hübner) (Lepidoptera: Pyralidae). Appl Environ Microbiol 59:1828–1837

Dow JAT (1984) Extremely high pH in biological systems: a model for carbonate transport. Am J Physiol 246:R633–R636

Dow JAT (1992) pH gradients in lepidopteran midgut. J Exp Biol 172:355–375

Dow JAT, Harvey WR (1988) Role of midgut electrogenic K^+ pump potential difference in regulating lumen K^+ and pH in larval Lepidoptera. J Exp Biol 140:455–463

Dow JAT, Peacock JM (1989) Microelectrode evidence for the electrical isolation of goblet cell cavities in *Manduca sexta* middle midgut. J Exp Biol 143:101–114

Dow JAT, Gupta BL, Hall TA, Harvey WR (1984) X-ray microanalysis of elements in frozen-hydrated sections of an electrogenic K^+ transport system: the posterior midgut of tobacco hornworm (*Manduca sexta*) in vivo and in vitro. J Membr Biol 77:223–241

Dulmage HT, Cooperators (1981) Insecticidal activity of isolates of *Bacillus thuringiensis* and their potential for pest control. In: Burges HD (ed) Microbial control of pests and plant diseases 1970–1980. Academic Press, London, pp 193–222

Endo Y, Nishiitsutsuji-Uwo J (1980) Mode of action of *Bacillus thuringiensis* δ-endotoxin: histopathological changes in the silkworm midgut. J Invertebr Pathol 36:90–103

Endo Y, Nishiitsutsuji-Uwo J (1981) Mode of action of *Bacillus thuringiensis* δ-endotoxin: ultrastructural changes of midgut epithelium of *Pieris*, *Lymantria* and *Ephestia* larvae. Appl Entomol Zool 16:231–241

English LH, Cantley LC (1985) Delta-endotoxin inhibits Rb^+ uptake, lowers cytoplasmic pH and inhibits a K^+-ATPase in *Manduca sexta* CHE cells. J Membr Biol 85:199–204

English LH, Readdy TL, Bastian AE (1991) Delta-endotoxin-induced leakage of ^{86}Rb$^+$-K$^+$ and H$_2$O from phospholipid vesicles is catalyzed by reconstituted midgut membrane. Insect Biochem 21:177–184

Escriche B, De Decker N, Van Rie J, Jansens S, Van Kerkhove E (1998) Changes in permeability of brush border membrane vesicles from *Spodoptera littoralis* midgut induced by insecticidal crystal proteins from *Bacillus thuringiensis*. Appl Environ Microbiol 64:1563–1565

Fast PG (1981) The crystal toxin of *Bacillus thuringiensis*. In: Burges HD (ed) Microbial control of pests and plant diseases 1970–1980. Academic Press, London, pp 223–248

Fast PG, Angus TA (1965) Effects of parasporal inclusions of *Bacillus thuringiensis* var. *sotto* Ishiwata on the permeability of the gut wall of *Bombyx mori* (Linnaeus) larvae. J Invertebr Pathol 7:29–32

Fast PG, Donaghue TP (1971) The δ-endotoxin of *Bacillus thuringiensis*. II. On the mode of action. J Invert Pathol 18:135–138

Fast PG, Morrison IK (1972) The δ-endotoxin of *Bacillus thuringiensis*. IV. The effect of δ-endotoxin on ion regulation by midgut tissue of *Bombyx mori* larvae. J Invertebr Pathol 20:208–211

Fast PG, Murphy DW, Sohi SS (1978) *Bacillus thuringiensis* δ-endotoxin: evidence that toxin acts at the surface of susceptible cells. Experientia 34:762–763

Feitelson JS, Payne J, Kim L (1992) *Bacillus thuringiensis*: Insects and beyond. Bio/Technology 10:271–275

Ferré J, Real MD, Van Rie J, Jansens S, Peferoen M (1991) Resistance to *Bacillus thuringiensis* bioinsecticide in a field population of *Plutella xylostella* is due to a change in a midgut membrane receptor. Proc Natl Acad Sci USA 88:5119–5123

Forcada C, Alcácer E, Garcerá MD, Tato A, Martínez R (1999) Resistance to *Bacillus thuringiensis* Cry1Ac toxin in three strains of *Heliothis virescens*: proteolytic and SEM study of the larval midgut. Arch Insect Biochem Physiol 42:51–63

Francis BR, Bulla LA Jr (1997) Further characterization of BT-R$_1$, the cadherin-like receptor for Cry1Ab toxin in tobacco hornworm (*Manduca sexta*) midguts. Insect Biochem Mol Biol 27:541–550

Garczynski SF, Adang MJ (1995) *Bacillus thuringiensis* CryIA(c) δ-endotoxin binding aminopeptidase in the *Manduca sexta* midgut has a glycosyl-phosphatidylinositol anchor. Insect Biochem Mol Biol 25:409–415

Garczynski SF, Crim JW, Adang MJ (1991) Identification of putative insect brush border membrane-binding molecules specific to *Bacillus thuringiensis* δ-endotoxin by protein blot analysis. Appl Environ Microbiol 57:2816–2820

Gazit E, Shai Y (1995) The assembly and organization of the α5 and α7 helices from the pore-forming domain of *Bacillus thuringiensis* δ-endotoxin. Relevance to a functional model. J Biol Chem 270:2571–2578

Gazit E, Bach D, Kerr ID, Sansom MSP, Chejanovsky N, Shai Y (1994) The α5 segment of *Bacillus thuringiensis* δ-endotoxin: in vitro activity, ion channel formation and molecular modelling. Biochem J 304:895–902

Gazit E, La Rocca P, Sansom MSP, Shai Y (1998) The structure and organization within the membrane of the helices composing the pore-forming domain of *Bacillus thuringiensis* δ-endotoxin are consistent with an "umbrella-like" structure of the pore. Proc Natl Acad Sci USA 95:12289–12294

Gill SS, Cowles EA, Francis V (1995) Identification, isolation, and cloning of a *Bacillus thuringiensis* CryIAc toxin-binding protein from the midgut of the lepidopteran insect *Heliothis virescens*. J Biol Chem 270:27277–27282

Giordana B, Sacchi FV, Hanozet GM (1982) Intestinal amino acid absorption in lepidopteran larvae. Biochim Biophys Acta 692:81–88

Giordana B, Sacchi VF, Parenti P, Hanozet GM (1989) Amino acid transport systems in intestinal brush-border membranes from lepidopteran larvae. Am J Physiol 257:R494–R500

Giordana B, Tasca M, Villa M, Chiantore C, Hanozet GM, Parenti P (1993) *Bacillus thuringiensis* subsp. *aizawai* δ-endotoxin inhibits the K$^+$/amino acid cotransporters of lepidopteran larval midgut. Comp Biochem Physiol 106C:403–407

Gould F, Martinez-Ramirez A, Anderson A, Ferre J, Silva FJ, Moar WJ (1992) Broad-spectrum resistance to *Bacillus thuringiensis* toxins in *Heliothis virescens*. Proc Natl Acad Sci USA 89: 7986–7990

Griego VM, Moffett D, Spence KD (1979) Inhibition of active K^+ transport in the tobacco hornworm (*Manduca sexta*) midgut after ingestion of *Bacillus thuringiensis* endotoxin. J Insect Physiol 25:283–288

Gringorten JL (1999) Ion regulation in the larval lepidopteran midgut and the response to *Bacillus thuringiensis* δ-endotoxin. Pestic Sci 55:604–606

Gringorten JL, Milne RE, Fast PG, Sohi SS, van Frankenhuyzen K (1992) Suppression of *Bacillus thuringiensis* δ-endotoxin activity by low alkaline pH. J Invertebr Pathol 60:47–52

Gringorten JL, Crawford DN, Harvey WR (1993) High pH in the ectoperitrophic space of the larval lepidopteran midgut. J Exp Biol 183:353–359

Gringorten JL, Sohi SS, Masson L (1999) Activity spectra of *Bacillus thuringiensis* δ-endotoxins against eight insect cell lines. In Vitro Cell Dev Biol 35A:299–303

Grochulski P, Masson L, Borisova S, Pusztai-Carey M, Schwartz JL, Brousseau R, Cygler M (1995) *Bacillus thuringiensis* CryIA(a) insecticidal toxin: crystal structure and channel formation. J Mol Biol 254:447–464

Gupta BL, Dow JAT, Hall TA, Harvey WR (1985) Electron probe X-ray microanalysis of the effects of *Bacillus thuringiensis* var. *kurstaki* crystal protein insecticide on ions in an electrogenic K^+-transporting epithelium of the larval midgut in the lepidopteran, *Manduca sexta*, in vitro. J Cell Sci 74:137–152

Haider MZ, Ellar DJ (1989) Mechanism of action of *Bacillus thuringiensis* insecticidal δ-endotoxin: interaction with phospholipid vesicles. Biochim Biophys Acta 978:216–222

Hannay CL (1953) Crystalline inclusions in aerobic spore-forming bacteria. Nature 172:1004

Hanozet GM, Giordana B, Sacchi VF, Parenti P (1989) Amino acid transport systems in brush-border membrane vesicles from lepidopteran enterocytes. J Exp Biol 143:87–100

Harris JG (1997) Microbial insecticides – an industry perspective. In: Evans HF (ed) Microbial insecticides: novelty or necessity? BCPC Symp Proc No 68. British Crop Protection Council, Farnham, Surrey, pp 41–50

Harvey WR (1980) Water and ions in the gut. In: Locke M, Smith DS (eds) Insect biology in the future – "VBW 80." Academic Press, New York, pp 105–124

Harvey WR, Nedergaard S (1964) Sodium-independent active transport of potassium in the isolated midgut of the cecropia silkworm. Proc Natl Acad Sci USA 51:757–764

Harvey WR, Wolfersberger MG (1979) Mechanism of inhibition of active potassium transport in isolated midgut of *Manduca sexta* by *Bacillus thuringiensis* endotoxin. J Exp Biol 83:293–304

Harvey WR, Zerahn K (1972) Active transport of potassium and other alkali metals by the isolated midgut of the silkworm. Curr Top Membr Transp 3:367–410

Harvey WR, Haskell JA, Zerahn K (1967) Active transport of potassium and oxygen consumption in the isolated midgut of *Hyalophora cecropia*. J Exp Biol 46:235–248

Harvey WR, Haskell JA, Nedergaard S (1968) Active transport by the cecropia midgut. III. Midgut potential generated directly by active K-transport. J Exp Biol 48:1–12

Harvey WR, Cioffi M, Wolfersberger MG (1986) Transport physiology of lepidopteran midgut in relation to the action of Bt delta-endotoxin. In: Samson RA, Vlak JM, Peters D (eds) Fundamental and applied aspects of invertebrate pathology. Foundation of the 4th International Colloquium of Invertebrate Pathology, Wageningen, Netherlands, pp 11–14

Heimpel AM, Angus TA (1959) The site of action of crystalliferous bacteria in Lepidoptera larvae. J Insect Pathol 1:152–170

Hendrickx K, De Loof A, Van Maellaert (1990) Effects of *Bacillus thuringiensis* delta-endotoxin on the permeability of brush border membrane vesicles from tobacco hornworm (*Manduca sexta*) midgut. Comp Biochem Physiol 95C:241–245

Himeno M (1987) Mechanism of *Bacillus thuringiensis* insecticidal δ-endotoxin action on insect cells in vitro. In: Maramorosch K (ed) Biotechnology in invertebrate pathology and cell culture. Academic Press, San Diego, pp 29–43

Hodgman TC, Ellar DJ (1990) Models for the structure and function of the *Bacillus thuringiensis* δ-endotoxins determined by compilational analysis. DNA Sequence 1:97–106

Hofmann C, Lüthy P, Hütter R, Pliska V (1988a) Binding of the delta-endotoxin from *Bacillus thuringiensis* to brush-border membrane vesicles of the cabbage butterfly (*Pieris brassicae*). Eur J Biochem 173:85–91

Hofmann C, Vanderbruggen H, Höfte H, Van Rie J, Jansens S, Van Mellaert H (1988b) Specificity of *Bacillus thuringiensis* δ-endotoxins is correlated with the presence of high-affinity binding sites in the brush border membrane of target insect midguts. Proc Natl Acad Sci USA 85:7844–7848

Huber HE, Lüthy P (1981) *Bacillus thuringiensis* delta-endotoxin: composition and activation. In: Davidson EW (ed) Pathogenesis of invertebrate microbial diseases. Allanheld Osmun, Totowa, NJ, pp 209–234

Ihara H, Kuroda E, Wadano A, Himeno M (1993) Specific toxicity of δ-endotoxins from *Bacillus thuringiensis* to *Bombyx mori*. Biosci Biotechnol Biochem 57:200–204

Keeton TP, Bulla LA Jr (1997) Ligand specificity and affinity of BT-R₁, the *Bacillus thuringiensis* Cry1A toxin receptor from *Manduca sexta*, expressed in mammalian and insect cell cultures. Appl Environ Microbiol 63:3419–3425

Keller B, Langenbruch GA (1993) Control of coleopteran pests by *Bacillus thuringiensis*. In: Entwistle PF, Cory JS, Bailey MJ, Higgs S (eds) *Bacillus thuringiensis*, an environmental biopesticide: theory and practice. Wiley, Chichester, pp 171–191

Klein U, Löffelmann G, Wieczorek H (1991) The midgut as a model system for insect K⁺-transporting epithelia: immunocytochemical localization of a vacuolar-type H⁺ pump. J Exp Biol 161:61–75

Klein U, Koch A, Moffett DF (1996) Ion transport in Lepidoptera. In: Lehane MJ, Billingsley PF (eds) Biology of the insect midgut. Chapman and Hall, London, pp 236–264

Knight PJK, Crickmore N, Ellar DJ (1994) The receptor for *Bacillus thuringiensis* CryIA(c) delta-endotoxin in the brush border membrane of the lepidopteran *Manduca sexta* is aminopeptidase N. Mol Microbiol 11:429–436

Knight PJK, Knowles BH, Ellar DJ (1995) Molecular cloning of an insect aminopeptidase N that serves as a receptor for *Bacillus thuringiensis* CryIA(c) toxin. J Biol Chem 270:17765–17770

Knowles BH (1994) Mechanism of action of *Bacillus thuringiensis* insecticidal δ-endotoxins. Adv Insect Physiol 24:275–308

Knowles BH, Dow JAT (1993) The crystal δ-endotoxin of *Bacillus thuringiensis*: models for their mechanism of action on the insect gut. BioEssays 15:469–476

Knowles BH, Ellar DJ (1987) Colloid-osmotic lysis is a general feature of the mechanism of action of *Bacillus thuringiensis* δ-endotoxins with different insect specificity. Biochim Biophys Acta 924:509–518

Knowles BH, Farndale RW (1988) Activation of insect cell adenylate cyclase by *Bacillus thuringiensis* δ-endotoxins and melittin. Biochem J 253:235–241

Kramer KJ, Muthukrishnan S (1997) Insect chitinases: molecular biology and potential use as biopesticides. Insect Biochem Mol Biol 27:887–900

Lecadet MM, Martouret D (1967) Enzymatic hydrolysis of the crystals of *Bacillus thuringiensis* by the proteases of *Pieris brassicae*. II. Toxicity of the different fractions of the hydrolysate for larvae of *Pieris brassicae*. J Invertebr Pathol 9:322–330

Lee MK, Milne RE, Ge AZ, Dean DH (1992) Location of a *Bombyx mori* receptor binding region on a *Bacillus thuringiensis* δ-endotoxin. J Biol Chem 267:3115–3121

Lee MK, Curtiss A, Alcantara E, Dean DH (1996a) Synergistic effect of the *Bacillus thuringiensis* toxins CryIAa and CryIAc on the gypsy moth, *Lymantria dispar*. Appl Environ Microbiol 62:583–586

Lee MK, You TH, Young BA, Cotrill JA, Valaitis AP, Dean DH (1996b) Aminopeptidase N purified from gypsy moth brush border membrane vesicles is a specific receptor for *Bacillus thuringiensis* CryIAc toxin. Appl Environ Microbiol 62:2845–2849

Leonardi MG, Parenti P, Casartelli M, Giordana B (1997) *Bacillus thuringiensis* Cry1Aa δ-endotoxin affects K⁺/amino acid symport in *Bombyx mori* larval midgut. J Membr Biol 159:209–217

Lepier A, Azuma M, Harvey WR, Wieczorek H (1994) K⁺/H⁺ antiport in the tobacco hornworm midgut: the K⁺-transporting component of the K⁺ pump. J Exp Biol 196:361–373

Lereclus D, Delécluse A, Lecadet MM (1993) Diversity of *Bacillus thuringiensis* toxins and genes. In: Entwistle PF, Cory JS, Bailey MJ, Higgs S (eds) *Bacillus thuringiensis*, an environmental biopesticide: theory and practice. Wiley, Chichester, pp 37–69

Li J, Carroll J, Ellar DJ (1991) Crystal structure of insecticidal δ-endotoxin from *Bacillus thuringiensis* at 2.5 Å resolution. Nature 353:815–821

Liang Y, Patel SS, Dean DH (1995) Irreversible binding kinetics of *Bacillus thuringiensis* CryIA δ-endotoxins to gypsy moth brush border membrane vesicles is directly correlated to toxicity. J Biol Chem 270:24719–24724

Liebig B, Stetson DL, Dean DH (1995) Quantification of the effect of *Bacillus thuringiensis* toxins on short-circuit current in the midgut of *Bombyx mori*. J Insect Physiol 41:17–22

Lorence A, Darszon A, Diaz C, Liévano A, Quintero R, Bravo A (1995) δ-Endotoxins induce cation channels in *Spodoptera frugiperda* brush border membranes in suspension and in planar lipid bilayers. FEBS Lett 360:217–222

Lorence A, Darzon A, Bravo A (1997) Aminopeptidase dependent pore formation of *Bacillus thuringiensis* Cry1Ac toxin on *Trichoplusia ni* membranes. FEBS Lett 414:303–307

Lu YJ, Adang MJ (1996) Conversion of *Bacillus thuringiensis* CryIAc-binding aminopeptidase to a soluble form by endogenous phosphatidylinositol phospholipase C. Insect Biochem Mol Biol 26:33–40

Luo K, Lu YJ, Adang MJ (1996) A 106 kDa form of aminopeptidase is a receptor for *Bacillus thuringiensis* CryIC δ-endotoxin in the brush border membrane of *Manduca sexta*. Insect Biochem Mol Biol 26:783–791

Luo K, Sangadala S, Masson L, Mazza A, Brousseau R, Adang MJ (1997) The *Heliothis virescens* 170 kDa aminopeptidase functions as "Receptor A" by mediating specific *Bacillus thuringiensis* Cry1A δ-endotoxin binding and pore formation. Insect Biochem Mol Biol 27:735–743

Luo K, Banks D, Adang MJ (1999) Toxicity, binding, and permeability analyses of four *Bacillus thuringiensis* Cry δ-endotoxins using brush border membrane vesicles of *Spodoptera exigua* and *Spodoptera frugiperda*. Appl Environ Microbiol 65:457–464

Lüthy P, Ebersold HR (1981) *Bacillus thuringiensis* delta-endotoxin: histopathology and molecular mode of action. In: Davidson EW (ed) Pathogenesis of invertebrate microbial diseases. Allanheld Osmun, Totowa, NJ, pp 235–267

MacIntosh SC, Stone TB, Jokerst RC, Fuchs RL (1991) Binding of *Bacillus thuringiensis* proteins to a laboratory-selected line of *Heliothis virescens*. Proc Natl Acad Sci USA 88:8930–8933

Martin PAW, Travers RS (1989) Worldwide abundance and distribution of *Bacillus thuringiensis* isolates. Appl Environ Microbiol 55:2437–2442

Martin F, Wolfersberger MG (1995) *Bacillus thuringiensis* δ-endotoxin and larval *Manduca sexta* brush-border membrane vesicles act synergistically to cause very large increases in the conductance of planar lipid bilayers. J Exp Biol 198:91–96

Martínez-Ramírez AC, Gould F, Ferré J (1999) Histopathological effects and growth reduction in a susceptible and a resistant strain of *Heliothis virescens* (Lepidoptera: Noctuidae) caused by sublethal doses of pure Cry1A crystal proteins from *Bacillus thuringiensis*. Biocontr Sci Technol 9:239–246

Masson L, Mazza A, Brousseau R, Tabashnik B (1995) Kinetics of *Bacillus thuringiensis* toxin binding with brush border membrane vesicles from susceptible and resistant larvae of *Plutella xylostella*. J Biol Chem 270:11887–11896

Masson L, Tabashnik BE, Liu YB, Brousseau R, Schwartz JL (1999) Helix 4 of the *Bacillus thuringiensis* Cry1Aa toxin lines the lumen of the ion channel. J Biol Chem 274:31996–32000

Milne R, Ge AZ, Rivers D, Dean DH (1990) Specificity of insecticidal crystal proteins: Implications for industrial standardization. In: Hickle LA, Fitch WL (eds) Analytical chemistry of *Bacillus thuringiensis*. ACS symposium series 432. American Chemical Society, Washington, DC, pp 22–35

Milne RE, Pang ASD, Kaplan H (1995) A protein complex from *Choristoneura fumiferana* gut-juice involved in the precipitation of δ-endotoxin from *Bacillus thuringiensis* subsp. *sotto*. Insect Biochem Mol Biol 25:1101–1114

Milne R, Wright T, Kaplan H, Dean D (1998) Spruce budworm elastase precipitates *Bacillus thuringiensis* δ-endotoxin by specifically recognizing the C-terminal region. Insect Biochem Mol Biol 28:1013–1023

Mitani K, Watarai J (1916) A new method to isolate the toxin of *Bacillus sotto* Ishiwata by passing through a bacterial filter and preliminary report on the toxic action of this toxin to the silkworm larva. Aichi Gensanshu Seizojo Hokoku (in Japanese; translation by T.B. Tsay, Canadian Forest Service)

Moffett DF, Cummings SA (1994) Transepithelial potential and alkalization in an in situ preparation of tobacco hornworm (*Manduca sexta*) midgut. J Exp Biol 194:341–345

Moffett DF, Koch AR (1988) Electrophysiology of K$^+$ transport by midgut epithelium of lepidopteran insect larvae. I. The transbasal electrochemical gradient. J Exp Biol 135:25–38

Moffett DF, Koch A, Woods R (1995) Electrophysiology of K$^+$ transport by midgut epithelium of lepidopteran insect larvae. III. Goblet valve patency. J Exp Biol 198:2103–2113

Monette R, Savaria D, Masson L, Brousseau R, Schwartz JL (1994) Calcium-activated potassium channels in the UCR-SE-1a lepidopteran cell line from the beet armyworm (*Spodoptera exigua*). J Insect Physiol 40:273–282

Monette R, Potvin L, Baines D, Laprade R, Schwartz JL (1997) Interaction between calcium ions and *Bacillus thuringiensis* toxin activity against Sf9 cells (*Spodoptera frugiperda*, Lepidoptera). Appl Environ Microbiol 63:440–447

Murphy DW, Sohi SS, Fast PG (1976) *Bacillus thuringiensis* enzyme-digested delta-endotoxin: effect on cultured insect cells. Science 194:954–956

Narayanan K, Jayaraj S (1974) The effect of *Bacillus thuringiensis* endotoxin on hemolymph cation levels in the citrus leaf caterpillar, *Papilio demoleus*. J Invertebr Pathol 23:125–126

Navon A (1993) Control of lepidopteran pests with *Bacillus thuringiensis*. In: Entwistle PF, Cory JS, Bailey MJ, Higgs S (eds) *Bacillus thuringiensis*, an environmental biopesticide: theory and practice. Wiley, Chichester, pp 125–146

Nishiitsutsuji-Uwo J, Endo Y (1980) Mode of action of *Bacillus thuringiensis* δ-endotoxin: general characteristics of intoxicated *Bombyx* larvae. J Invertebr Pathol 35:219–228

Nishiitsutsuji-Uwo J, Endo Y (1981a) Mode of action of *Bacillus thuringiensis* δ-endotoxin: effect on *Galleria mellonella* (Lepidoptera: Pyralidae). Appl Entomol Zool 16:79–87

Nishiitsutsuji-Uwo J, Endo Y (1981b) Mode of action of *Bacillus thuringiensis* δ-endotoxin: changes in hemolymph pH and ions of *Pieris*, *Lymantria* and *Ephestia* larvae. Appl Entomol Zool 16:225–230

Nishiitsutsuji-Uwo J, Endo Y, Himeno M (1979) Mode of action of *Bacillus thuringiensis* δ-endotoxin: effect on TN-368 cells. J Invertebr Pathol 34:267–275

Norris JR (1971) The protein crystal toxin of *Bacillus thuringiensis*: biosynthesis and physical structure. In: Burges HD, Hussey NW (eds) Microbial control of insects and mites. Academic Press, London, pp 229–246

Pang ASD, Gringorten JL (1998) Degradation of *Bacillus thuringiensis* δ-endotoxin in host insect gut juice. FEMS Microbiol Lett 167:281–285

Parenti P, Villa M, Hanozet GM, Tasca M, Giordana B (1995) Interaction of the insecticidal crystal protein Cry1A from *Bacillus thuringiensis* with amino acid transport into brush border membranes from *Bombyx mori* larval midgut. J Invertebr Pathol 65:35–42

Pendleton IR (1970) Sodium and potassium fluxes in *Philosamia ricini* during *Bacillus thuringiensis* protein crystal intoxication. J Invertebr Pathol 16:313–314

Percy J, Fast PG (1983) *Bacillus thuringiensis* crystal toxin: ultrastructural studies of its effect on silkworm midgut cells. J Invertebr Pathol 41:86–98

Peyronnet O, Vachon V, Brousseau R, Baines D, Schwartz JL, Laprade R (1997) Effect of *Bacillus thuringiensis* toxins on the membrane potential of lepidopteran insect midgut cells. Appl Environ Microbiol 63:1679–1684

Pietrantonio PV, Gill SS (1996) *Bacillus thuringiensis* endotoxins: action on the insect midgut. In: Lehane MJ, Billingsley PF (eds) Biology of the insect midgut. Chapman and Hall, London, pp 345–372

Potvin L, Laprade R, Schwartz JL (1998) Cry1Ac, a *Bacillus thuringiensis* toxin, triggers extracellular Ca^{2+} influx and Ca^{2+} release from intracellular stores in Cf1 cells (*Choristoneura fumiferana*, Lepidoptera). J Exp Biol 201:1851–1858

Rajamohan F, Lee MK, Dean DH (1998) *Bacillus thuringiensis* insecticidal proteins: molecular mode of action. Progr Nucl Acid Res Mol Biol 60:1–27

Ramakrishnan N (1968) Observations on the toxicity of *Bacillus thuringiensis* for the silkworm, *Bombyx mori*. J Invertebr Pathol 10:449–450

Reuveni M, Dunn PE (1991) Differential inhibition by *Bacillus thuringiensis* δ-endotoxin of leucine and aspartic acid uptake into BBMV from midgut of *Manduca sexta*. Biochem Biophys Res Commun 181:1089–1093

Ridgway RL, Moffett DF (1986) Regional differences in the histochemical localization of carbonic anhydrase in the midgut of tobacco hornworm (*Manduca sexta*). J Exp Zool 237:407–412

Sacchi VF, Wolfersberger MG (1996) Amino acid absorption. In: Lehane MJ, Billingsley PF (eds) Biology of the insect midgut. Chapman and Hall, London, pp 265–292

Sacchi VF, Parenti P, Hanozet GM, Giordana B, Lüthy P, Wolfersberger MG (1986) *Bacillus thuringiensis* toxin inhibits K^+-gradient-dependent amino acid transport across the brush border membrane of *Pieris brassicae* midgut cells. FEBS Lett 204:213–218

Sangadala S, Walters FS, English LH, Adang MJ (1994) A mixture of *Manduca sexta* aminopeptidase and phosphatase enhances *Bacillus thuringiensis* insecticidal CryIA(c) toxin binding and $^{86}Rb^+$-K^+ efflux in vitro. J Biol Chem 269:10088–10092

Schnepf E, Crickmore N, Van Rie J, Lereclus D, Baum J, Feitelson J, Zeigler DR, Dean DH (1998) *Bacillus thuringiensis* and its pesticidal crystal proteins. Microbiol Mol Biol Rev 62:775–806

Schultz JC, Lechowicz MJ (1986) Hostplant, larval age, and feeding behavior influence midgut pH in the gypsy moth (*Lymantria dispar*). Oecologia 71:133–137

Schwab GE, Culver P (1990) In vitro analyses of *Bacillus thuringiensis* δ-endotoxin action. In: Hickle LA, Fitch WL (eds) Analytical chemistry of *Bacillus thuringiensis*. ACS symposium series 432. American Chemical Society, Washington, DC, pp 36–45

Schwartz JL, Laprade R (2000) Membrane permeabilization by *Bacillus thuringiensis* toxins: protein insertion and pore formation. In: Charles JF, Delécluse A, Nielsen-LeRoux C (eds) Entomopathogenic bacteria: from laboratory to field application. Kluwer, Dordrecht, Netherlands, pp 199–218

Schwartz JL, Garneau L, Masson L, Brousseau R (1991) Early response of cultured lepidopteran cells to exposure to δ-endotoxin from *Bacillus thuringiensis*: involvement of calcium and anionic channels. Biochim Biophys Acta 1065:250–260

Schwartz JL, Garneau L, Savaria D, Masson L, Brousseau R, Brousseau E (1993) Lepidopteran-specific crystal toxins from *Bacillus thuringiensis* form cation- and anion-selective channels in planar lipid bilayers. J Membr Biol 132:53–62

Schwartz JL, Lu YJ, Söhnlein P, Brousseau R, Laprade R, Masson L, Adang M (1997a) Ion channels formed in planar lipid bilayers by *Bacillus thuringiensis* toxins in the presence of *Manduca sexta* midgut receptors. FEBS Lett 412:270–276

Schwartz JL, Juteau M, Grochulski P, Cygler M, Préfontaine G, Brousseau R, Masson L (1997b) Restriction of intramolecular movements within the Cry1Aa toxin molecule of *Bacillus thuringiensis* through disulfide bond engineering. FEBS Lett 410:397–402

Schwartz JL, Potvin L, Chen XJ, Brousseau R, Laprade R, Dean DH (1997c) Single-site mutations in the conserved alternating-arginine region affect ionic channels formed by CryIAa, a *Bacillus thuringiensis* toxin. Appl Environ Microbiol 63:3978–3984

Schweikl H, Klein U, Schindlbeck M, Wieczorek H (1989) A vacuolar-type ATPase, partially purified from potassium transporting plasma membranes of tobacco hornworm midgut. J Biol Chem 264:11136–11142

Skibbe U, Christeller JT, Callaghan PT, Eccles CD, Laing WA (1996) Visualization of pH gradients in the larval midgut of *Spodoptera litura* using ^{31}P-NMR microscopy. J Insect Physiol 42:777–790

Slatin SL, Abrams CK, English L (1990) Delta-endotoxins form cation-selective channels in planar lipid bilayers. Biochem Biophys Res Commun 169:765–772

Smith RA, Couche GA (1991) The Phylloplane as a source of *Bacillus thuringiensis* variants. Appl Environ Microbiol 57:311–315

Steinhaus EA (1961) On the correct author of *Bacillus sotto*. J Insect Pathol 3:97–100

Tabashnik BE, Finson N, Groeters FR, Moar WJ, Johnson MW, Luo K, Adang MJ (1994) Reversal of resistance to *Bacillus thuringiensis* in *Plutella xylostella*. Proc Natl Acad Sci USA 91:4120–4124

Travers RS, Faust RM, Reichelderfer CF (1976) Effects of *Bacillus thuringiensis* var. *kurstaki* δ-endotoxin on isolated lepidopteran mitochondria. J Invertebr Pathol 28:351–356

Vachon V, Paradis MJ, Marsolais M, Schwartz JL, Laprade R (1995) Ionic permeabilities induced by insecticidal toxins of *Bacillus thuringiensis* in Sf9 cells. J Membr Biol 148:57–63

Vadlamudi RK, Ji TH, Bulla LA Jr (1993) A specific binding protein from *Manduca sexta* for the insecticidal toxin of *Bacillus thuringiensis* subsp. *berliner*. J Biol Chem 268:12334–12340

Vadlamudi RK, Weber E, Ji I, Ji TH, Bulla LA Jr (1995) Cloning and expression of a receptor for an insecticidal toxin of *Bacillus thuringiensis*. J Biol Chem 270:5490–5494

Valaitis AP, Lee MK, Rajamohan F, Dean DH (1995) Brush border membrane aminopeptidase-N in the midgut of the gypsy moth serves as the receptor for the CryIA(c) δ-endotoxin of *Bacillus thuringiensis*. Insect Biochem Mol Biol 25:1143–1151

van Frankenhuyzen K (1993) The challenge of *Bacillus thuringiensis*. In: Entwistle PF, Cory JS, Bailey MJ, Higgs S (eds) *Bacillus thuringiensis*, an environmental biopesticide: theory and practice. Wiley, Chichester, pp 1–35

van Frankenhuyzen K (2000) Application of *Bacillus thuringiensis* in forestry. In: Charles JF, Délécluse A, Nielsen-LeRoux C (eds) Entomopathogenic bacteria: from laboratory to field application. Kluwer, Dordrecht, Netherlands, pp 371–382

van Frankenhuyzen K, Gringorten JL, Milne RE, Gauthier D, Pusztai M, Brousseau R, Masson L (1991) Specificity of activated CryIA proteins from *Bacillus thuringiensis* subsp. *kurstaki* HD-1 for defoliating forest Lepidoptera. Appl Environ Microbiol 57:1650–1655

van Frankenhuyzen K, Gringorten JL, Gauthier D, Milne RE, Masson L, Peferoen M (1993) Toxicity of activated CryI proteins from *Bacillus thuringiensis* to six forest Lepidoptera and *Bombyx mori*. J Invrebr Pathol 62:295–301

Van Rie J, Jansens S, Höfte H, Degheele D, Van Mellaert H (1989) Specificity of *Bacillus thuringiensis* δ-endotoxins: importance of specific receptors on the brush border membrane of the midgut of target insects. Eur J Biochem 186:239–247

Van Rie J, Jansens S, Höfte H, Degheele D, Van Mellaert H (1990a) Receptors on the brush border membrane of the insect midgut as determinants of the specificity of *Bacillus thuringiensis* delta-endotoxins. Appl Environ Microbiol 56:1378–1385

Van Rie J, McGaughey WH, Johnson DE, Barnett BD, Van Mellaert H (1990b) Mechanism of insect resistance to the microbial insecticide *Bacillus thuringiensis*. Science 247:72–74

Villalon M, Vachon V, Brousseau R, Schwartz JL, Laprade R (1998) Video imaging analysis of the plasma membrane permeabilizing effects of *Bacillus thuringiensis* insecticidal toxins in Sf9 cells. Biochim Biophys Acta 1368:27–34

Walters FS, Slatin SL, Kulesza CA, English LH (1993) Ion channel activity of N-terminal fragments from CryIA(c) delta-endotoxin. Biochem Biophys Research Commun 196:921–926

Walters FS, Kulesza CA, Phillips AT, English LH (1994) A stable oligomer of *Bacillus thuringiensis* delta-endotoxin, CryIIIA. Insect Biochem Mol Biol 24:963–968

Wieczorek H (1992) The insect V-ATPase, a plasma membrane proton pump energizing secondary active transport: molecular analysis of the electrogenic potassium transport in the tobacco hornworm midgut. J Exp Biol 172:335–343

Wieczorek H, Wolfersberger MG, Cioffi M, Harvey WR (1986) Cation-stimulated ATPase activity in purified plasma membranes from tobacco hornworm midgut. Biochim Biophys Acta 857:271–281

Wieczorek H, Weerth S, Schindlbeck M, Klein U (1989) A vacuolar-type proton pump in a vesicle fraction enriched with potassium transporting plasma membranes from tobacco hornworm midgut. J Biol Chem 264:11143–11148

Wieczorek H, Putzenlechner M, Zeiske W, Klein U (1991) A vacuolar-type proton pump energizes K^+/H^+ antiport in an animal plasma membrane. J Biol Chem 266:15340–15347

Wieczorek H, Brown D, Grinstein S, Ehrenfeld J, Harvey WR (1999a) Animal plasma membrane energization by proton-motive V-ATPases. BioEssays 21:637–648

Wieczorek H, Grüber G, Harvey WR, Huss M, Merzendorfer H (1999b) The plasma membrane H^+-V-ATPase from tobacco hornworm midgut. J Bioenerget Biomembr 31:67–74

Wilson GR, Benoit TG (1990) Activation and germination of *Bacillus thuringiensis* spores in *Manduca sexta* larval gut fluid. J Invertebr Pathol 56:233–236

Witt DP, Carson H, Hodgdon JC (1986) Cytotoxicity of *Bacillus thuringiensis* δ-endotoxins to cultured Cf-1 cells does not correlate with in vivo activity toward spruce budworm larvae. In: Samson RA, Vlak JM, Peters D (eds) Fundamental and applied aspects of invertebrate pathology. Foundation of the 4th International Colloquium of Invertebrate Pathology, Wageningen, Netherlands, pp 3–6

Wolfersberger MG (1989) Neither barium nor calcium prevents the inhibition by *Bacillus thuringiensis* δ-endotoxin of sodium- or potassium gradient-dependent amino acid accumulation by tobacco hornworm midgut brush border membrane vesicles. Arch Insect Biochem Physiol 12:267–277

Wolfersberger MG (1990) The toxicity of two *Bacillus thuringiensis* δ-endotoxins to gypsy moth larvae is inversely related to the affinity of binding sites on midgut brush border membranes for the toxins. Experientia 46:475–477

Wolfersberger MG (1991) Inhibition of potassium-gradient-driven phenylalanine uptake in larval *Lymantria dispar* midgut by two *Bacillus thuringiensis* delta-endotoxins correlates with the activity of the toxins as gypsy moth larvicides. J Exp Biol 161:519–525

Wolfersberger MG (1992) V-ATPase-energized epithelia and biological insect control. J Exp Biol 172:377–386

Wolfersberger MG (1996) Localization of amino acid absorption systems in the larval midgut of the tobacco hornworm *Manduca sexta*. J Insect Physiol 42:975–982

Wolfersberger MG, Spaeth DD (1987) Activity of spore-crystal preparations from twenty serotypes of *Bacillus thuringiensis* toward *Manduca sexta* larvae in vivo and in vitro. J Appl Entomol 103:138–141

Wolfersberger MG, Harvey WR, Cioffi M (1982) Transepithelial potassium transport in insect midgut by an electrogenic alkali metal ion pump. Curr Top Membr Transp 16:109–133

Wolfersberger MG, Luethy P, Maurer A, Parenti P, Sacchi FV, Giordana B, Hanozet GM (1987) Preparation and partial characterization of amino acid transporting brush border membrane vesicles from the larval midgut of the cabbage butterfly (*Pieris brassicae*). Comp Biochem Physiol 86A:301–308

Wolfersberger MG, Chen XJ, Dean DH (1996) Site-directed mutations in the third domain of *Bacillus thuringiensis* δ-endotoxin CryIAa affect its ability to increase the permeability of *Bombyx mori* midgut brush border membrane vesicles. Appl Environ Microbiol 62:279–282

Wood JL, Farrand PS, Harvey WR (1969) Active transport of potassium by the cecropia midgut. VI. Microelectrode potential profile. J Exp Biol 50:169–178

Yaoi K, Kadotani T, Kuwana H, Shinkawa A, Takahashi T, Iwahana H, Sato R (1997) Aminopeptidase N from *Bombyx mori* as a candidate for the receptor of *Bacillus thuringiensis* Cry1Aa toxin. Eur J Biochem 246:652–657

Yi S, Pang ASD, van Frankenhuyzen K (1996) Immunocytochemical localization of *Bacillus thuringiensis* CryI toxins in the midguts of three forest insects and *Bombyx mori*. Can J Microbiol 42:634–641

Evolution of Amplified Esterase Genes as a Mode of Insecticide Resistance in Aphids

L. M. Field[1], R. L. Blackman[2], and A. L. Devonshire[1]

1
Introduction

It is now 20 years since the development of insecticide resistance in the aphid *Myzus persicae* (Sulzer) was proposed as an example of evolution by gene duplication (Devonshire and Sawicki 1979). This hypothesis was based on the observation that the molar amount of an insecticide-detoxifying esterase increased geometrically, doubling between a series of aphid clones having increasing levels of insecticide resistance. This was best explained by a succession of tandem duplications of the esterase structural genes (gene amplification) selected by the presence of the insecticide. The idea was in line with concurrent studies on the development of resistance to chemotherapeutic drugs in tumours and cell cultures, where amplification of genes encoding target proteins was beginning to be demonstrated (Alt et al. 1978; Padgett et al. 1979). Intensive research to test the hypothesis followed, notably in aphids (by the present authors) and in *Culex* mosquitoes, where a similar increase in esterase activity was associated with insecticide resistance (recently reviewed by Hemingway and Karunaratne 1998). The first molecular genetic evidence supporting the hypothesis was with mosquitoes, when a large increase (ca. 250-fold) in esterase gene copy number was found in a strongly resistant strain (Mouches et al. 1986). Two years later, amplification of the esterase genes in *M. persicae* was confirmed (Field et al. 1988), and a further 5 years on, the genes were fully cloned and sequenced (Field et al. 1993).

Since then, the story itself has "evolved" to reveal that the gene amplification mechanism conferring resistance to aphicides involves a complex interaction between genome evolution, aphid migration and selection, with both copy number and gene regulation contributing to the resistant phenotype.

Identical amplified esterase genes were found in a closely related, interbreeding species, *Myzus nicotianae* (Field et al. 1994), and a similar mechanism has recently been detected in another aphid species, *Schizaphis graminum* (Rider et al. 1998; Ono et al. 1999).

[1] IACR-Rothamsted, Harpenden, Herts AL5 2JQ, UK
[2] Natural History Museum, Cromwell Road, London, SW7 5BD, UK

Isaac Ishaaya (Ed.): Biochemical Sites of
Insecticide Action and Resistance
© Springer-Verlag Berlin, Heidelberg 2001

2
Biochemistry of Esterase-Based Resistance in *M. persicae*

The story began to unfold with the demonstration by Needham and Sawicki (1971) of a positive correlation between insecticide resistance in *M. persicae* clones (note that aphids are reared in the laboratory as asexually-reproducing clonal cultures established from individual aphids collected in the field) and their total carboxylesterase activity, as measured by the in vitro hydrolysis of 1-naphthyl acetate. Beranek (1974) and Devonshire (1975) then showed that only one of several esterases present in *M. persicae* was more active in resistant aphids. This esterase, designated E4, was purified from extremely resistant aphids and used to study the kinetics of insecticide hydrolysis (Devonshire 1977). This showed that E4 was responsible for the hydrolysis of both naphthyl esters and insecticides, and demonstrated unequivocally that this enzyme was the cause of resistance. The enzyme acylated by organophosphorus (OP) or carbamate insecticides is either stable (e.g. with *O,S*-dialkyl phosphates) or is hydrolysed extremely slowly (with carbamates, diethyl phosphates and dimethyl phosphates having progressively faster hydrolysis rates). Resistance levels reflect these rates, being highest towards those insecticides that are hydrolysed fastest. Measurements of the catalytic centre activity (k_3) of E4 showed that its increased activity in resistant aphids results from their producing more E4 protein rather than from a series of more efficient mutant enzymes (Devonshire and Moores 1982). The extremely large amount of enzyme present (10 pmol in the most resistant aphid variants, or 1–2% of total protein) both hydrolyses and sequesters insecticide molecules, so conferring resistance. Two very closely related variant forms of the enzyme, E4 and FE4, were subsequently distinguished, the latter having a slightly higher catalytic centre activity (1.5-fold) towards OPs (Devonshire et al. 1983) and a molecular weight of 66 kDa compared to 65 kDa for E4. The difference was in the primary structure rather than resulting from different post-translational modification (Devonshire et al. 1986).

3
Molecular Genetics of Esterase Overproduction

The initial molecular evidence for esterase gene amplification in *Myzus periscae* came from using partial cDNAs encoding E4 as probes for DNA dot blots and Southern blots (Field et al. 1988). When the full-length cDNAs for the genes encoding the E4 and FE4 esterases were cloned and sequenced (Field et al. 1993), the derived amino acid sequences predicted enzymes with extensive similarity to other serine hydrolases. The enzymes have a signal peptide, 2 potential cysteine bridges and the characteristic catalytic triad involving an active site nucleophilic serine activated by a charge relay from glutamate through histidine.

3.1
Esterase Genes in Susceptible Aphids

For many years, despite extensive studies of the amplified esterases, our understanding of the esterase genes in wild-type, susceptible aphids was very limited. The 98% sequence identity between *E4* and *FE4* (Field et al. 1993) suggested that they had arisen by a recent duplication event, but the sequence that could be identified by polymerase chain reaction (PCR) amplification of esterase sequences from susceptible aphids was neither *E4* nor *FE4*. Then, in 1998, studies of the genes and their flanking DNA produced evidence that susceptible aphids do have both the *E4* and *FE4* genes adjacent to each other in a head-to-tail arrangement, with *E4* approx. 19 kb upstream of *FE4* (Field and Devonshire 1998). The same study identified two other closely related sequences, suggesting the presence of an esterase gene family.

The duplication and divergence of genes to give families is well established and depends on the need for the duplicated gene to acquire a useful function and therefore be retained by natural selection (Fryxell 1996). Indeed, this concept (Ohno 1970) contributed to the formulation of the original hypothesis for aphid esterase gene duplication (Devonshire and Sawicki 1979). In the case of *E4* and *FE4* this criterion might be fulfilled by the consequent increase in the capacity to detoxify xenobiotic esters. There are precedents for such families of insect genes being associated with detoxification, such as the clusters of *P450* genes in housefly (*Musca domestica*), thought to have arisen by duplication and inversion events (Cohen and Feyereisen 1995), and those associated with insecticide resistance in *Drosophila melanogaster* (Maitra et al. 1996).

The duplication event leading to the divergence of *E4* and *FE4* coding sequences in *M. persicae* also led to differences upstream from the 5′ ends of the genes (Field and Devonshire 1998). Approximately 1.7 kb of sequence, present in amplified *E4* genes, is absent in *FE4*, creating a junction (FJ) in the *FE4* sequence, as shown in Fig. 1. Transcription of *FE4* starts 94 bp upstream of its ATG initiation codon; there is a CAP site and a putative TATA box, suggesting that expression of *FE4* genes is controlled by this conventional promoter region. For *E4* genes there is approx. 300 bp of untranslated leader sequence interrupted by two introns; this additional region contains a CpG island analogous with promoters of vertebrate housekeeping genes (see below).

Recent PCR-based analysis has identified another amplified sequence, apparently from a deleted *E4* gene, suggesting a fifth member of the esterase gene family (L.M. Field, unpubl.).

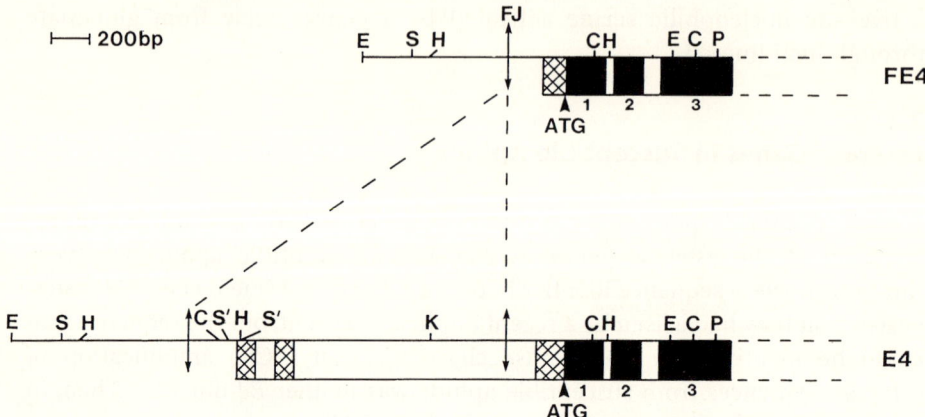

Fig. 1. Structure of the 5′ ends of amplified *E4* and *FE4* genes. introns; 5 untranslated regions; n translated regions. 1, 2 and 3 indicate the first three exons and FJ denotes the junction upstream of *FE4* resulting from the absence of approx. 1.7 kb of sequence present in the *E4* gene. *Letters* show position of restriction sites. C *Cla*1; E *Eco*R1; H *Hpa*1; K *Kpa*1; S *Spe*1; S′ *Sal*1. (Reproduced with permission from Field and Devonshire 1998 and the Biochemical Society)

3.2
Organization of Amplified Esterase Genes

The amplified *E4* genes and flanking DNA in one aphid clone (794J) are present at a single site on multiple 24 kb units (amplicons) arranged as a tandem array of head-to-tail direct repeats (Field et al. 1996). These have apparently been created by multiple breakages just upstream of the *E4* gene at exactly the same site as that generating the initial E4/FE4 duplication, followed by re-assembly to generate multiple 'novel joints' (NJ) between copies (Field and Devonshire 1997, 1998), as shown in Fig. 2. This arrangement is consistent with a model for the generation of amplified DNA by unequal sister chromatid exchange, although there are no obvious homologies or adenine-rich regions usually associated with such events (Triglia et al. 1991).

The same novel joint sequence as that detected in 794J aphids has been found between the amplicons of 13 other resistant aphid clones of diverse geographical origin (Field and Devonshire 1997), showing that either the same event has occurred many times or, more likely, that the *E4* amplification arose only once and then spread throughout the world. This could have occurred through migration, coupled with world-wide selection by insecticides of the beneficial trait conferred by the amplified genes, as has been suggested for the esterase genes in mosquitoes (Raymond et al. 1991). In aphids, spread could have occurred naturally by long-range aerial movement (Loxdale et al. 1993) or, perhaps more likely, as a consequence of the international trade in plants and produce harboring insect pests (Frey 1993).

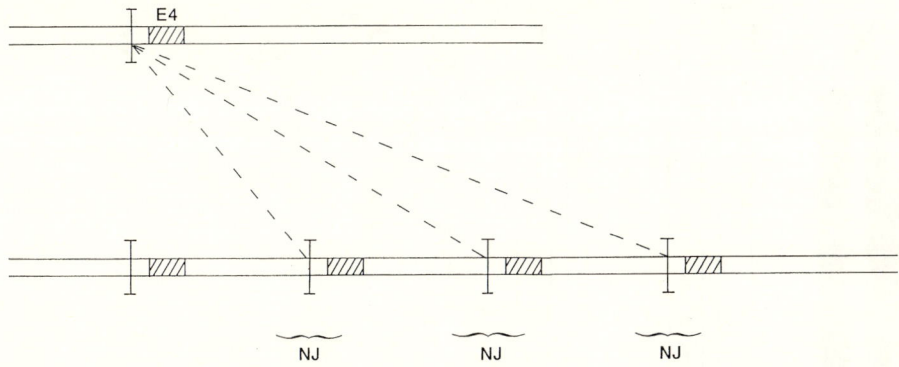

Fig. 2. Model (not to scale) for the creation of a novel joint (*NJ*) during amplification of the *E4* gene and flanking DNA. *Top* Wild-type single-copy gene; *bottom* E4 amplicons in tandem array. (Reproduced with permission from Field and Devonshire 1997 and the Biochemical Society)

Earlier work indicated that, in at least one aphid clone, the number of *E4* gene copies was insufficient to account for the increase in enzyme content (Field et al. 1996). However, the recent use of quantitative competitive PCR, together with pulsed-field gel electrophoresis, has allowed a more accurate assessment of the relative gene copy number in aphid clones with different levels of esterase enzyme (Field et al. 1999). This has shown that a wide range of copy number can exist for both *E4* and *FE4*, giving a proportionate increase in esterase level and consequent resistance, and reaching a maximum of 80 gene copies. Furthermore, crossing studies enabled the partial resolution of multiple E4 sites, and their measurement, in the one aphid clone known to have loci other than that on 3^T.

3.3
Cytogenetic Studies of Amplified Esterases

Fluorescence in situ hybridization (FISH) has shown that the amplified genes in most E4 over-producing aphid clones are located at a single site on autosome 3 near the breakpoint of an autosomal 1,3 translocation, which previous work had shown to be genetically linked to insecticide resistance (Blackman et al. 1995). This site could have been moved to the truncated chromosome 3 (3^T) by reciprocal exchange from a site on chromosome 1 (Fig. 3) thought to be the locus of the unamplified esterase genes. This site (1I) is situated near subtelomeric repetitive DNA (Blackman et al. 1996) and is sometimes seen by FISH with susceptible aphid chromosomes (R.L. Blackman, unpublished). Only one aphid clone has so far been detected with multiple sites of amplified *E4* genes having the same 3^T site but additional sites on autosomes 5 and 2 (Blackman et al. 1995). Despite this dispersion of the amplicons in this partic-

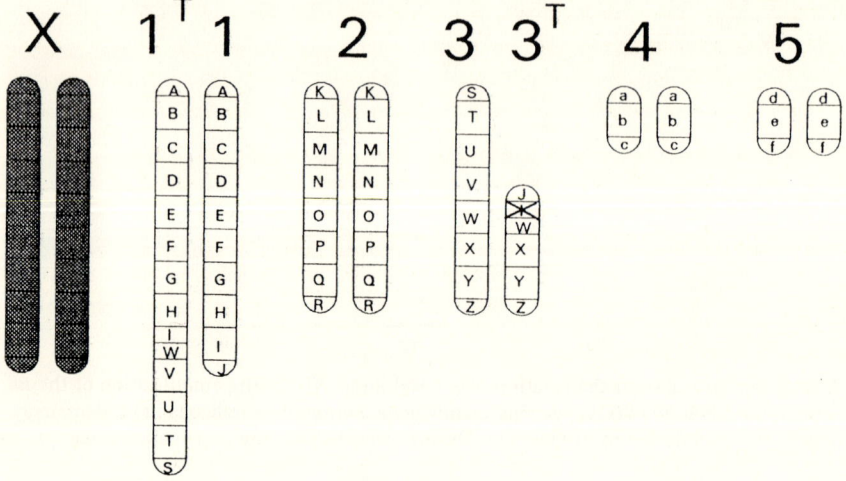

Fig. 3. Diagram of the chromosomes in aphids with amplified *E4* genes and a reciprocal A1, 3 translocation. ⊠ site of amplified *E4* genes

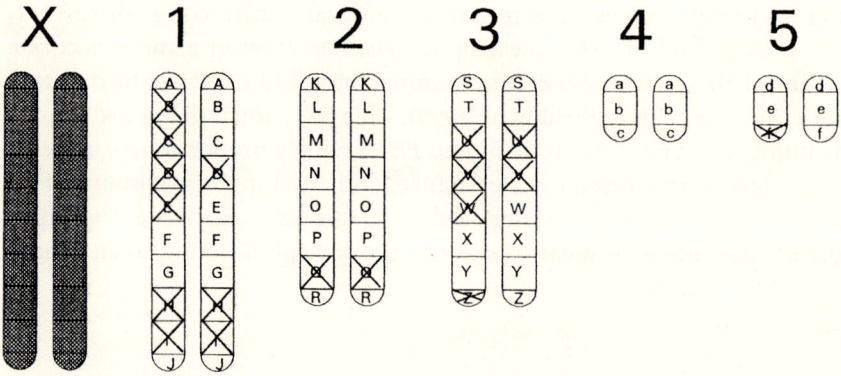

Fig. 4. Diagram of the chromosomes in aphids with amplified *FE4* genes and a normal karyotype. 5 ⊠ sites where amplified *FE4* genes have been detected (up to eight sites in one genome)

ular clone, their structure and that of the DNA immediately flanking the genes are indistinguishable from that in clones with all copies of the amplicon at a single locus.

In contrast to the studies of amplified *E4* genes, much less is known about *FE4*. The genes are again arranged as blocks, this time of 18 kb repeat units (Field and Devonshire 1998), but the nature of the novel joints has not yet been resolved. There are no visible chromosome abnormalities associated with *FE4* amplification (Blackman et al. 1995) but a striking feature is that the genes are scattered around the genome (Fig. 4), with as many as eight sites per nucleus involving several different chromosomes (Blackman et al. 1995, 1999).

Contrary to our initial findings with laboratory clones (Blackman et al. 1995), it now seems that in field populations, at least those of Greek origin, other locations of *FE4* amplicon clusters occur most commonly at various sites on the same chromosome (autosome 1) as the putative ancestral esterase gene locus (Blackman et al. 1999). Subsequent relocation of *FE4*-containing amplicons to other sites on autosome 1 may have occurred by inversion, and to other autosomes by reciprocal interchange. The postulated ancestral locus is situated near subtelomeric heterochromatin that contains numerous copies of a 169 bp repeat that also occurs subtelomerically on all the other autosomes of *M. persicae* (Spence et al. 1998), suggesting that interchange of subtelomeric sections may occur relatively frequently.

Transposable elements might be involved in any or all of such rearrangements. In this context, a non-LTR retrotransposon-like sequence associated with the telomeric DNA of *M. persicae* has recently been isolated (J.M. Spence, unpubl. results), and inverted repeat sequences have been found within the *E4* amplicon (L.M. Field, unpubl.).

Overall, the emerging picture is one of two independent amplification events occurring to generate either amplified *E4* or *FE4*, followed by chromosome relocations and changes in copy number, particularly for *FE4*. This extensive introgression of multiple *FE4* loci into aphid populations might reflect the more sexual orientation of the non-translocated genotype of aphids with amplified *FE4* genes. The translocation appears to impose a genetic load in that it predisposes its carrier to asexual reproduction.

4
Expression of Esterase Genes

In addition to the complex genome rearrangements giving rise to amplified esterase genes in aphids, there can also be transcriptional down-regulation of *E4* genes. Very occasionally, the offspring of aphids in a highly resistant, *E4*-producing clone (reproducing parthenogenetically) will spontaneously lose their elevated enzyme and hence their resistant phenotype (revertants). This is accompanied by a loss of *E4* mRNA but the amplified genes are retained (Field et al. 1989). Rather surprisingly, it has been shown that the transcribed *E4* genes contain 5-methylcytosine (5mC) at *Msp*I sites (CCGG) in and around the genes and that this methylation is lost concurrently with the loss of elevated esterase (Field et al. 1989; Hick et al. 1996). This positive association between DNA methylation and gene expression is in direct contrast to extensive data from vertebrates, which link DNA methylation with suppression of gene activity (Holliday 1996). It is also one of very few examples of the presence of 5mC in an invertebrate genome (Tweedie et al. 1997) and the only one where changes in methylation have been linked to changes in transcription.

Clearly, aphids are able to methylate their DNA at CpG doublets and the pattern is generally stably inherited during both sexual and asexual reproduction (Blackman et al. 1996). However, the extent to which it occurs in the aphid genome and its general effect on gene expression are unknown. It may be specific to the amplified *E4* genes, in line with reports that multicopy and foreign DNA are methylated as part of a genomic defence mechanism (Yoder et al. 1997).

As mentioned earlier, the initial duplication to give *E4* and *FE4* generated distinct sequences at the 5' end of each gene. *E4* has an unusual leader sequence not preceded by conventional promoter sequences (Field et al. 1996) and this region is very rich in the dinucleotide CpG (L.M. Field, unpublished) analogous to the CpG islands present in vertebrate housekeeping genes, where DNA methylation appears to play a direct role in controlling transcription. Preliminary studies of this *E4* CpG island upstream of the coding region have shown no methylation for either the expressed or silent genes (L.M. Field, unpubl.). The methylated sites originally identified by differential cleavage by *Msp*I and *Hpa*II are located within the gene (in exons 3 and 6) and in the 3' flanking DNA. Clearly the role of 5mC in aphid gene expression warrants further investigation, especially in the light of the ongoing controversy over the role of methylation in genome evolution (Martienssen 1998).

5
Wider Implications

The amplification of esterase genes in *M. persicae* and *Culex* mosquitoes are some of the best-understood examples of a resistance mechanism at the molecular level, but to what extent is this relevant to other species?

Amplified genes, identical to *E4* and *FE4*, have been found in a closely related tobacco-feeding form of the aphid, *M. nicotianae*, which is probably best explained by the selection of the amplified genes in *M. persicae*, transfer to hybrids between the species by sexual reproduction and subsequent spread through *M. nicotianae* populations (Field et al. 1994). Such a transfer of resistance genes is obviously limited to species which can form hybrids. However, an unrelated aphid, *Schizaphis graminum*, has clearly developed resistance independently by the amplification of an esterase gene with 73% identity to *E4/FE4* (Ono et al. 1999). Not only have the genes amplified, but they also appear to be methylated, as judged by their differential digestion by *Msp*I and *Hpa*II.

There is potential for further changes in copy number of *M. persicae* esterase genes and for further sequence divergence. Consequently, selection of new types of esterase-based resistance to insecticides may evolve. Two forms of target-site resistance also occur in this species, involving changes in the acetylcholinesterase (*AChE*) and sodium channel (*kdr*) genes (Devonshire et al.

1998), which together give *M. persicae* a wide range of defence mechanisms against our attempts to control their populations.

Acknowledgements. IACR-Rothamsted receives grant-aided support from the Biotechnology and Biological Sciences Research Council of the UK.

References

Alt W, Kellems RE, Bertino JR, Schimke RT (1978) Selective multiplication of dihydrofolate reductase genes in methotrexate-resistant variants of cultured murine cells. J Biol Chem 253:1357–1370

Beranek AP (1974) Esterase variation and organophosphate resistance in populations of *Aphis fabae* and *Myzus persicae*. Ent Exp Appl 17:129–142

Blackman RL, Spence JM, Field LM, Devonshire AL (1995) Chromosomal location of the amplified esterase genes conferring resistance to insecticides in *Myzus persicae* (Homoptera: Aphididae). Heredity 75:297–302

Blackman RL, Spence JM, Field LM, Javed N, Devine GJ, Devonshire AL (1996) Inheritance of the amplified esterase genes responsible for insecticide resistance in *Myzus persicae* (Homoptera: Aphididae). Heredity 77:154–167

Blackman RL, Spence JM, Field LM, Devonshire AL (1999) Variation in the chromosomal distribution of amplified esterase (*FE4*) genes in Greek field populations of *Myzus persicae* (Sulzer). Heredity 82:180–186

Cohen MB, Feyereisen R (1995) A cluster of cytochrome-P450 genes of the CYP6 family in the house-fly. DNA Cell Biol 14:73–82

Devonshire AL (1975) Studies of the carboxylesterases of *Myzus persicae* resistant and susceptible to organophosphorus insecticides. In: Proceedings 8th British Insecticide and Fungicide Conference, vol 1, pp 67–73

Devonshire AL (1977) The properties of a carboxylesterase from the peach-potato aphid *Myzus persicae* (Sulz.) and its role in conferring insecticide resistance. Biochem J 167:675–683

Devonshire AL, Moores GD (1982) A carboxylesterase with broad substrate specificity causes organophosphorus, carbamate and pyrethroid resistance in peach-potato aphids (*Myzus persicae*). Pestic Biochem Physiol 18:235–246

Devonshire AL, Sawicki RM (1979) Insecticide-resistant *Myzus persicae* as an example of evolution by gene duplication. Nature 280:140–141

Devonshire AL, Moores GD, Chiang C-L (1983) The biochemistry of insecticide resistance in the peach-potato aphid *Myzus persicae*. In: Miyamoto J (ed) IUPAC pesticide chemistry, human welfare and the environment. Pergamon Press, Oxford, pp 191–196

Devonshire AL, Searle LM, Moores GD (1986) Quantitative and qualitative variation in the mRNA for carboxylesterases in insecticide-susceptible and resistant *Myzus persicae* (Sulz.). Insect Biochem 16:659–665

Devonshire AL, Field LM, Foster SP, Moores GD, Williamson MS, Blackman RL (1998) The evolution of insecticide resistance in the peach-potato aphid, *Myzus persicae*. Philos Trans R Soc Lond B 353:1677–1684

Field LM, Devonshire AL (1997) Structure and organisation of amplicons containing the *E4* esterase genes responsible for insecticide resistance in the aphid *Myzus persicae* (Sulzer). Biochem J 322:867–871

Field LM, Devonshire AL (1998) Evidence that the *E4* and *FE4* esterase genes responsible for insecticide resistance in the aphid *Myzus persicae* (Sulzer) are part of a gene family. Biochem J 330:169–173

Field LM, Devonshire AL, Forde BG (1988) Molecular evidence that insecticide resistance in peach-potato aphids (*Myzus persicae* Sulz.) results from amplification of an esterase gene. Biochem J 251:309–312

Field LM, Devonshire AL, Ffrench-Constant RH, Forde BG (1989) Changes in DNA methylation are associated with loss of insecticide resistance in the peach-potato aphid *Myzus persicae* (Sulz.). FEBS Lett 243:323–327

Field LM, Williamson MS, Moores GD, Devonshire AL (1993) Cloning and analysis of the esterase genes conferring insecticide resistance in the peach-potato aphid *Myzus persicae* (Sulzer). Biochem J 294:569–574

Field LM, Javed N, Stribley MF, Devonshire AL (1994) The peach-potato aphid *Myzus persicae* and the tobacco aphid *Myzus nicotianae* have the same esterase-based mechanisms of insecticide resistance. Insect Mol Biol 3:143–148

Field LM, Devonshire AL, Tyler-Smith C (1996) Analysis of amplicons containing the esterase genes responsible for insecticide resistance in the peach-potato aphid *Myzus persicae* (Sulzer). Biochem J 313:543–547

Field LM, Blackman RL, Tyler-Smith C, Devonshire AL (1999) Relationship between amount of esterase and gene copy number in insecticide resistant *Myzus persicae* (Sulzer). Biochem J 339:737–742

Frey JE (1993) The analysis of anthropod pest movement through trade in ornamental plants. Brighton Crop Protect Conf Monogr 54:157–165

Fryxell KJ (1996) The coevolution of gene family trees. Trends Genet 12:364–369

Hemingway J, Karunaratne SHPP (1998) Mosquito carboxylesterases: a review of the molecular biology and biochemistry of a major insecticide resistance mechanism. Med Vet Entomol 12:1–12

Hick CA, Field LM, Devonshire AL (1996) Changes in the methylation of amplified esterase DNA during loss and reselection of insecticide resistance in peach-potato aphids, *Myzus persicae*. Insect Biochem Mol Biol 26:41–47

Holliday R (1996) DNA methylation in eukaryotes: 20 years on. In: Russo VEA, Martienssen RA, Riggs AD (eds) Epigenetic mechanisms of gene regulation. Cold Spring Harbor Laboratory Press, Cold Spring Harbor, pp 5–27

Loxdale HD, Hardie J, Halbert S, Footit R, Kidd NAC, Carter CI (1993) The relative importance of short-range and long-range movement of flying aphids. Biol Rev 68:291–311

Maitra S, Dombrowski SM, Waters LC, Ganguly R (1996) Three second chromosome-linked clustered Cyp6 genes show differential constitutive and barbital-induced expression in DDT-resistant and susceptible strains of *Drosophila melanogaster*. Gene 180:165–171

Martienssen R (1998) Transposons, DNA methylation and gene control. Trends Genet 14:263–264

Mouches C, Pasteur N, Berge JB, Hyrien O, Raymond M, de Saint Vincent BR, de Silvestri M, Georghiou GP (1986) Amplification of an esterase gene is responsible for insecticide resistance in a Californian *Culex* mosquito. Science 233:778–780

Needham PH, Sawicki RM (1971) Diagnosis of resistance to organophosphorus insecticides in *Myzus persicae* Sulz. Nature 230:126–127

Ohno S (1970) Evolution by gene duplication. Allen and Unwin, London and Springer, Berlin Heidelberg New York

Ono M, Swanson JJ, Field LM, Devonshire AL, Siegfried BD (1999) Amplification and methylation of an esterase gene associated with insecticide resistance in greenbugs, *Schizaphis graminum* (Rondani) (Homoptera: Aphididae). Insect Biochem Mol Biol 29: 1065–1073

Padgett RA, Wahl GM, Coleman PF, Stark GR (1979) N-(Phosphonacetyl)-L-aspartate-resistant hamster cells overaccumulate a single mRNA coding for the multifunctional protein that catalyzes the first steps of UMP synthesis. J Biol Chem 254:974–980

Raymond M, Callaghan A, Font P, Pasteur N (1991) Worldwide migration of amplified insecticide resistance genes in mosquitoes. Nature 350:151–153

Rider SD, Wilde GE, Kambhampati S (1998) Genetics of esterase-mediated insecticide resistance in the aphid *Schizaphis gramiminum*. Heredity 81:14–19

Spence JM, Blackman RL, Testa JM, Ready PD (1998) A 169-base pair tandem repeat DNA marker for subtelomeric heterochromatin and chromosomal rearrangements in aphids of the *Myzus persicae* group. Chromosome Res 6:167–175

Triglia T, Foster SJ, Kemp DJ, Cowman AF (1991) Implication of the multidrug resistance gene PFMDR1 in *Plasmodium falciparum* has arisen as multiple independent events. Mol Cell Biol 11:5244–5250

Tweedie S, Charlton J, Clark V, Bird A (1997) Methylation of genomes and genes at the inverte-
    brate-vertebrate boundary. Mol Cell Biol 17:1469–1475
Yoder JA, Walsh CP, Bestor (1997) Cytosine methylation and the ecology of intragenomic para-
    sites. Trends Genet 13:335–340

Insensitive Acetylcholinesterase as Sites for Resistance to Organophosphates and Carbamates in Insects: Insensitive Acetylcholinesterase Confers Resistance in Lepidoptera

ROBIN V. GUNNING[1] and GRAHAM D. MOORES[2]

1
Introduction

Acetylcholinesterase (EC 3.1.1.7) is the enzyme responsible for hydrolysing the neurotransmitter acetylcholine at the nerve synapse. If this hydrolysis does not take place, build-up of acetylcholine occurs leading to repeated firing of neurones and ultimately death by exhaustion. Vertebrate species possess acetylcholinesterase and butyrylcholinesterase (EC 3.1.1.8), differentiated by substrate preferences, whilst arthropods have only one enzyme which, whilst containing characteristics of both, is generally considered to be AChE (Toutant 1989).

AChE belongs to the class of proteins characterized by the (hydrolase fold, which is shared by several hydrolytic enzymes of different origins and function. The active site of AChE contains two subsites, the esteratic and the anionic site, corresponding to the catalytic site and the choline-binding site, respectively. There is also one other peripheral anionic binding site for acetylcholine and other quaternary ligands.

The mode of action of AChE resembles other serine hydrolases, involving the formation of an acyl-enzyme intermediate. The acylation step occurs as follows:

$$SerOH + (CH_3)_3 N^+ CH_2 CH_2 COOCH_3$$

$$SerOCOCH_3 + (CH_3)_3 N^+ CH_2 CH_2 OH$$

where SerOH represents the serine in the active site (see below).

In the presence of water the acylated enzyme is readily hydrolysed to release the free enzyme:

$$SerOCOCH_3 + H_2O \quad SerOH + CH_3COO^- H^+$$

[1] NSW Agriculture, Tamworth Centre for Crop Improvement, RMB 944 Calala Lane, Tamworth, NSW, Australia 2340
[2] Department of Biological and Ecological Chemistry, IACR-Rothamsted, Harpenden, Herts. AL5 2JQ, UK

Isaac Ishaaya (Ed.): Biochemical Sites of Insecticide Action and Resistance
© Springer-Verlag Berlin, Heidelberg 2001

AChE contains a catalytic triad of amino acid residues. Studies on the enzyme isolated from the electric ray, *Torpedo californica*, have shown that during hydrolysis a negatively charged glutamate (E^{327}) draws a hydrogen atom from an adjacent histidine (H^{440}) which, in turn, draws a hydrogen atom from the active site serine (S^{200}). This activates the serine residue enabling a powerful nucleophilic attack on the substrate, resulting in an intermediate acylated enzyme. The covalent bond thus formed is subsequently hydrolysed and acetate or, in the case of butyrylcholinesterase, butyrate, is released. The last stage of catalysis requires a residue in the vicinity of the active site to activate a water molecule, which attacks the bound acid moiety, activates it, and transforms the resultant complex into a transition state with a new tetrahedral conformation and high energy level. This leads to hydrolysis, and a reduction in the level of energy in the system. The key water molecule involved in catalysis is held in place by E^{443} and E^{199} (Soreq et al. 1992).

Further studies on the AChE from *Torpedo* have revealed that the active site is located within a deep and narrow gorge, which penetrates halfway into the enzyme and widens out close to the base (Sussman and Silman 1992). This gorge is often referred to as the "aromatic gorge" because a substantial part of its lining is composed of the rings of aromatic amino acid residues (Axelson et al. 1994). It is thought that some of the aromatic residues facilitate the movement of the substrate to the active site. Docking studies show that the primary interaction of the quaternary group of acetylcholine is with the indole of a conserved tryptophan, Trp^{84}. It has been suggested that the substrate may then move along to the active site by a mechanism of electrostatic guidance (Sussman and Silman 1992). This is made possible because the enzyme possesses a strong electrostatic dipole that is aligned with the gorge, so that a positively charged substrate would be drawn to the active site by the electrostatic field. The affinity of the quaternary ammonium compounds for aromatic rings, combined with the electrostatic force, create an efficient guidance mechanism for the substrate (Ripoll et al. 1993).

Insect AChE is found in one main form, a globular disulphide-linked dimer of 150 kDa. It is glycosylated and linked to the membrane via a glycosyl-phosphatidylinositol (GPI) anchor (Gnagey et al. 1987). In *Drosophila*, each subunit is composed of two non-covalently linked peptides of 18 and 55 kDa. The 55-kDa peptide contains the active-site serine and the GPI anchor (Fournier et al. 1992b).

2
Acetylcholinesterase as a Resistance Mechanism

The reaction of AChE with organophosphate inhibitors follows pseudo-first-order kinetics with respect to inhibitor concentration. The serine residue in the active site is progressively phosphorylated as follows:

$$K_d \quad k_2 \quad k_3$$

$$E + AX \quad E - AX \quad EA + X \quad E+A$$

$$k_i$$

Where E is the free enzyme, AX is the organophosphate, E–AX the Michaelis complex and EA the phosphorylated enzyme. K_d is the dissociation constant of the enzyme inhibitor complex and k_2 the phosphorylation rate constant. When researching AChE sensitivity and its role in conferring resistance, the three most relevant kinetic parameters to study are the bimolecular rate constant (k_i), which defines the sensitivity of the enzyme to a given inhibitor, and may be defined as k_2/K_d from the above equation; the Michaelis constant, K_m, a measure of the affinity of the enzyme for its substrate; and the reactivation constant (k_3).

The lower the value of k_i, the more inherent insensitivity the enzyme possesses to the inhibitor. Measurement of this constant is the best technique for comparing the sensitivity of AChE in susceptible and resistant variants. The lower the value of K_m the higher the affinity of the enzyme for its substrate. A high affinity of the AChE for its substrate can protect the enzyme against an insecticide (substrate protection) because of the competition between substrate and inhibitor for binding to the catalytic centre. Thus, if two AChE variants had the same k_i but differing values for K_m, the variant with the lower K_m would exhibit the greater insensitivity. The reactivation rate constant of inhibited enzyme (k_3) may be an important resistance factor after inhibition by carbamate insecticides, where recovery of AChE activity is usually faster than after inhibition by organophosphates.

The existence of an AChE exhibiting insensitivity to an organophosphate insecticide was first discovered in a paraoxon-resistant strain of red spider mite, *Tetranychus urticae* (Smissaert 1964). The resistance to paraoxon was associated with a decreased ability by the AChE to hydrolyse acetylcholine. This decreased activity can be considered as a disadvantage to the organism that can be tolerated, and is caused by the same modification of the enzyme as that which confers the insensitivity to the insecticide.

Insect species may possess more than one insensitive form of AChE, each showing different insensitivity profiles, and each having its own characteristic kinetic parameters. For example, two insensitive AChE variants differing in their insensitivity to paraoxon have been described in the tobacco whitefly, *Bemisia tabaci* (Byrne and Devonshire 1993). Similarly, insensitive variants were reported in Colorado potato beetle, *L. decemlineata*, where the greatest differences between the two insensitive enzyme forms were found in response to carbofuran and azinphosmethyl, the two insecticides used to select for resistance (Wierenga and Hollingworth 1993). A further variant discovered in the mosquito, *C. tritaeniorhynchus*, gave an extremely high insensitivity factor, 7000-fold, against dichlorvos (Mamiya et al. 1997).

In the case of the housefly, multiple insensitive forms of the enzyme have been found. Some of these insensitive forms had a higher maximal rate of reaction (V_{max}) than the original, susceptible form of the AChE. Thus, the modifications that conferred insensitivity in these cases also bestowed an increase in the enzymatic activity (Devonshire and Moores 1984).

It is commonly the case that the mutation conferring insensitivity also decreases the affinity of the protein for its natural substrate. This decrease in affinity is not surprising, since resistance often results in a decreased affinity for organophosphates or carbamates, which are substrate analogues. However, insensitive forms have also been found that have a lower K_m (higher affinity for ATChI) than the susceptible form, enhancing the resistance conferred by a lower k_i. These have been reported in housefly (Devonshire and Moores 1984; Price 1988) and in *C. tritaeniorhynchus*, where the K_m for ACh in the resistant strain was 0.65 that of the susceptible, although the affinity for ATChI was reduced (Mamiya et al. 1997).

This resistance mechanism has since been reported in a variety of insect and acarine species including cattle tick, *Boophilus microplus* (Roulston et al. 1968), the green rice leafhopper, *Nephotettix cincticeps* (Hama and Iwata 1971), the diamondback moth, *Plutella xylostella* (Noppun et al. 1987), and several species of mosquito, *Anopheles albimanus*, *A. atroparvus*, *A. sacharovi*, *A. nigerimus*, *Culex pipiens*, *C. tritaeniorhynchus* and *C. quinquefasciatus* (cited in ffrench-Constant and Bonning 1989).

Modification of insect AChEs as a mechanism of resistance to insecticides has formed the subject of several reviews (e.g., Hama 1983; Fournier and Mutero 1994) and it is not our intention to duplicate these works. Instead, we will concentrate on the status of insensitive AChE as a resistance mechanism in lepidopteran species.

3
Insensitive AChE in Lepidopteran Species

Some Lepidoptera, and in particular Noctuidae, are major pests of crops throughout the world. These insects are routinely exposed to organophosphate and carbamate insecticides and it is therefore not surprising that a number of species have evolved target-site resistance against these xenobiotics. Organophosphate-insensitive AChE has been recorded in the following lepidopteran pests: *Plutella xylostella* (Yponomeutidea), *Spodoptera littoralis* (Noctuidae), *Heliothis virescens* (Noctuidae) and *Helicoverpa armigera* (Noctuidae). Carbamate-insensitive AChE has been reported in: *S. litoralis*, *Spodoptera frugiperda* (Noctuidae), *H. virescens*, *H. armigera*, *Phyllonorycter blancardella* (Gracillariidae). Table 1 shows details of the resistance status of these insects.

Table 1. Resistant AChEs in Lepidoptera

Species	Insecticide	Reference
Plutella xylostella	Paraoxon	Tang and Zhou (1992)
Plutella xylostella	Fenitroxon	Konno and Shishido (1994)
	Fenthion-oxon	
	Pyridaphenthion-oxon	
	Isoathion-oxon	
	Pyrimiphosmethyl-oxon	
	Chlorpyrifosmethyl-oxon	
	Tetachlorovinphos	
	Dimethylvinphos	
	Monocrotophos	
	Dichlorvos	
Spodoptera littoralis	Methyl paraoxon	Zaazou et al. (1973)
Spodoptera littoralis	Paraoxon	Dittrich et al. (1979)
	Dichlorvos	
	Moncrotophos	
	Profenofos	
	Carbaryl	
	Aldicarb	
Spodoptera frugiperda	Carbaryl	Yu (1991, 1992)
Phyllonorycter blancardella	Methomyl	Pree et al. (1990)
	Azinphosmethyl-oxon	
Grapholita molesta	Carbfuran	Kanga et al. (1997)
	Carbaryl	
	Methomyl	
Heliothis virescens	Methyl paraoxon	Brown and Bryson (1992)
	Monocrotophos	Brown et al. (1996)
	Propoxur	
	Tetrachlorovinphos	Kanga and Plapp (1995)
	Methomyl	
Helicoverpa armigera	Methomyl	Gunning et al. (1996)
	Thiodicarb	
	Pirimcarb	Gunning (unpubl.)
	Methyl paraoxon	Gunning et al. (1998)
Helicoverpa punctigera	Methomyl	Gunning (unpubl.)

4
Insensitive AChE in *H. punctigera*

Toxicological and biochemical studies of methomyl-resistant and susceptible *Helicoverpa punctigera* larvae collected from Australian cotton were performed, using methods described fully by Gunning et al. (1996). Kinetic parameters were calculated as described by Gunning et al. (1998).

New data reporting methomyl resistance in *H. punctigera* and AChE insensitivity to methomyl in the resistant strain are presented in Tables 1–5 and Fig. 1. Bioassay data (Table 5) show that the resistant strain of *H. punctigera* was

Table 2. Molecular rate constants of AChE from susceptible and resistant Lepidoptera

Species	Insecticide	k_i S-AChE ($M^{-1} min^{-1}$)	k_i R-AChE ($M^{-1} min^{-1}$)	Reference
H. armigera (2nd-instar larvae)	Methyl paraoxon	3.8×10^5	3.6×10^3	Gunning et al. (1998)
H. armigera (2nd-instar larvae)	Methomyl	5.3×10^4	1.3×10^3	Gunning (unpubl.)
H. punctigera (2nd-instar larvae)	Methomyl	1.2×10^4	4.5×10^2	Gunning (unpubl.)
H. virescens (larval heads)	Paraoxon	6.89×10^5	–	Rose and Sparks (1984)
	Acephate	5.19×10^1	–	
	Methamidophos	3.26×10^2	–	
H. virescens (brain and ganglia)	Methyl paraoxon	9.3×10^1	3.8×10^1	Brown and Bryson (1992)
H. virescens (adult heads)	Tetrachlorvinphos	1.6×10^1	–	Kanga and Plapp (1995)
	Methomyl	1.7×10^1	–	
	Carbaryl	3.7×10^2	–	
S. littoralis (1st-instar larvae)	Paraoxon	2.3×10^6	4.3×10^5	Dittrich et al. 1979
	Dichlorvos	8.1×10^5	1.3×10^5	
	Monocrotophos	3.2×10^4	3.5×10^3	
	Profenfos	5.4×10^3	1.1×10^3	
	Carbaryl	3.0×10^5	2.1×10^5	
	Aldicarb	3.1×10^2	7.5×10^2	

approximately 12 times more resistant to methomyl compared with the susceptible strain. The effects of methomyl on AChE from mass homogenates of the susceptible and resistant strains are shown in Fig. 1. AChE from the resistant strain was clearly less sensitive to inhibition than the susceptible strain. Based on the bimolecular rate constants (k_i) for the susceptible and resistant strains (Table 2), AChE from the resistant strain was approximately 23 times less sensitive to inhibition than the susceptible. This is the first time that insensitive AChE has been reported in *H. punctigera*.

5
Forms of AChE in Lepidoptera

As stated previously, insect AChE is found in one main form, a globular disulphide-linked dimer of 150 kDa. However, modified forms may be distinguished by differential mobility in non-denaturing polyacrylamide gel electrophoresis (Bonning et al. 1991), thus suggesting the existence of different allelic variations. For example, in *H. armigera*, there are at least three forms of AChE which correspond to susceptibility, specific insensitivity to carbamates and specific

Table 3. Relationship between resistance ratio and AChE insensitivity to organophosphates and carbamates in Lepidoptera

Species	Insecticide	RR	IR	RR (k_i)	Reference
H. armigera	Methyl paraoxon	52	6.1	10.5	Gunning et al. (1998)
H. armigera	Methomyl	75	>500	40.8	Gunning (unpubl.)
	Thiodicarb	35	>500	–	
H. punctigera	Methomyl	12.3	23	26	Gunning (unpubl.)
H. virescens	Methyl paraoxon	65	–	21.6	Brown and Bryson (1992) (brain and ganglion)
H. virescens	Tetrachlorvinfos	–	2.99–4.06	–	Kanga and Plapp
	Methomyl	–	3.1–7.59	–	(1995) (adult heads)
	Carbaryl	–	4.99–7.37	–	
S. littoralis	Monocrotophos	137	–	4.9	Dittrich et al. (1979)
	Profenfos	6.1	–	9.1	(1st–instar larvae)
	Paraoxon	–	–	5.3	
	Dichlorvos	–	–	6.2	
	Carbaryl	–	–	1.4	
	Aldicarb	–	–	0.41	
S. littoralis	Methyl paraoxon	–	–	12	Zaazou et al. (1973)
S. frugiperda	Carbaryl	–	–	3.8	Yu (1991)
G. molesta	Carbofuran	–	33.0	–	Kanga et al. (1997)
	Carbaryl	–	43.8	–	(3rd instar larvae)
	Methomyl	–	3.3	–	
	Gluthoxon	–	6.3	–	
	Carbofuran	25	53	–	Kanga et al. (1997)
	Carbaryl	–	1757	–	(adult heads)
	Methomyl	–	8.0	–	
	Gluthoxon	–	10.5	–	
P. xylostella	Fenitroxon	87	5.4	–	Konno and Shishido
	Fenthion–oxon	34	2.8	–	(1994) (4 th instar
	Pyridaphenthion–oxon	52	7.3	–	larvae)
	Isoxathion–oxon	61	1.2	–	
	Pirimphosmethyl–oxon	6.6	1.1	–	
	Chlorpyrifosmethyl–oxon	1168	1.6	–	
	Tetrachlorovinphos	249	32	–	
	Monocrotophos	17	29	–	
	Dichlorvos	9.3	8.8	–	
P. xylostella	Paraoxon	–	–	127	Tang and Zhou (1992)

Table 4. Effects of AChE insensitivity on acetylcholine hydrolysis

Species	Insensitivity	AChE activity[a]	Reference
H. armigera	Methomyl	Increase	Gunning et al. (1996)
	Methyl paraoxon	Decrease	Gunning et al. (1998)
H. punctigera	Methomyl	Increase	Gunning (unpubl.)
G. molesta	Methomyl	Unaffected	Kanga et al. (1997)
	Guthoxon	Unaffected	
	Carbofuran	Unaffected	
H. virescens	Methyl paraoxon	Increase	Brown and Bryson (1992)
P. blancardella	Methomyl	Unaffected	Pree et al. (1990)
	Azinphosmethyl-oxon	Unaffected	

[a] Uninhibited resistant AChE activity with respect to susceptible AChE activity

Table 5. Toxicity of methomyl to 3rd-instar *Helicoverpa punctigera*

Strain	Slope	LD_{50} (µg/larva)	Fiducial limits	X^2	RF
Susceptible	3.4	0.048	0.032–0.069	0.34	–
Resistant (sel)	2.8	0.59	0.46–0.77	0.46	12.3

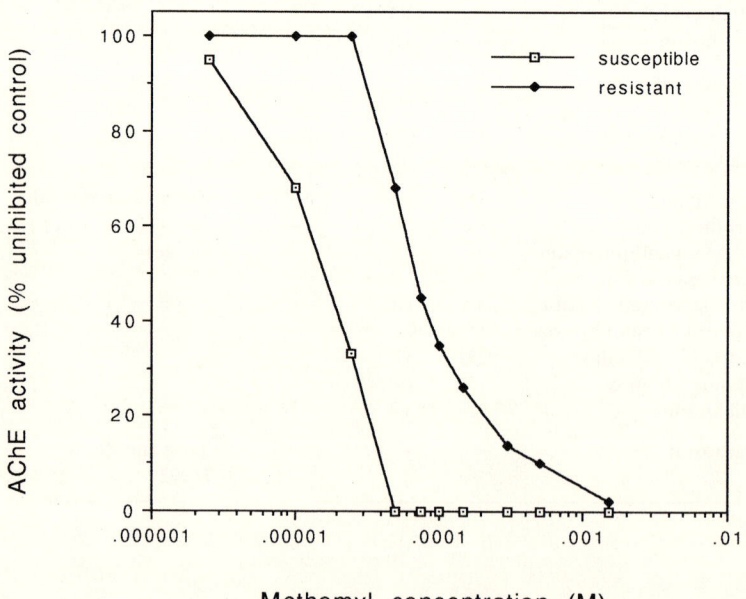

Fig. 1. Inhibition of acetylcholinesterase by methomyl-susceptible and methomyl-resistant strains of *Helicoverpa punctigera*

insensitivity to methyl paraoxon. Forms of AChE from *H. armigera* resistant and susceptible to methomyl are electrophoretically indistinguishable, whilst AChE insensitive to methyl paraoxon is clearly separated from the susceptible form (Gunning, unpubl.).

Calculations of the bimolecular rate constant provide a reliable measure of AChE sensitivity to inhibition. Differences in k_i or other kinetic properties from a resistant strain would indicate differing forms of the enzyme (Fournier and Mutero 1994). The bimolecular rate constants for a number of susceptible and resistant lepidopteran species are summarized in Table 2. Reduced AChE sensitivity to inhibition by organophosphates, indicated by lower k_i values when compared with a corresponding susceptible enzyme, was found in *H. armigera* (methyl paraoxon), *H. virescens* (methyl paraoxon) and *S. littoralis* (paraoxon, dichlorvos, monocrotophos and profenofos) (Dittrich et al. 1979; Brown and Bryson 1992; Gunning et al. 1998). Values of k_i against organophosphates within and between species (Table 2) are variable but generally indicate a high sensitivity to compounds such as paraoxon and methyl paraoxon (see also Rose and Sparks 1984). In *H. virescens*, k_i values for paraoxon and methyl paraoxon (Rose and Sparks 1984; Brown and Bryson 1992) differed by many orders of magnitude, possibly because of contrasting means of enzyme preparation.

Bimolecular rate constants for inhibition of AChE by carbamates indicated reduced sensitivity in resistant *H. armigera* (methomyl), *H. punctigera* (methomyl) and *S. littoralis* (aldicarb) (Dittrich et al. 1979; Gunning et al. 1998; Gunning, unpubl.). Susceptible AChE from *S. littoralis* shows a greater sensitivity for carbaryl compared with aldicarb, whilst carbaryl remains a potent inhibitor against susceptible AChE from *H. virescens*.

6
Effects of Altered AChE on Acetylcholine Hydrolysis

Uninhibited AChE activity (or acetylcholine hydrolysis) is commonly reduced in modified AChE (Fournier and Mutero 1994), although this is not the case in *M. persicae* nor several resistant strains of *M. domestica* (Devonshire and Moores 1984; Moores et al. 1994). In the case of the housefly, multiple insensitive forms of the enzyme have been found; some of these insensitive forms had a higher maximal rate of reaction (V_{max}) than the original, susceptible form of the AChE. Thus, the modifications that conferred insensitivity in these cases, also bestowed an increase in the enzymatic activity (Devonshire and Moores 1984).

In organophosphate-resistant *H. armigera*, activity of the insensitive enzyme was somewhat decreased, compared with the susceptible (Gunning et al. 1996, 1998; Gunning, unpubl.), but in carbamate-resistant *H. armigera*, *H. punctigera* and paraoxon-resistant *H. virescens*, there was an increase in activity in the resistant AChE (Brown and Bryson 1992). Furthermore, in *C. molesta*, *and P. blancardella,* resistant AChE activity was found to be unaltered, compared with the susceptible (Pree et al. 1990; Kanga et al. 1997).

Quantitative changes in AChE from field populations have never been reported, although experiments have shown that quantitative changes can affect the resistant status of individuals (Fournier et al. 1992a). Should enhanced AChE activity be found in strains which possess the same bimolecular rate constant values and the same K_m value as a corresponding susceptible strain, it may be conceivable that overproduction of the enzyme has occurred, and that this could result in resistance.

7
Inhibition Ratios and Toxicity in Lepidoptera

The relationship between AChE insensitivity and insecticide resistance is a difficult issue to resolve (Table 3). Any correlation between the resistance ratio and inhibition ratio (calculated either from the IC_{50} or k_i) is difficult to make in organophosphate-resistant *P. xylostella* and *S. littoralis* (Dittrich et al. 1979; Konno and Schishido 1994) and it is very probable that altered AChE may have been accompanied by other mechanisms of resistance. Similarly, in *H. virescens,* it is known that other mechanisms of resistance to organophosphates are present (Konno et al. 1989, 1990; Harold and Ottea 1997).

Methomyl and thiodicarb resistance in *H. armigera* was clearly correlated to high levels of AChE insensitivity (Gunning et al. 1996); moreover, assays showed no significant differences between the frequency of resistant *H. armigera* detected by insensitive AChE or by discriminating dose bioassay (Gunning et al. 1997, 1999). Methyl parathion resistance (52x) in *H. armigera* was associated with somewhat lower IR and RR_{ki} values; however, this can be attributed to the K_m (Table 6), where the greater affinity of the modified AChE for its natural substrate is conferring greater resistance than would be expected from the k_i values alone. Furthermore, the frequency of individuals carrying insensitive AChE correlated well to resistance frequency calculated from discriminating dose bioassay (Gunning 1999). In *H. punctigera,* the low level of methomyl resistance (12.5x) was associated with commensurate IR and RR_{ki} values.

Table 6. K_m of AChE from organophosphate- and carbamate-susceptible and -resistant *H. armigera* and carbamate-resistant and -susceptible *H. punctigera*[a]

Species	Insecticide	K_m S-AChE	K_m R-AChE (μM)	Reference
H. armigera	Methyl paraoxon	180	45	Gunning et al. (1998)
	Methomyl	180	30	Gunning (unpubl.)
H. punctigera	Methomyl	10	4	Gunning (unpubl.)

[a] Substrate acetylthiocholine iodide

8
Cross Resistance Between Organophosphates and Carbamates in Lepidoptera

While in some insects resistance due to an altered AChE can produce a broad range of resistance amongst organophosphates and carbamates (Fournier and Mutero 1994), in most Lepidoptera results suggest that insensitivity is rather more specific to either carbamates or organophosphates.

In *H. virescens*, Kanga and Plapp (1995) showed insensitivity to both organophosphate and carbamate insecticides; however, it is not known if these findings were a result of cross-resistance or multi-resistance.

For methomyl-resistant *H. armigera* and *H. punctigera*, altered AChE insensitivity is restricted to carbamates, with no cross-resistance to organophosphates (Gunning et al. 1996, 1998). Similarly, *H. armigera* strains demonstrating organophosphate insensitivity are fully sensitive to carbamates. Furthermore, insensitivity appears to be specific to methyl paraoxon and profenofos, with complete susceptibility to chlorpyrifos. Thus, tolerance of AChE for some organophosphates does not necessarily confer insensitivity across the whole class of compounds.

In *S. littoralis* it is interesting that the modified AChE that confers significant insensitivity to several organophosphates confers little insensitivity to carbaryl and appears to be four-fold hypersensitive to aldicarb.

9
Genetics of Resistance in Lepidoptera

Studies have shown that a single gene appears to be responsible for AChE insensitivity to insecticides (Fournier and Mutero 1994). In biochemical assays, altered AChE seems to be "incompletely dominant", although heterozygotes can appear to be as resistant as homozygotes in insecticide bioassays (Hama and Iwata 1978). This "effective dominance" may have been as a result of insects needing only a fraction of their AChE for viability (Smissaert et al. 1975). In Lepidoptera, information on the genetics of insensitive AChE is limited, but data for *Heliothis* and *Helicoverpa* spp. also suggest a single, incompletely dominant gene which may or may not be effectively dominant under exposure to insecticide.

In *H. virescens*, AChE insensitivity to methyl paraoxon appears to be a single, dominant gene (Brown et al. 1996), since individuals can be assigned to one of three genotypes, susceptible homozygote, heterozygote and resistant homozygote. In *H. virescens*, there appears to be only one resistant AChE allele, but molecular genetic studies (Brown et al. 1996) suggest that several mutations may be required in the structural gene to produce increasingly resistant forms. Heckel et al. (1998) characterized an autosomal locus controlling AChE insensitivity in paraoxon-insensitive *H. virescens* and mapped it to linkage group 2.

In *H. armigera*, two different forms of AChE, insensitive to organophosphates and carbamates, respectively, seem to be controlled by alternative alleles at the same locus (Gunning et al. 1996, 1998). Carbamate resistance is autosomal, semi-dominant and heterozygotes have a reduced resistance factor compared with homozygotes. Heterozygotes are, therefore, considerably easier to control in the field. Organophosphate-insensitive AChE is currently being investigated. However, it also appears to be autosomal, semi-dominant, and heterozygotes are more susceptible to high doses of insecticide than homozygotes (Gunning, unpubl.). Methomyl-insensitive AChE in *H. punctigera*, which is presently being researched, also appears semi-dominant and sex-linked.

10
Fitness of Resistance in Lepidoptera

Reduced fitness is commonly associated with resistance genes. However, in the Lepidoptera studied so far, there is little evidence of any fitness deficit associated with insensitive AChE, at least in heterozygotes, since mutant AChE seems to have quickly replaced the wild-type enzyme in natural populations.

In *H. virescens*, methyl-paraoxon-insensitive AChE was found to be a fully functional enzyme and apparently not deleterious to the resistant strain (Brown and Bryson 1992). Field monitoring demonstrated its widespread nature and, furthermore, the resistance mechanism persisted in the field long after withdrawal of the use of methyl parathion against *H. virescens* in the USA (Brown et al. 1996). Nevertheless, these authors also noted a lack of resistant homozygotes when compared with frequencies predicted from Hardy-Weinberg statistics and this may indicate some lack of viability in homozygotes.

Studies of carbamate-resistant *H. armigera* also indicate that altered AChE, at least in the heterozygous form, does not appear to confer any fitness deficit in the field. Since the identification of altered AChE in this species in 1994, frequency of the mechanism in field populations has increased in spite of restrictions on carbamate use (Gunning et al. 1997, 1999); however, there is a lack of resistant homozygotes compared with frequencies predicted by Hardy-Weinberg statistics (Gunning 1999). There also seems to be no reduction in fitness of *H. punctigera* carrying carbamate-insensitive AChE in Australia. The frequency of insensitive AChE is widespread in field populations and is increasing (Gunning 1999). Conversely, methyl-paraoxon-insensitive AChE, in Australian *H. armigera*, appears to have a real fitness deficit resulting in slower larval growth (Gunning et al. 1998). In Australia, despite greatly expanded use of organophosphates against *H. armigera* on cotton (due to poor control with other insecticides and variable performance of transgenic cotton). the frequency of modified AChE able to confer organophosphate resistance remains low overall. This may be as a consequence of the lack of fitness in individuals carrying this form of insensitive AChE.

11
Evolution

Carbamates have been registered for the control of *H. armigera* in Australia since the mid-1970s, but were rarely used on cotton because more effective insecticides soon became available. However, as resistance to other chemicals increased, so did the use of alternative chemicals such as thiodicarb and methomyl. Carbamate resistance in *H. armigera* due to altered AChE first arose in sweet-corn crops after intensive selection with methomyl and thiodicarb (Gunning et al. 1996). Resistance was further exacerbated by the use of ovicidal rates of methomyl and a lack of interest in resistance management. *H. punctigera* has also been subjected to considerable selection pressure, both direct and indirect, from carbamates on cotton and other crops in Australia. While it was generally assumed that ecological factors would prevent the development of resistance in *H. punctigera* (Fitt 1989), pyrethroid, endosulfan and carbamate resistance have all been detected in field populations (Gunning et al. 1997; Gunning 1999). Organophosphate-insensitive AChE in *H. armigera* was first diagnosed in 1997 (Gunning et al. 1998). This occurred after increased use of profenofos and methyl parathion due to poor control with other insecticides and the mediocre performance of transgenic cotton. However, low levels of resistance to profenofos had been known for some years previously (Gunning and Easton 1993).

In *H. virescens*, methyl parathion resistance resulted from significant selection pressure over many years until the cessation of methyl parathion use on cotton (Brown et al. 1996).

12
Control of Altered AChE in Lepidoptera

Examples from Lepidoptera show that selection pressure can rapidly lead to an increase in the frequency of individuals carrying mutant AChE, and that such mechanisms can result in significant control problems. Nonetheless, an insensitive AChE is not beyond the scope of resistance management, nor are the ways to circumvent this resistance mechanism.

Insecticide tolerance in Australian *H. armigera* has been limited by an insecticide resistance management strategy (Forrester et al. 1993); however, management tactics limiting methomyl and thiodicarb use were introduced too late to halt the escalation of resistant AChE in field populations (Gunning et al.

1997). Pree et al. (1990), however, reported success in controlling carbamate-resistant *P. blancardella*, by accurately timing sprays to the most sensitive life stage and by restricting carbamate use.

The apparent lack of resistant homozygotes in field populations of methyl-parathion-resistant *H. virescens* (Brown et al. 1996) and carbamate-resistant *H. armigera* (Gunning 1999) suggests that the RR mutations may be less viable. If this were true, field control would be much facilitated. In Australia, reduced fitness in methyl-parathion- and profenofos-resistant *H. armigera* may help reduce the numbers of individuals carrying this insensitive AChE. Furthermore, this AChE is fully susceptible to another widely used organophosphate, chlorpyrifos (Gunning et al. 1998).

The idea of designing compounds to overcome insensitive AChE is not new, and deserves further consideration for Lepidoptera. This, however, is difficult without knowledge of the structure of the enzyme and the mutations involved in resistance. Attempts at this approach have already been pursued in *P. xylostella* and *H. virescens* by Konno and Shishido (1994), who explored the relationship between chemical structure of organophosphate insecticides and AChE insensitivity. Results indicated that the altered AChE was only specifically insensitive to P-O-aliphatic organophosphates. In *H. virescens*, organophosphate-resistant AChE did not seem to be based on reduced vulnerability to phosphorylation, as resistance could be reduced by altering the R moiety specifically, so that it contained a 4-nitrophenyl leaving group (Brown and Bryson 1992). Clearly, knowledge of the spatial structure of the enzyme surrounding the active site, and mutations involved in resistance, will enhance understanding of structure-activity relationships and could lead to a more directed effort at designing chemicals to overcome insensitivity.

13
Population Genetics and Monitoring

Resistance monitoring and population genetics studies are a prerequisite for appraising the efficacy of any resistance management strategy. Sensitive biochemical tests for specific insecticide resistance mechanisms in single insects can provide a powerful means of monitoring the effect of control treatments and assessing resistance management tactics.

The Ellman reaction (Ellman et al. 1961), using an artificial substrate acetylthiocholine iodide and 5,5'-dithio-bis(2-nitrobenzoic acid), produces a brilliant yellow coloration in the presence of a small amount of AChE. This technique can be used to detect an insensitive AChE if duplicate samples from an insect are used, one containing a diagnostic concentration of insecticide in the reaction mix, and the other containing only substrate. Even visual assessments can sometimes differentiate sensitivity variants and this technique has been automated for resistance detection in several insect species (Raymond et al. 1985; Hemmingway et al. 1986; Brogden and Barber 1987). This approach was further improved and adapted for a microplate reader system utilizing kinetic software (Moores et al. 1988; ffrench-Constant and Bonning 1989), allowing clear resolution of resistant genotypes. Dot-blot tests, which can be used in the field, have also been developed for the identification of resistant AChE in single insects (Dary et al. 1991).

In Lepidoptera, microplate-based techniques which identify insensitive AChE have been successfully used in resistance monitoring programmes for *H. virescens* in the USA and for *Helicoverpa* spp. in Australia (Brown et al. 1996; Gunning et al. 1997, 1999). These methods have also been further adapted in Australia to produce field-based resistance detection kits for use with individual *H. armigera* and *H. punctigera* (Gunning et al. 1999). Results from the kits, which can be read by eye, provide satisfactory resolution of genotypes, and promise to be a very useful tool for resistance management.

14
Conclusions

Insecticide resistance, mediated by insensitive AChE, has yielded much interesting information about Lepidoptera. It is clear that altered AChE has evolved independently in a number of species and especially in major noctuid crop pests. Examination of the patterns of cross resistance, kinetics of insensitive enzymes, fitness of mechanisms and its contribution to toxicity, reveal many differences in the properties of altered AChE, even amongst closely related species. The evidence discussed here certainly does not support the concept of a common, single mutation giving rise to insensitive AChE in various species of Lepidoptera. Data instead indicate that insensitive AChE is largely species specific, and the profiles of the sensitive enzymes sometimes contrast sharply to the findings of Voss (1981) who concluded that susceptible AChE in Lepidoptera was very uniform, at least with respect to the inhibition patterns obtained. Fournier and Mutero (1994) postulated that there may be several weak mutations of an underlying AChE gene manifesting in various combinations to produce differing forms of insensitive AChE, depending on the insect species. This existence of multiple forms of insensitive AChE has been well documented in some insects, e.g. *Musca domestica*, and the underlying mutations characterized (Williamson et al. 1992) which lends strong support to the hypothesis. Clearly, this is an area of research that requires further study in the Lepidoptera.

Acknowledgements. The authors acknowledge the technical assistance of M.E. Balfe and N.A. Coleman (both of NSW Agriculture). This work was partially funded by the Cotton Research and Development Corporation, Australia. IACR-Rothamsted receives grant-aided support from the Biological Sciences Research Council of the United Kingdom.

References

Axelson PH, Harel M, Silman I, Sussman JL (1994) Structure and dynamics of the active site gorge of AChE. Protein Sci 3:188–197

Bonning BC, Hemmingway J, Romi R, Majori G (1991) Interaction of insecticide resistant genes in field populations of *Culex pipiens* (Diptera: Culicidae) from Italy in response to changing insecticide selection pressure. Bull Entomol Res 88:5–10

Brogdon WG, Barber AM (1987) Microplate assay of acetylcholinestase inhibition kinetics in single mosquito homogenates. Pestic Biochem Physiol 29:252–259

Brown TM, Bryson PK (1992) Selective inhibitors of methyl parathion resistant acetyl-cholinesterase from *Heliothis virescens*. Pestic Biochem Physiol 44:155–164

Brown TM, Bryson PK, Arnette MR, Mallett JBG, Nemec SJ (1996) Surveillance of resistant acetyl-cholinesterase in *Heliothis virescens*. In: Brown TM (ed) Molecular genetics and pesticide resistance. ACS symposium series no 645. American Chemical Society, Washington, DC, pp 149–157

Byrne FJ, Devonshire AL (1993) Insensitive acetylcholinesterase and esterase polymorphism in susceptible and resistant populations of the tobacco whitefly *Bemisia tabaci* (Genn.). Pestic Biochem Physiol 45:34–42

Dary O, Georghiou GP, Parsons E, Pasteur N (1991) Dot-blot test for identification of insecticide resistant acetylcholinesterase in single insects. J Econ Entomol 84:28–33

Devonshire AL, Moores GD (1984) Different forms of insensitive acetylcholinesterase in insecti-cide-resistant houseflies (*Musca domestica*). Pestic Biochem Physiol 21:336–340

Dittrich V, Luetkemeier N, Voss G (1979) Monocrotophos and profenofos: two organophosphates with a different mechanism of action in resistant races of *Spodoptera littoralis*. J Econ Entomol 72:380–384

Ellman GL, Courtney KD, Andres JV, Featherstone RM (1961) A new and rapid colorimetric deter-mination of acetylcholinesterase activity. Biochem Pharmacol 7:88–95

ffrench-Constant RH, Bonning BC (1989) Rapid microplate test distinguishes insecticide resis-tant acetylcholinesterase genotypes in the mosquitoes *Anopheles albimanus*, *An. niggerimus* and *Culex pipiens*. Med Vet Entomol 3:9–13

Fitt GP (1989) The ecology of *Heliothis* species in relation to agrosystems. Annu Rev Entomol 34:17–52

Forrester NW, Cahill M, Bird LJ, Layland JK (1993) Management of pyrethroid and endosulfan resistance in *Helicoverpa armigera* (Lepidoptera: Noctuidae) in Australia. Bull Entomol Res [Suppl 1]:132 pp

Fournier D, Bride JM, Hoffman F, Karch F (1992) Acetylcholinesterase: two types of modifications confer resistance to insecticide. J Biol Chem 267:14270–14274

Fournier D, Mutero A, Rungger D (1992) Drosophila acetylcholinesterase: expression of a func-tional precursor in *Xenopus* oocytes. Eur J Biochem 203:513–519

Fournier D, Mutero A (1994) Modification of acetylcholinesterase as a mechanism of resistance to insecticides. Comp Biochem Physiol 108C:19–31

Gnagey AL, Forte M, Rosenberry TL (1987) Isolation and characterisation of acetylcholinesterase from *Drosophila*. J Biol Chem 262:13290–13296

Gunning RV (1999) Organophosphate and carbamate resistance in *Helicoverpa* spp. Final report to Cotton Research and Development Corporation, Australia. NSW Agriculture, Australia

Gunning RV, Easton CS (1993) Organophosphate resistance in *Helicoverpa armigera* (Lepi-doptera: Noctuidae) in Australia. Gen Appl Entomol 25:27–34

Gunning RV, Moores GD, Devonshire AL (1996) Insensitive acetylcholinesterase and resistance to thiodicarb in Australian *Helicoverpa armigera* Hübner (Lepidoptera: Noctuidae). Pestic Biochem Physiol 55:21–28

Gunning RV, Moores GD, Devonshire AL (1997) Biochemical resistance detection in *Helicoverpa armigera* in Australia. Rec Res Dev Entomol 1:203–214

Gunning RV, Moores GD, Devonshire AL (1998) Insensitive acetylcholinesterase and resistance to organophosphates in Australian *Helicoverpa armigera* Hübner (Lepidoptera: Noctuidae). Pestic Biochem Physiol 62:147–151

Gunning RV, Moores GD, Devonshire AL (1999) Biochemical resistance detection: a tool for resistance management in *Helicoverpa armigera*. Proceedings of regional consultation on insecticide resistance management in cotton, Multan, Pakistan, June 1999 (in press)

Hama H (1983) Resistance to insecticides due to reduced sensitivity of acetylcholinesterase. In: Georghiou GP, Saito T (eds) Pest resistance to pesticides. Plenum Press, New York, pp 299–331

Hama H, Iwata T (1971) Insensitive cholinesterase in the Nagawaka strain of the green rice leafhopper, *Nephotettix cincticeps* Uhler, as a cause of resistance to carbamate insecticides. Appl Entomol Zool 6:183–191

Hama H, Iwata T (1978) Studies on the inheritance of carbamate-resistance in the green rice leafhopper, *Nephotettix cincteps* Uhler (Hemiptera: Cicadellidae). Relationships between insensitivity of acetylcholinesterase and cross resistance to carbamate and organophosphorous insecticides. Appl Entomol Zool 15: 249–261

Harold JA, Ottea JA (1997) Toxicological significance of enzyme activities in profenofos-resistant tobacco budworms, *Heliothis virescens* (F.). Pestic Biochem Physiol 58:23–33

Heckel DG, Bryson PK, Brown TM (1998) Linkage analysis of insecticide-resistant acetylcholinesterase in *Heliothis virescens*. J Hered 89:71–78

Hemmingway J, Smith C, Jaywardena KGI, Herath C (1886) Field and laboratory detection of the altered acetylcholinesterase resistance genes which confer organophosphate and carbamate resistance in mosquitos (Diptera: Culicidae). Bull Entomol Res 76:559–565

Kanga LHB, Plapp FW Jr (1995) Target-site insensitivity as the mechanism of resistance to organophosphorus, carbamate and cyclodiene insecticides in tobacco budworms. J Econ Entomol 88:1150–1157

Kanga LHB, Pree DJ, van Lier JL, Whitty KJ (1997) Mechanisms of resistance to organophorus and carbamate insecticides in oriental fruit moth populations (*Grapholita molesta* Busck). Pestic Biochem Physiol 59:11–23

Konno Y, Shishido T (1994) A relationship between the chemical structure of acetylcholinesterase in the diamondback moth, *Plutella xylostella* L. (Lepidoptera: Yponomeutidae). Appl Entomol Zool 29:595–597

Konno T, Hodgson E, Dauterman WC (1989) Studies on methyl parathion resistance in *Helithis virescens*. Pestic Biochem Physiol 33:189–199

Konno T, Kassai Y, Rose RL, Hodgson EL, Dauterman WC (1990) Purification and characterisation of a phosphotriester hydrolase from methyl parathion resistant *Heliothis virescens*. Pestic Biochem Physiol 36:1–12

Mamiya A, Ishikawa Y, Kono Y (1997) Acetylcholinesterase in insecticide resistant *Culex tritaeniorhynchus*: characteristics accompanying insensitivity to inhibitors. Appl Entomol Zool 32:37–44

Moores GD, Devonshire AL, Denholm I (1988) A microtitre plate assay for characterising insensitive acetylcholinesterase genotypes of insecticide resistant insects. Bull Entomol Res 78:537–544

Moores GD, Devine GJ, Devonshire AL (1994) Insecticide-insensitive acetylcholinesterase can enhance esterase-based resistance in *Myzus persicae* and *Myzus nicotianae*. Pestic Biochem Physiol 49:114–120

Moores GD, Gao X, Denholm I, Devonshire AL (1996) Characterisation of insensitive acetylcholinesterase in insecticide-resistant cotton aphids, *Aphis gossypii* Glover (Homoptera: Aphididae). Pestic Biochem Physiol 56:102–110

Noppun V, Miyata T, Saito T (1987) Insensitivity of acetylcholinesterase in phenthoate resistant diamondback moth, *Plutella xylostella* (Lepidoptera: Yponomentidae). Appl Entomol Zool 22:116–118

Pree DJ, Archibald DE, Cole KJ (1990) Insecticide resistance in spotted tentiform leafminer (Lepidoptera: Gracillariidae): mechanisms and management. J Econ Entomol 83:678–685

Price NR (1988) Insecticide-insensitive acetylcholinesterase from a laboratory selected and a field strain of housefly (*Musca domestica* L.). Comp Biochem Physiol 90C:221–224

Raymond M, Fournier D, Bergé JB, Cuany A, Bride JM, Pasteur N (1985) Single mosquito test to determine genotypes with an acetylesterase insensitive to inhibition to propoxur insecticide. J Am Mosq Control Assoc 1:425–427

Ripoll DR, Faerman CH, Axelsen PH, Silman I, Sussman JL (1993) An electrostatic mechanism for substrate guidance down the aromatic gorge of acetylcholinesterase. Proc Natl Acad Sci USA 90:5128–5132

Rose RL, Sparks TM (1984) Acephate toxicity, metabolism and anticholinesterase activity in *Heliothis virescens* (F.) and *Anthonomus grandis grandis* (Boheman). Pestic Biochem Physiol 22:69–77

Roulston WJ, Schnitzerling HJ, Schuntner CA (1968) Acetylcholinesterase insensitivity in the Biarra strain of the cattle tick, *Boophilus microplus*, as a cause of resistance to organophosphate and carbamate insecticides. Aust J Biol Sci 21:759–767

Smissaert HR (1964) Cholinesterase inhibition in spider mites susceptible and resistant to organophosphate. Science 143:129–131

Smissert RH, Abd EI, Hamid FM, Overmeer WP (1975) The minimum acetylcholinesterase (AChE) fraction compatible with life derived by the aid of a simple model explaining the degree of dominance of resistance to inhibitors in AChE "mutants". Biochem Pharmacol 24:1043–1047

Soreq H, Gnatt A, Loewenstein Y, Neville LF (1992) Excavations into the active-site gorge of cholinesterases. Trends Biochem Sci 17:353–358

Sussman JL, Silman I (1992) Acetylcholinesterase: structure and use as a model for specific cation-protein interactions. Curr Opin Struct Biol 2:721–729

Tang ZH, Zhou CL (1992) Acetylcholinesterase sensitivity in resistant *Plutella xylostella* (L.). Acta Entomol Sin 35:385–446

Toutant J-P (1989) Insect acetylcholinesterase: catalytic properties, tissue distribution and molecular forms. Prog Neurobiol 32:423–446

Voss G (1981) Taxonomy-related cholinesterase patterns in insects. J Econ Entomol 74:555–557

Wierenga JM, Hollingworth RM (1993) Inhibition of altered acetylcholinesterases from insecticide-resistant Colorado potato beetles (Coleoptera: Chrysomelidae). J Econ Entomol 86:673–679

Williamson MS, Moores GD, Devonshire AL (1992) Altered forms of acetylcholinesterase in insecticide-resistant houseflies (*Musca domestica*). In: Shafferman A, Velan B (eds) Multidisciplinary approaches to cholinesterase functions. Plenum Press, New York, pp 83–86

Yu SJ (1991) Insecticide resistance in the fall armyworm *Spodoptera frugiperda* (J.E. Smith). Pestic Biochem Physiol 39:84–91

Yu SJ (1992) Detection and biochemical characterisation of insecticide resistance in the fall armyworm (Lepidoptera: Noctuidae). J Econ Entomol 85:675–682

Zaazou MH, Ali AM, Abdallah MD, Rizkallah MR (1973) In vivo and in vitro inhibition of cholinesterase and aliesterase in susceptible strains of *Spodoptera littoralis* (Biosd.). Bull Entomol Soc Egypt Econ Ser VII:25–30

Glutathione *S*-Transferases and Insect Resistance to Insecticides

Chih-Ning Sun[1], Shin-Yi Huang[1], Nien-Tai Hu[2] and Wei-Yuan Chung[1]

1
Introduction

Booth et al. (1961) were the first to report an enzyme that catalyzed the conjugation of 1,2-dichloro-4-nitrobenzene (DCNB) with glutathione in a cytosolic extract of rat liver. Since then, this group of enzymes, glutathione *S*-transferases (GSTs, EC 2.5.1.18), has been widely observed in many aerobic organisms. The roles of GSTs, and their biochemical and physiological characteristics have been well investigated, most intensively in mammals. Molecular biology studies since the mid-1970s have resulted in significant understanding of the structure of GST genes and regulation of their expression. The three-dimensional structures of GST proteins, established since the early 1990s, have helped further elucidate the evolution and function of these enzymes in biological systems.

This chapter will deal with insect GSTs as one of the major mechanisms that confer insecticide resistance, and discuss relevant studies in several insects, with a brief review of the general features of this group of enzymes.

In order to shorten the list of references, review articles are referred to, rather than the original papers, for the discussion on the general features of GSTs.

2
General Features of Glutathione *S*-Transferases (GSTs)

2.1
Roles

Glutathione *S*-transferases (GSTs) are a family of multi-functional enzymes that catalyze the nucleophilic addition of reduced glutathione (GSH) (*r*-glutamyl-cysteine-glycine) to electrophilic substrates (RX):

$$RX + GSH \rightarrow GSR + HX$$

---

[1] Department of Entomology, National Chung-Hsing University, Taichung, Taiwan 40227, Republic of China
[2] Graduate Institute of Biochemistry, National Chung-Hsing University, Taichung, Taiwan 40227, Republic of China

Isaac Ishaaya (Ed.): Biochemical Sites of Insecticide Action and Resistance
© Springer-Verlag Berlin, Heidelberg 2001

Formation of the more water-soluble conjugates (GSR), with further processing when necessary, facilitates the detoxification and eventual excretion of many hydrophobic endogenous and foreign substances from organisms (Mannervik and Danielson 1988; Clark 1989). An ATP-dependent efflux pump that mediates the export of glutathione conjugates from cells has been described (Ishikawa 1992). The GSTs also bind, with high affinity, a variety of hydrophobic compounds and serve as intracellular carrier proteins for the transport of these compounds. The covalent binding of certain ligands by GSTs, with consequent inactivation and immobilization, has been proposed as an additional protective role of this enzyme (Mannervik and Danielson 1988).

2.2
Biochemical and Physiological Characteristics

Multiple forms of GSTs that are able to catalyze reactions towards a large number of structurally diverse substrates have been observed in almost all kinds of aerobic organisms. An early attempt to classify different forms of GSTs was based on the enzymatic properties and/or physicochemical properties of the protein subunits (Mannervik and Danielson 1988). More recently, human cytosolic GSTs, on account of their substrate and inhibitor specificity, antibody cross-reactivity and primary structures, have been grouped into five classes, i.e., alpha, mu, pi, kappa and theta (Armstrong 1997). Squid lens proteins identified as GSTs have been proposed to constitute a separate class, sigma, which has not been identified in vertebrates (Wilce and Parker 1994). Members within each class exhibit similar monomer sizes (24–28 kDa, 210–223 amino acids), share high amino acid sequence identity (60–80%), and have distinctive but overlapping substrate specificity. This nomenclature has been extended to GSTs of all vertebrates and other organisms.

The cytosolic GSTs are either homodimers or heterodimers of subunits belonging to the same class, with two active sites per dimer that behave independently of each other (Wilce and Parker 1994). Three-dimensional structure studies have revealed that all GST subunits have the same basic protein folding, which consists of two domains.

Domain I, covering roughly the N-terminal one-third of the protein, has the primary function to provide the binding site for GSH. Domain II, at the C-terminal two-thirds of the protein, provides structural elements for the recognition of xenobiotic substrates and helps define the substrate selectivity of the various isozymes (Armstrong 1997). Therefore, the sequence of domain I is rather conserved and that of domain II is relatively diversified. While most members of class-alpha GSTs have chemically blocked N-termini, no post-translational modifications have been noted (Mannervik and Danielson 1988).

Toxic electrophiles that are produced intracellularly, such as quinones, organic hydroperoxides, epoxides, and alkenes, may function as "natural" substrates for GSTs (Mannervik and Danielson 1988). In addition to catalyzing the reactions towards a large number of structurally diverse model substrates and xenobiotics, GSTs bind reversibly non-substrate ligands in a passive detoxification process and as an intracellular transport system. These ligands may bind at the catalytic site or separate binding site not far away from the GSH and substrate binding sites, causing inhibition of the enzyme (Mannervik and Danielson 1988; Clark 1989).

Three microsomal GSTs are known, one of which appears to be involved in xenobiotic metabolism. The three proteins are about the same size (17 kDa), membrane-bound, and share some sequence identity. Their sequences bear no relationship to those of cytosolic GSTs (Armstrong 1997).

Despite the ubiquitous occurrence of GSTs, there is a high degree of variation of activity of this detoxifying enzyme among organisms. Cytosolic GSTs are most abundant in human, rat and mouse tissues. The occurrence of the different forms of GSTs varies greatly in an organ- or tissue-specific manner during the life cycle of an organism; and the expression of a specific GST form may be controlled by hormones or induced by foreign compounds (Mannervik and Danielson 1988). For example, the pi-class GST is the most widely distributed isozyme and usually the most abundant, and is predominant in placenta, erythrocytes, breast, lung and prostate. On the other hand, the alpha-class GST is the major isozyme in human liver and kidney (Wilce and Parker 1994). The expression of some GSTs is developmentally regulated, e.g., relatively high levels of pi-class GST observed in the fetal liver decreased rapidly after the first week of postnatal development to become undetectable in the adult liver (Daniel 1993).

The cytosolic GSTs are inducible by drugs, xenobiotics, food additives and natural dietary components in animals (Daniel 1993). The induction of rat liver class-alpha and -mu GSTs by xenobiotics is due to the transcriptional activation of the respective GST genes (Ding and Pickett 1985). Both constitutive and novel GSTs may be induced (Daniel 1993).

Induction of GSTs has been considered a major mechanism of protection against chemical stress and carcinogenesis (Daniel 1993). On the other hand, certain forms of GSTs, particularly the pi-class, are expressed at high levels in mammalian tumor cells while their over-expression is implicated in carcinogenesis and the development of resistance to drugs of cells and organisms (Daniel 1993).

2.3
Structure, Regulation, and Evolution of GST Genes

Sequence analysis of cDNA clones has revealed that alpha- and mu-class GST subunits are encoded by multigene families, and pi-class subunits are expressed from a single gene (Daniel 1993). cDNA clones of rat alpha-class

GSTs are >90% homologous, encode 221–223 amino acids, and the structural genes span a region of about 11 kb and contain seven exons. The multiple mu GST genes of rat span about 5 kb, encode subunits of 210 amino acids, and contain eight exons. A human mu-class GST gene is a polymorphic locus for which three alleles have been described. The pi-class GST genes of both rat and human span about 3 kb, contain seven exons, and encode a 210 amino acid protein.

A single *cis*-regulatory element, composed of two adjacent AP-1-like binding sites, and a *trans*-acting factor, Fos/Jun complex, has been identified for the basal and xenobiotics-inducible expression for rat and mouse alpha-class GSTs and a rat pi class GST (Daniel 1993).

Armstrong (1997) proposed that cytosolic and membrane-bound GSTs arose by convergent evolution, and the mammalian cytosolic GST superfamily was probably derived from a common ancestral gene by divergence. The possible evolutionary relationship of the six gene families has been proposed through an analysis of cDNA sequences, intron-exon boundaries of the genes, as well as the three-dimensional structures of the proteins (Armstrong 1997).

3
Insect GSTs

Insect GSTs have received much attention mainly because of their relevance to the metabolism of insecticides and the development of resistance to these control agents. Using mammalian GSTs as a basis for comparison, Clark (1989) reviewed in detail the enzymology of GSTs from non-vertebrate organisms, and concluded that non-vertebrate GSTs generally resemble very closely those from mammals. However, the amino acid sequences of most invertebrate GSTs show <40% identity to those of mammalian GSTs (Ranson et al. 1998).

3.1
Roles

The involvement of GSTs in the metabolism of organophosphorus (OP) insecticides was first proposed by Fukami and Shishido (1966). They found that, in rat and insect tissues, methyl parathion was demethylated by a soluble enzyme that required glutathione. Lewis (1969) reported deethylation of diazinon by housefly homogenate fortified with GSH. Dearylation of diazinon by GST in rat and insects was subsequently observed by Shishido et al. (1972). Many OP insecticides can be metabolized and detoxified by this enzyme (Motoyama and Dauterman 1980).

DDT dehydrochlorination is a major route of detoxification in some insects, notably the housefly and several species of mosquito (Oppenoorth 1985). The reaction is catalyzed by a soluble, GSH-requiring DDT-dehydrochlorinase

(DDTase). Clark and Shamaan (1984) provided the evidence to show that DDTase from the housefly is a GST. Now it is well established that some forms of GSTs possess DDTase activity (Tang and Tu 1994; Prapanthadara et al. 1995). In addition, GSTs are able to metabolize plant allelochemicals, and thus play an important role in the feeding strategy of phytophagous insects (Yu 1992).

A protein synthesized and stored primarily at the fat body of diapausing 2nd-instar larvae has been isolated from the spruce budworm, *Choristoneura fumiferana*, and identified by Feng et al. (1999) as a GST. It is absent in the midgut of active feeding 2nd- and 6th-instar larvae. They proposed that this GST might be involved in the metabolism of potentially toxic substances accumulated during diapause. A GST isozyme found in the fully developed ovary of *Aedes aegypti* was suggested to have either an unknown in vivo function in oocytes or a function of protecting the developing oocytes from a large amount of toxic compounds formed due to aerobic respiration (Grant 1991).

3.2
Biochemical and Physiological Characteristics

Yu (1996) compiled a list of insects from which GSTs have been purified. The list includes 13 species of lepidopterans, 6 dipterans, 2 coleoptrans and 1 each of dictyopteran and hymenopteran. In all these preparations, GSTs are either homo- or heterodimers with subunit masses from 20 to 35 kDa, and apparent pI from 3.8 to 9.7. While one form of GST is observed in some species, e.g., *Trichoplusiani* and *Ceratitis captitata*, several isoforms are detected in others, e.g., *Periplaneta america* and *Spodoptera frugiperda*. Yu (1992) observed multiple forms and higher activity of GSTs in highly polyphagous lepidopterous insects as compared with more specialized lepidopterans. Varied numbers of GST forms have been observed for the same insect species, e.g., one and four isoforms, respectively, for *Helicoverpa zea* (Chien and Dauterman 1991; Yu 1989), and this might reflect the differences between strains or result from incomplete characterization. The expression of some GST forms is tissue-specific. Yu (1999) observed six heterodimeric GSTs in the midgut, three homodimeric GSTs in the fat body, and five heterodimeric GSTs in the Malpighian tubules of larvae of *S. frugiperda*. The GST recently identified from the spruce budworm, *C. fumiferana*, is abundant in the fat body and absent in the midgut of 6th-instar larvae; and the diapausing 2nd-instar larvae contains higher levels of this GST than the feeding 2nd- and 6th-instar larvae (Feng et al. 1999).

Housefly GSTs have been grouped, according to their physicochemical and biochemical properties, into two immunologically distinct classes: class I GSTs with pI > 6.5 and apparent activity toward DCNB and several insecticides, and class II GSTs with pI < 6.5 and little or no activity toward DCNB (Clark et al.

1984, 1986). Members of the same class show at least 30–40% identity in amino acid sequences. GSTs of other insects have often been categorized into one of these two classes (Feng et al. 1999).

GSTs can be induced to varied extents by insecticides, drugs, host plants and allelochemicals, and species differences in GST inducibility have been observed in lepidopterous insects (Yu 1992). Although it was once thought that induction resulted in increased levels of the existing isoforms only, it is now evident that new GST isozymes can be induced by allelochemicals. Yu (1999) studied the induction by xanthotoxin of GSTs in different tissues of the fall armyworm, *S. frugiperda*. He found that the treatment resulted in 32-, 59-, and 76-fold increase in GST activity toward DCNB in Malpighian tubules, fat body, and midgut, respectively; and two new GST isoforms were produced in fat body while the isozyme composition in the other two tissues remained the same.

3.3
GSTs and Insecticide Resistance

Yang et al. (1971) found that GST-dependent cleavage of diazinon in resistant housefly was considerably higher than in the susceptible strain. Lewis and Sawicki (1971) showed that several OP insecticides were deethylated in resistant housefly with enhanced GST activity. Since then, much work has been carried out on the association of GSTs with insect resistance to insecticides, primarily OP insecticides. Higher GST activity towards model substrates, such as 1-chloro-2,4-dinitrobenzene (CDNB) and especially DCNB, is found in resistant strains as compared with susceptible strains (Dauterman 1985). In some other cases, increased GST metabolism of insecticides has been observed in resistant strains (Oppenoorth et al. 1977; Motoyama and Dauterman 1980; Kao and Sun 1991).

Some forms of GSTs also possess DDTase activity (Clark and Shamaan 1984), and thus contribute to DDT resistance in insects. In a DDT-resistant strain of *Anopheles gambiae*, all seven partially purified GSTs showed DDTase activity, one possessed 60% of total DDTase activity; and the DDTase activity of the GSTs in both resistant and susceptible strains of *A. gambiae* was correlated with the GST activity toward DCNB (Prapanthadara et al. 1995).

As stated above, GST activity can be induced to varied extents by many drugs, insecticides that are either substrates or non-substrates, host plants and some allelochemicals (Yu 1992), and both constitutive and new GST isoforms can be induced (Yu 1999). This induction may increase the tolerance of insects to some insecticides (Yu 1996).

3.4
Molecular Biology Studies

A total of 37 cDNAs of insect GSTs have been cloned since 1990 (Feng et al. 1999). Among them 31 are GSTs of dipterans, mostly *Anopheles gambiae, Drosophila melanogaster, and Musca domstica*, and 5 are GSTs of lepidopterans. Two distinct classes can be established by comparing the deduced amino acid sequences and they complement the classification already in use, i.e., class I and class II. While some members of the same class show a high degree of similarity (>90%), others may share <40% similarity. Class I insect GSTs appear to be closer to mammalian theta-class GSTs (Fig. 1) while class II GSTs are closer to sigma-class GSTs (Feng et al. 1999). Wilce et al. (1995) reported the crystal structure of a GST from the Australian sheep blowfly, *Lucilia cuprina*, the first theta-class GST for which a crystal structure was ever elucidated. While a tyrosine near the N-terminus strictly conserved in alpha-, mu-, and pi-class GSTs is considered to play a critical role in the reaction mechanism, the equivalent residue in the theta-class structure is not in the active site (Wilce et al. 1995).

Class I GSTs were once considered to be products of intronless genes (Toung et al. 1991, 1993). This conclusion has since been proved to be oversimplified (Huang 1998; Ranson et al. 1998; Lougarre et al. 1999). Current data suggest

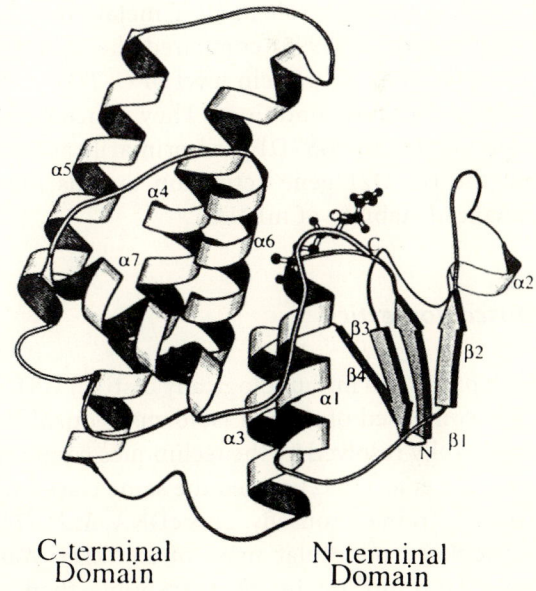

Fig. 1. Ribbon representation of crystal structure of a theta-class GST from *Lucilia cuprina*, the Australian sheep blowfly. (With permission from Wilcer et al. 1995)

that class II GSTs are products of single-copy genes with no closely related sequences present in the genome, and the better-studied class I GSTs are encoded by members of a large gene family (Ranson et al. 1998). This will be discussed in more detail below.

4
GST Studies of Several Insects

4.1
Drosophila melanogaster

A class II GST gene, *DmGST2*, mapped to chromosome 2 at 53F and with three introns, encodes a protein of 249 amino acids and has no close relatives within the genome (Beall et al. 1992).

A supergene family of class I GSTs, *DmGSTD* genes, with a cluster of at least eight intronless genes organized in divergent orientations within a ca. 60-kb DNA segment, has been located to chromosome 3R at 87B of the polytene chromosome (Toung et al. 1993). Two of the eight genes are probably GST pseudogenes, and the amino acid sequence identity among the functional genes ranges from 53 to 75%. The *E. coli* expressed recombinant *DmGSTD*1 (209 amino acids) and D21 (215 amino acids) are immunologically distinct from each other, and functionally active with different substrate preferences (Tang and Tu 1994). One of them, *DmGSTD*1, displays very low DDTase activity, and is implied to play a role in DDT metabolism in this insect.

Tang and Tu (1995) compared the steady-state level of mRNA, the transcription rate and protein level of *GSTD*1 and D21 under control and pentobarbital-induced conditions. They concluded that the inducer enhanced the expression of the *GSTD*1 gene primarily at the level of transcription initiation, but the *GSTD21* gene activation by this inducer was primarily due to the enhanced stability of mRNA.

4.2
Musca domestica

Each of the two immunologically distinct GST classes purified from the housefly is composed of several isoforms (Fournier et al. 1992). These forms could not be fully resolved by the techniques commonly used. Wang et al. (1991) and Fournier et al. (1992) cloned the same class I GST cDNA from the OP-resistant Cornell strain of housefly. This cDNA, *MdGST*1, encodes 208 amino acids with a calculated molecular mass of 23.6 kDa. Wang et al. (1991) detected more *MdGST*1 transcript in the susceptible than in the highly resistant Cornell strains, and concluded that this GST was not responsible for resistance. Yet, the over-transcription of the same GST gene was proposed to contribute at least partly to the insecticide resistance in the Cornell strain by Fournier et al. (1992). Syvanen et al. (1994) cloned, in addition to *MdGST1*, three new GST genes, *MdGST2, 3* and *4*, and their recombinant proteins expressed in *E. coli* exhibited CDNB conjugating activity. While all four GSTs were over-produced in Cornell-R as compared with susceptible strains, only *MdGST4* with greater activity toward DCNB was proposed as one of the genes involved in OP resistance. A subsequent report (Syvanen et al. 1996) suggested that *MdGST3* was the major factor contributing to OP resistance in the Cornell strain, and a total of 12 different *MdGST3* genes were detected in this strain. They proposed that these 12 genes had resulted from the amplification of the *MdGST3* gene and the sequence divergence between the amplified copies during the course of selection for insecticide resistance.

4.3
Anopheles gambiae

The major mechanism of DDT resistance in this insect has been attributed to increased levels of DDT-dehydrochlorination by GSTs (Hemingway et al. 1985), and the DDT-resistant ZANDS strain has eightfold higher DDTase activity than the susceptible G3 strain (Prapanthadara et al. 1995). All seven partially purified GSTs from this insect have shown varied DDTase activity, and one of them has 60% of total DDTase activity (Prapanthadara et al. 1995). The DDTase activity of the GSTs in both strains is correlated with the GST activity toward DCNB; and different subsets of GSTs are responsible for DDT resistance in larvae and adults of this mosquito. Each of the above-mentioned GSTs represents a fraction eluted from the column in sequential column chromatography, and presumably contains multiple GSTs. Thus, there are probably more forms of GST than the number known.

AgGST2-1, a cDNA isolated from the susceptible adults of this malaria-vector mosquito, was identified as a member of the class II GST gene family (Reiss and James 1993). Class I GSTs have been extensively studied in this insect vector primarily because of their involvement in DDT resistance. Partial or full-length cDNA sequences of six class I GSTs so far reported are localized to chromosome 2R division 18A and arranged sequentially in the genome (Ranson et al. 1998). Among them, *AgGST1-2*, an intronless gene from larvae, encodes a 209 amino acid protein with >40% sequence similarity to class I GSTs of *D. melanogaster* and *M. domestica*. The recombinant protein from this gene exhibits catalytic activity toward the model substrate CDNB (Ranson et al. 1997). Two additional GST genes, *AgGST1α* and *AgGST1β*, are sequentially arranged with *AgGST1-2* on the chromosome in divergent orientations (Ranson et al. 1998). The *AgGST1α* contains five coding exons that are alternatively spliced to produce four mature GST transcripts, AgGST1-3, 1-4, 1-5 and 1-6. Each of these four GSTs contains a common 5' exon encoding the N-terminus of the protein spliced to one of four distinct 3' exons encoding the C-terminus. A full-length GST transcript, AgGST1-7, identified from an adult female mosquito, is a product of splicing of exon 1 of *AgGST*1β to one or more 3' exons. The recombinant proteins expressed in *E. coli* from cDNAs of AgGST1-5 and AgGST1-6 isolated from a DDT-resistant strain of this mosquito are able to catalyze the dehydrochlorination of DDT.

As mentioned above, while the highly conserved N-terminal domain of GST provides the GSH binding site, the remaining two-thirds C-terminus provides xenobiotic substrate binding site (Armstrong 1997). The divergence in the C-terminal domain confers the variation in substrate specificity of GSTs, and this variation is obviously essential to the development of insecticide resistance. The genomic organization of GST genes and the presence of alternative RNA splicing in generating multiple functional transcripts in *A. gambiae* provide a mechanism of increasing the diversity of this detoxifying enzyme produced with minimal increase in the length of genome in this mosquito (Ranson et al. 1998).

In view of the complexity of the GST profiles in *A. gambiae*, more GST genes or alternative splice variants of the genes remain to be cloned.

4.4
Plutella xylostella

The diamondback moth (DBM), *Plutella xylostella*, an insect pest of cruciferous vegetables (Talekar and Shelton 1995), has evolved high levels of resistance to almost all chemical insecticides (Sun 1992), as well as the microbial agent *Bacillus thuringiensis* (Tabashnik 1994).

Kao and Sun (1991) first demonstrated that glutathione conjugation was a major detoxifying reaction for parathion and methyl parathion in the DBM, and a considerably higher degradation of both insecticides was found in the resistant than in the susceptible strains. Subsequently, four GST isozymes have been isolated and purified from DBM larvae (Chiang and Sun 1993; Ku et al. 1994). They are homodimers with subunits of molecular mass ranging from 23.6 to 27.1 kDa. All, except GST-1, are basic proteins with pIs > 8.0; and GST-3 and GST-4 are highly immuno-related, with similar biochemical and toxicological properties. In contrast to purified housefly GSTs which prefer conjugation with the methyl group (Motoyama 1982), all four GSTs of DBM appear to conjugate only with the aryl group of both parathion and methyl parathion (Chiang and Sun 1993). They are unable to catalyze the conjugation of some other OPs known as GST substrates, such as fenitrothion, diazinon, tetrachlorvinphos and azinphosmethyl (Kao and Sun 1991).

An important finding regarding these GSTs is their distinct substrate preferences (Fig. 2). While GST-2 shows considerably higher preference toward CDNB than the others, GST-3 and GST-4 exhibit 8- to 20-fold higher activity

Fig. 2. Substrate preference of purified GST isozymes from DBM. CDNB 1-Chloro-2,4-dinitrobenzene; DCNB 1,2-dichloro-4-nitrobenzene; PA parathion; MPA methyl parathion; PAO paraoxon. (Adapted from Ku et al. 1994)

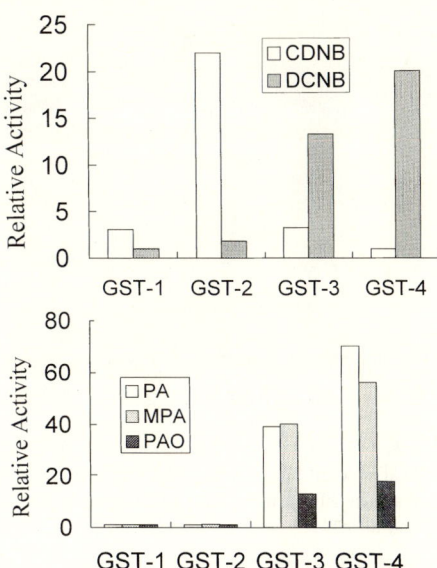

toward DCNB. The latter two GST isozymes degrade parathion, methyl parathion and paraoxon 13- to 70-fold more effectively than GST-1 or GST-2 (Ku et al. 1994).

A study of the GST profiles of several strains of the DBM further reveals the roles these isozymes play in the insecticide resistance of this insect (Fig. 3). While no GST-3 is seen in the susceptible MT strain or fenvalerate-resistant FEN strain, this isoform is present in a methyl-parathion-resistant MPA strain, a field-collected TC strain, a mixed-field MD strain as well as a teflubenzuron (TFB)-resistant CME strain. The absence of GST-3 in the FEN strain is in accordance with the lack of cross-resistance to these OPs (Sun et al. 1992). The somewhat lower proportion of GST-3 in the MD strain could be related to the fact that this strain was reared in the laboratory without any insecticide exposure. Among the six DBM strains examined, the CME strain is unique in terms of GST isozyme profile. Unexpectedly high GST activity toward both DCNB and methyl parathion has been observed (Sun et al. 1992). Its GST-3 content is low, yet it contains an additional form, GST-4. This isozyme displays even stronger preference for the three OPs than GST-3, yet why the CME strain does not exhibit a level of methyl parathion resistance comparable to that of the MPA strain (>450-fold) is unclear. Preliminary results suggest that GST-4 might bind TFB to render it ineffective as a chitin synthesis inhibitor and this binding does not affect its catalytic activity toward DCNB or CDNB (Chung and Sun, unpubl. data). The presence of a higher amount of more efficient OP-degrading GST isozymes, GST-3 and GST-4, contributes to the resistance to some OPs in this notorious insect pest of crucifers.

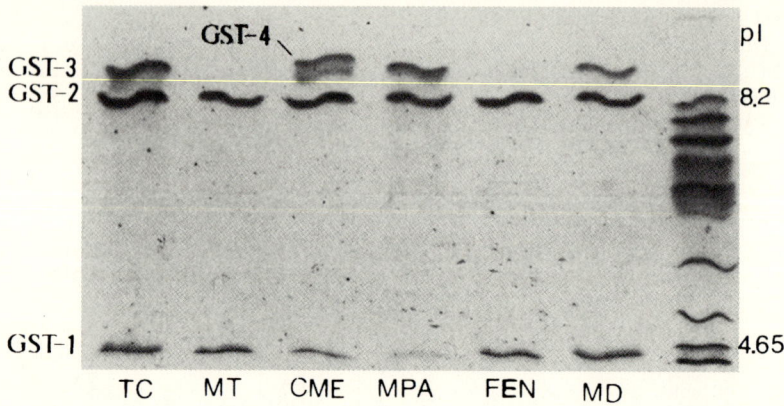

Fig. 3. GST isoform profiles of several strains of DBM. TC A recently collected field strain, MT a local susceptible strain; CME a teflubenzuron-resistant strain; MPA a methyl-parathion-resistant strain; FEN a fenvalerate-resistant strain; MD a field strain reared in the laboratory for years. (Ku et al. 1994)

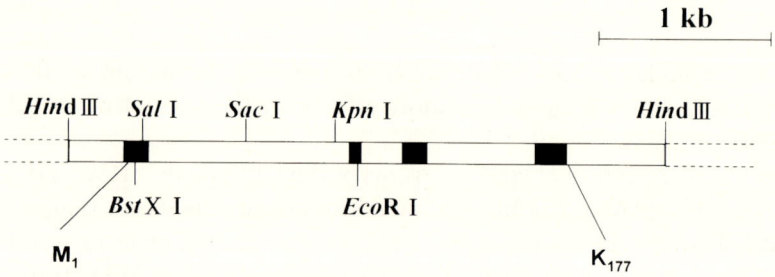

Fig. 4. Partial map of the GST 3 gene of DBM. (Huang 1998)

A cDNA of 809 pb, *PxGST3*, obtained by screening a midgut cDNA library of the MPA strain, encodes a GST-3 protein of 216 amino acids (calculated mass of 24,083 and pI 8.50) (Huang et al. 1998). Enzymatically active PxGST3 heterologously expressed in *E. coli* exhibits biochemical and toxicological properties similar to those of GST-3 purified from DBM larvae. This protein shares >35% identity to class I insect GSTs and has the highest (46.3%) amino acid sequence identity to MsGST1 of *Manduca sexta*, the first lepidopterous insect for which a GST has been cloned. A partially constructed map (Fig. 4) reveals at least five exons in this gene, *PxGST3* (Huang 1998). The higher amount of GST-3 found in the OP-resistant DBM probably results from an increased transcription of this gene. It is the first cloned GST with a well-defined role in insecticide resistance.

The cDNA of GST-4 has been cloned from the midgut cDNA library of a CME strain, and the GST-4 subunit (calculated mass of 24,120 and pI of 9.63) is different from GST-3 only in one amino acid, glutamine$_{207}$ → lysine (Chung 1999). The location of this difference is significant in view of the finding that the C-terminal region of GSTs is associated with substrate selectivity (Wilce and Parker 1994). The accumulation of single amino acid replacements has been proposed as an important mechanism for generating diversity in human alpha-class GSTs with various xenobiotic substrates (Chow et al. 1988). The relationship between GST-3 and GST-4 genes remains to be elucidated.

At least two more GST isozymes, both basic proteins, have been observed in the DBM. One of them, with pI between that of GST-3 and GST-4, may be related to DBM resistance to another OP, mevinphos (Chung and Sun, unpubl. data).

Lolah et al. (1995) observed some carboxylesterase activity in purified GSTs from housefly and rat. At least one GST isozyme, GST3, in the DBM shows apparent carboxylesterase activity (Chung, He and Sun, unpubl. data), and the role of this hydrolytic activity in DBM resistance to insecticides remains to be studied.

The induction of GSTs in the DBM by some potent inducers, such as indole-3-acetonitrile and xanthotoxin, has been unstable and quite limited (Hung and Sun, unpubl. data). Among the seven lepidopterous species studied, the DBM was the least inducible when indole-3-acetonitrile was used as an inducer; the inducibility of GSTs in these species was also independent of their levels of constitutive enzyme activity (Yu and Hsu 1993).

5
Concluding Remarks

Although GSTs are multiple in forms, they are nevertheless much easier to study than cytochrome P450 dependent monooxygenases, another group of detoxifying enzymes that confer insect resistance to many classes of insecticides, including OPs, carbamates, pyrethroids, and chitin synthesis inhibitors (Scott 1999). GSTs are soluble proteins, relatively easy to characterize, and a general lack of post-translational modification of the protein makes the heterologous expression of functional enzyme less troublesome. If the GST isoform profile of an insect is not too complicated, such as that of the DBM, one can try to resolve the proteins first and then continue onto the molecular work. Otherwise, cloning cDNAs may be taken as an alternative approach, with subsequent over-expression in *E. coli* or other systems to obtain pertinent biochemical and toxicological data. With the accumulation of molecular data on GSTs, it is not as difficult a task as it used to be to clone their cDNAs from insects with significance in agriculture or public health. Thus, many more insect GSTs will be characterized in time. Then the real challenges lie in the elucidation of the structure and the regulation of these multiple genes.

While some progress has been made in understanding insect GSTs, much remains to be studied, notably in the relationship between these GSTs and insecticide resistance. Although elevated GST activity is known to be related to insecticide resistance in many insects, identification and characterization of the relevant isoforms, especially for agricultural insect pests, is still much in demand. This could be done either through protein purification or by cDNA cloning. Although the cDNAs of several dozens of insect GST genes have been cloned and more would be cloned in the future, the roles of these GSTs in insecticide metabolism and resistance have yet to be confirmed with data at protein level. While a thorough understanding of the structure, induction and regulation of the relevant genes will add valuable information to the overall GST database, it is essential to the elucidation of the selection course of insecticide resistance as well as the eventual solutions to this problem.

References

Armstrong RN (1997) Structure, catalytic mechanism and evolution of the glutathione transferases. Chem Res Toxicol 10:2–18

Booth J, Boyland E, Sims P (1961) An enzyme from rat liver catalysing conjugations with glutathione. Biochem J 79:516–524

Beall C, Fyrberg C, Song S, Fyrberg E (1992) Isolation of a *Drosophila* gene encoding glutathione S-transferase. Biochem Genet 30:515–527

Chiang FM, Sun CN (1993) Glutathione transferase isozymes of diamondback moth larvae and their role in the degradation of some organophosphorus insecticides. Pestic Biochem Physiol 45:7–14

Chien C, Dauterman WC (1991) Studies on glutathione S-transferase in *Helicoverpa* (=*Heliothis*) *zea*. Insect Biochem 21:857–864

Chow NW, Whang-Peng J, Kao-Shan CS, Tam MF, Lai HCJ, Tu CPD (1988) Human glutathione S-transferases. The Ha multigene family encodes products of different but overlapping substrate specificities. J Biol Chem 263:12797–12800

Chung WY (1999) The molecular cloning of a glutathione S-transferase, GST-4, involved in insecticide resistance from the diamondback moth, *Plutella xylostella*. MS Thesis, National Chung-Hsing University, Taichung, 44pp

Clark AG (1989) The comparative enzymology of the glutathione S-transferases from non-vertebrate organisms. Comp Biochem Physiol 92B:419–446

Clark AG, Shamaan NA (1984) Evidence that DDT-dehydrochlorinase from the housefly is a glutathione S-transferase. Pestic Biochem Physiol 22:249–261

Clark AG, Shamaan NA, Dauterman WC, Hayaoka T (1984) Characterization of multiple glutathione transferases from houseflies, *Musca domestica* (L.). Pestic Biochem Physiol 22:51–59

Clark AG, Shamaan NA, Sinclair MD, Dauterman WC (1986) Insecticide metabolism by multiple glutathione S-transferases in two strains of the housefly, *Musca domestica* (L.). Pestic Biochem Physiol 24:169–175

Daniel V (1993) Glutathione S-transferases: gene structure and regulation of expression. CRC Crit Rev Biochem Mol Biol 28:173–207

Dauterman WC (1985) Insect metabolism: extramicrosomal. In: Kerkut GA, Gilbert LI (eds) Comprehensive insect physiology, biochemistry and pharmacology, vol 12. Pergamon Press, New York, pp 713–730

Ding VDH, Pickett CB (1985) Transcriptional regulation of rat liver glutathione S-transferase genes by phenobarbital and 3-methylcholanthrene. Arch Biochem Biophys 240:553–559

Feng QL, Davey KG, Pang ASD, Primavera M, Ladd TR, Zheng SC, Sohi SS, Retnakara A, Palli SR (1999) Glutathione S-transferase from the spruce budworm, *Choristoneura fumiferana*: identification, characterization, localization, cDNA cloning, and expression. Insect Biochem Mol Biol 29:779–793

Fournier D, Bride JM, Poirie M, Berge JB, Plapp FW Jr (1992) Insect glutathione S-transferases: biochemical characteristics of the major forms from housefly susceptible and resistant to insecticides. J Biochem Chem 167:1840–1845

Fukami J, Shishido T (1966) Nature of a soluble glutathione-dependent enzyme system active in cleavage of methyl parathion to desmethyl parathion. J Econ Entomol 59:1338–1352

Grant DF (1991) Evolution of glutathione S-transferase subunits in culicidae and related nematocera: electrophoretic and immunological evidence for conserved enzyme structure and expression. Insect Biochem 21:435–445

Hemingway J, Malcolm CA, Kissoon KE, Boddington RB, Curtis CF, Hill N (1985) The biochemistry of insecticide resistance in *Anopheles sacharovi*: comparative studies with a range of insecticide susceptible and resistant *Anopheles* and *Culex* species. Pestic Biochem Physiol 24:68–76

Huang SY (1998) Molecular cloning of a gluathione S-transferase gene involved in insecticide resistance in the diamondback moth, *Plutella xylostella* (L.). MS Thesis, National Chung-Hsing University, Taichung, 57 pp

Huang HS, Hu NT, Yao YE, Wu CY, Chiang SW, Sun CN (1998) Molecular cloning and heterologous expression of a glutathione S-transferase involved in insecticide resistance from the diamondback moth, *Plutella xylostella*. Insect Biochem Mol Biol 28:651–658

Ishikawa T (1992) The ATP-dependent glutathione S-conjugate export pump. Trends Biochem Sci 17:463–468

Kao CN, Sun CN (1991) *In vitro* degradation of some organophosphorus insecticides by susceptible and resistant diamondback moths. Pestic Biochem Physiol 41:132–141

Ku CC, Chiang FM, Hsin CY, Yao YE, Sun CN (1994) Glutathione transferase isozymes involved in insecticide resistance of diamondback moth larvae. Pestic Biochem Physiol 50:191–197

Lewis JB (1969) Detoxication of diazinon by subcellular fractions of diazinon-resistant and susceptible houseflies. Nature 22:917–919

Lewis JB, Sawicki RM (1971) Characterization of the resistance mechanism to diazinon, parathion and diazoxon in the organophosphorus-resistant SKA strain of houseflies (*Musca domestica* L.). Pestic Biochem Physiol 1:275–285

Lolah JO, Chien CI, Motoyama N, Dauterman WC (1995) Glutathione S-transferases: α-naphthyl acetate activity and possible role in insecticide resistance. J Econ Entomol 88:768–779

Lougarre A, Bride JM, Fournier D (1999) Is the insect glutathione S-transferase I gene family intronless? Insect Mol Biol 8:141–143

Mannervik B, Danielson UH (1988) Glutathione transferase: structure and catalytic activity. CRC Crit Rev Biochem 23:281–334

Motoyama N (1982) Characterization of glutathione S-transferases in relation to azinphosmethyl resistance. J Pestic Sci 7:415–425

Motoyama N, Dauterman WC (1980) Glutathione S-transferases: their role in the metabolism of organophosphorus insecticides. Rev Biochem Toxicol 2:49–69

Oppenoorth FJ (1985) Biochemistry and genetics of insecticide resistance. In: Kerkut GA, Gilbert LI (eds) Comprehensive insect physiology, biochemistry and pharmacology, vol 12. Pergamon Press, New York, pp 731–773

Oppenoorth FJ, Smissaert HR, Welling W, van der Pass LJT, Hitman KT (1977) Insensitive acetylcholinesterase, high glutathione S-transferase, and hydrolytic activity as resistance factors in a tetrachlorvinphos-resistant strain of housefly. Pestic Biochem Physiol 11:176–188

Prapanthadara L, Hemingway J, Ketterman AJ (1995) DDT-resistance in *Anopheles gambiae* (Diptera: Culicidae) from Zanzibar, Tanzania, based on increased DDT-dehydrochlorinase activity of glutathione S-transferases. Bull Entomol Res 85:267–274

Ranson H, Prapanthadara, L, Hemingway J (1997) Cloning and characterization of two glutathione S-transferases from a DDT-resistant strain of *Anopheles gambiae*. Biochem J 324:97–102

Ranson H, Collins, R, Hemingway J (1998) The role of alternative mRNA splicing in generating heterogeneity within the *Anopheles gambiae* class I glutathione S-transferase family. Proc Natl Acad Sci USA 95:14284–14289

Reiss RA, James AA (1993) A glutathione S-transferase gene of the vector mosquito, *Anopheles gambiae*. Insect Mol Biol 2:25–32

Scott JG (1999) Cytochrome P450 and insecticide resistance. Insect Biochem Mol Biol 29:757–777

Shishido T, Usui K, Sat M, Fukami J (1972) Enzymatic conjugation of diazinon with glutathione in rat and American cockroach. Pestic Biochem Physiol 2:51–63

Sun CN (1992) Insecticide resistance in diamondback moth. In: Talekar NS (ed) Management of diamondback moth and other cruciferous pests. Proceedings of the 2nd International Workshop, Asian Vegetable Research and Development Center, Shanghua, pp 419–426

Sun CN, Tsai YC, Chiang FM (1992) Resistance in the diamondback moth to pyrethroids and benzoylphenylureas. In: Mullin CA, Scott JG (eds) Molecular mechanisms of insecticide resistance. Diversity among insects. American Chemical Society, Washington, DC, pp 149–167

Synanen M, Zhou ZH, Wang JY (1994) Glutathione transferase gene family from the housefly *Musca domestica*. Mol Gen Genet 245:25–31

Syvanen M, Zhou Z, Wharton J, Goldsbury C, Clark A (1996) Heterogeneity of the glutathione transferase genes encoding enzymes responsible for insecticide degradation in the housefly. J Mol Evol 43:236–240

Tabashnik BE (1994) Evolution of resistance to *Bacillus thuringiensis*. Annu Rev Entomol 39:47–80

Talekar NS, Shelton AM (1995) Biology, ecology and management of the diamondback moth. Annu Rev Entomol 38:275–302

Tang AN, Tu CDP (1994) Biochemical characterization of *Drosophila* glutathione S-transferases D1 and D21. J Biol Chem 269:27876–27884

Tang AH, Tu CDP (1995) Pentobarbital-induced changes in *Drosophila* glutathione S-transferase D21 mRNA stability. J Biol Chem 270:13819–13825

Toung YPS, Hsieh TS, Tu CPD (1991) The Drosophila glutathione S-transferase 1-1 is encoded by an intronless gene at 87B. Biochem Biophys Res Commun 178:1205–1211

Toung YPS, Hsieh TS, Tu CPD (1993) The glutathione S-transferase D genes. A divergently organized, intronless gene family in *Drosophila melanogaster*. J Biol Chem 268:9737–9746

Wang JY, McCommas S, Syvanen M (1991) Molecular cloning of a glutathione S-transferase overproduced in an insecticide-resistant strain of the housefly (*Musca domestica*). Mol Gen Genet 227:260–266

Wilce MCJ, Parker MW (1994) Structure and function of glutathione S-transferases. Biochim Biophys Acta 1205:1–18

Wilce MCJ, Board PG, Feil SC, Parker MW (1995) Crystal structure of a theta-class glutathione transferase. EMBO J 14:2133–2143

Yang RSH, Hodgson E, Dauterman WC (1971) Metabolism in vitro of diazinon and diazoxon in susceptible and resistant houseflies. J Agric Food Chem 19:14–19

Yu SJ (1982) Host plant induction of glutathione S-transferase in the fall armyworm. Pestic Biochem Physiol 18:101–106

Yu SJ (1989) Purification and characterization of glutathione transferases from five phytophagous Lepidoptera. Pestic Biochem Physiol 35:97–105

Yu SJ (1992) Plant-allelochemical-adapted glutathione transferases in Lepidoptera. In: Mullin CA, Scott JG (eds) Molecular mechanisms of insect resistance. Diversity among insects. American Chemical Society, Washington, DC, pp 174–190

Yu SJ (1996) Insect glutathione S-transferases. Zool Stud 35:9–19

Yu SJ (1999) Induction of new glutathione S-transferase isozymes by allelochemicals in the fall armyworm. Pestic Biochem Physiol 63:163–171

Yu SJ, Hsu EL (1993) Induction of detoxification enzymes in phytophagous insects: roles of insecticide synergists, larvae age and species. Arch Insect Biochem Physiol 24:21–32

Cytochrome P450 Monooxygenases and Insecticide Resistance: Lessons from *CYP6D1*

Jeffrey G. Scott[1]

1
Cytochrome P450 Monooxygenases

The cytochrome P450-dependent monooxygenases (monooxygenases) are a vital biochemical system because they metabolize xenobiotics such as drugs, pesticides and plant toxins, and because they regulate the titers of endogenous compounds such as hormones, fatty acids and steroids. Cytochrome P450 (P450) is a hemoprotein which acts as the terminal oxidase in monooxygenase systems, and there are multiple P450 isoforms in eukaryotic species. Monooxygenases are remarkable in that they can oxidize widely diverse substrates and are capable of producing a bewildering array of reactions (Kulkarni and Hodgson 1980; Rendic and Di Carlo 1997; Mansuy 1998).

Monooxygenase-mediated oxidation requires (besides oxygen and substrate) P450, NADPH cytochrome P450 oxidoreductase (P450 reductase), NADPH and phospholipid (Lu and Coon 1968). P450 reductase is required to transfer reducing equivalents from NADPH to P450. Cytochrome b_5 is involved in certain monooxygenase reactions, depending upon the P450 and/or substrate involved (Vatsis et al. 1980; Peterson and Prough 1986; Pompon 1987; Epstein et al. 1989; Zhang and Scott 1996).

Cytochrome P450s are named *CYP* (for *CY*tochrome *P*450), followed by a number, a letter and a number indicating the family, subfamily and isoform, respectively (Nelson et al. 1996). Alleles are designated *v1*, *v2*, etc. This nomenclature is based on overall sequence similarity, and no information regarding the function of a P450 should be assumed by its classification within this system (Nelson 1998).

One remarkable feature of the monooxygenases is the large variation in substrate specificity of different P450s. For example, CYP1A1 can metabolize more than 20 substrates, while CYP7A1 has only one known substrate (Rendic and Di Carlo 1997). Certain P450s have overlapping substrate specificity (e.g. CYP2C subfamily in humans) (Rendic and Di Carlo 1997), so that a single compound may be subject to metabolism by multiple P450s. In addition, some P450s produce only a single metabolite from a given substrate, while other P450s can produce multiple metabolites. Studies of P450s are further compli-

[1] Department of Entomology, Comstock Hall, Cornell University, Ithaca, NY 14853-0901, USA

Isaac Ishaaya (Ed.): Biochemical Sites of
Insecticide Action and Resistance
© Springer-Verlag Berlin, Heidelberg 2001

cated by the observation that change of a single amino acid (i.e. Cyp2a4) can alter its substrate specificity (Lindberg and Negishi 1989). Information about known monooxygenase substrates can be found in several reviews (Kulkarni and Hodgson 1980; Wilkinson 1983, 1985; Agosin 1985; Ahmad 1986; Guengerich 1995, 1997; Rendic and Di Carlo 1997).

2
Insecticide Resistance

Resistance has been defined as "the inherited ability of a strain of some organism to survive doses of a toxicant that would kill the majority of individuals in a normal population of the same species" (World Health Organization 1957). Insecticide resistance is arguably the major obstacle in the control of medically and agriculturally significant arthropod pests (World Health Organization 1976; Georghiou 1980). The importance of insecticide resistance has been well documented (Brown 1958; Georghiou and Saito 1983; Oppenoorth 1985; Brattsten et al. 1986; Roush and Tabashnik 1990; Georghiou and Lagunes-Tejeda 1991; Scott 1991) and will not be elaborated on here.

3
Monooxygenase-Mediated Insecticide Resistance

The most common types of resistance found in insects are increased enzymatic detoxification and target site insensitivity (Agosin 1985; Oppenoorth 1985; Scott 1991). It has been postulated (Wilkinson 1983; Oppenoorth 1985; Scott 1991) that increased enzymatic detoxification is the most frequently occurring resistance mechanism, but altered target sites (ffrench-Constant 1999) are also very common. Monooxygenase-mediated resistance may be the most common type of metabolism-based insecticide resistance (Hodgson and Kulkarni 1983; Oppenoorth 1985; Brattsten et al. 1986; Scott 1991), although esterases are also important (Hemingway and Karunarantne 1998), and glutathione S-transferases are significant in some cases (Yu 1996). In addition, monooxygenase-mediated detoxification in susceptible strains of certain species substantially limits the toxicity and usefulness of some insecticides, such as pyrethrins (Sawicki 1962), imidacloprid (Wen and Scott 1997) and carbaryl (Wilkinson 1967). Furthermore, P450s are responsible for the activation of many organophosphate insecticides (Hodgson et al. 1991).

Most cases of monooxygenase-mediated resistance result from an increase in detoxification. However, in cases where the parent insecticide must undergo monooxygenase-mediated bioactivation, as is the case for many organophosphates, it is also possible that resistance could be achieved through decreased activation. Although this has been reported once (Konno et al. 1989), it does not appear to be a common mechanism of resistance. This may explain why esterases are relatively more common than monooxygenases in resistance to organophosphates in several species (Oppenoorth 1985; Scott 1991).

Early studies of insect monooxygenases frequently identified increases in total cytochrome P450s in resistant strains (references in Hodgson 1985). However, total P450 levels were usually increased only about three-fold, while monooxygenase activities were increased to much higher levels. The discovery of multiple P450 isoforms dictated that the individual P450s responsible for resistance would have to be isolated before our understanding of insecticide resistance could be improved.

Increased levels of P450 reductase (Vincent et al. 1985) and b_5 (Scott and Georghiou 1986b) were first associated with monooxygenase-mediated insecticide resistance in houseflies, and this observation has been confirmed in other species (Sun et al. 1992; Kotze 1993; Kotze and Wallbank 1996; Valles and Yu 1996). The discovery of elevated levels of P450 reductase and b_5 associated with monooxygenase-mediated resistance led to the suggestion that changes in one or both of these proteins could be involved in resistance (Scott and Georghiou 1986b).

The levels of certain P450s can be induced following exposure to drugs and other xenobiotics (Conney 1967, 1982). Many P450 inducers are known in insects (Terriere and Yu 1974; Hodgson 1983; Agosin 1985; Scott et al. 1998), and it appears that several of these inducers alter the expression of different P450s (references in Scott et al. 1998). In addition, induction has been associated with increased tolerance to insecticides (Brattsten and Wilkinson 1973). In 1983 Terriere proposed that "the same regulatory genes may be involved in both induction and biochemical resistance". Unfortunately, the lack of isolated P450s hampered investigations into this intriguing area for many years.

4
CYP6D1 and Insecticide Resistance

Research in my laboratory over the past several years has focused on the molecular basis of monooxygenase-mediated resistance to pyrethroid insecticides in the housefly. We have identified the P450 involved in this resistance (CYP6D1) and what is known about CYP6D1-mediated resistance will be elaborated on below. A recent review covers information on what is known about the role of other P450s in insecticide resistance (Scott 1999), and that information will not be repeated here.

The Learn pyrethroid resistant (LPR) strain of the housefly was collected in 1982 from a dairy in New York following the introduction of permethrin for fly control. After permethrin selection the LPR strain became homozygous for the major mechanisms of resistance and attained extremely high levels of resistance to commonly used pyrethroid insecticides (Scott and Georghiou 1986b).

The highest levels of resistance occur toward pyrethroids with a phenoxybenzyl moiety (e.g. >5000-fold resistance to cypermethrin; Scott and Georghiou 1985, 1986b).

Monooxygenase-mediated pyrethroid resistance in the LPR strain has been demonstrated by in vivo and in vitro studies (Scott and Georghiou 1986b; Wheelock and Scott 1992b; Liu and Scott 1995, 1996, 1997a). The two other mechanisms of resistance to pyrethroid insecticides in the LPR strain are knockdown resistance (*kdr*) and decreased cuticular penetration (*pen*) (Scott and Georghiou 1986a,b; Liu and Scott 1995). Monooxygenase-mediated detoxification appears to be the major mechanism of pyrethroid resistance in this strain (Scott and Georghiou 1986b).

Detailed genetic analysis of permethrin resistance revealed that the high level of PBO suppressible resistance (i.e., P450-mediated detoxification) was associated with autosomes 1 and 2 in combination, neither autosome having a substantial effect by itself (Liu and Scott 1995). Both *kdr* and *pen* were found on autosome 3 (Scott and Georghiou 1986a; Liu and Scott 1995).

In 1989 a single P450, CYP6D1 (a.k.a. $P450_{lpr}$), was purified from LPR houseflies using two high performance liquid chromatography (HPLC) steps (Wheelock and Scott 1989). Electrophoresis of housefly microsomes indicated that a protein band corresponding to CYP6D1 was expressed at elevated levels in LPR compared with susceptible houseflies (Scott and Lee 1993a). An isoform-specific antiserum was raised in rabbits using purified CYP6D1 protein as the antigen (Wheelock and Scott 1990) and this antiserum was used to characterize the expression of CYP6D1 (see below). The *CYP6D1* gene has been sequenced (Tomita and Scott 1995) and cloned (Smith and Scott 1997).

To ascertain the role of CYP6D1 in pyrethroid resistance in the LPR strain, the P450-dependent metabolism of deltamethrin was investigated. Monooxygenase-dependent deltamethrin metabolism occurred at increased levels in LPR microsomes compared with a susceptible strain, and this enhanced metabolism of deltamethrin was eliminated by the addition of anti-CYP6D1 antiserum. Thus, CYP6D1 is the major P450 responsible for deltamethrin metabolism in LPR flies (Wheelock and Scott 1992b). Similar results were observed for cypermethrin, and the primary CYP6D1-specific metabolite formed in vitro was identified using gas chromatography/mass spectrometry (GC/MS) as 4′-OH cypermethrin (Zhang and Scott 1994). Thus, CYP6D1 carries out metabolism at a single site on the pyrethroid phenoxybenzyl moiety. This helps to explain the reduced levels of resistance in LPR to pyrethroids lacking this functional group (Scott and Georghiou 1986b).

Using a b_5 antiserum it was demonstrated that b_5 is required for CYP6D1-mediated metabolism of cypermethrin in microsomes from LPR houseflies (Zhang and Scott 1996). Therefore, b_5 appears to be directly involved in CYP6D1-mediated pyrethroid resistance. The higher level of b_5 found in the

LPR strain (Scott and Georghiou 1986a), and the linkage of this trait to the same autosomes as monooxygenase-mediated resistance (Liu and Scott 1996), suggests that the elevated levels of b_5 may be a requisite for the enhanced metabolism of pyrethroids by CYP6D1 in LPR houseflies.

Northern blots of RNA from adults of the LPR and an insecticide-susceptible strain revealed that a CYP6D1 mRNA was expressed at about a 10-fold higher level in LPR flies compared with susceptible flies. This agreed with previous results of an eight-fold higher level of CYP6D1 protein in microsomes from LPR relative to susceptible flies (Wheelock and Scott 1990; Liu and Scott 1996).

Southern blots of genomic DNA hybridized with a *CYP6D1* cDNA probe revealed similar hybridization intensities between LPR and a susceptible strain, indicating that the elevated level of CYP6D1 mRNA in LPR flies is not due to gene amplification (Tomita and Scott 1995). Using an allele-specific polymerase chain reaction(PCR), *CYP6D1* was mapped to chromosome 1 (Liu et al. 1995).

To investigate the role of mRNA stability in the high level expression of CYP6D1 in the LPR strain, mRNA was isolated and quantified from aabys and LPR houseflies at different times following injection of a transcription inhibitor (actinomycin D). The same pattern of decrease in CYP6D1 mRNA abundance (approximate half-life of ~10h) was detected in both the LPR and aabys strains, indicating that the increased expression of *CYP6D1* in LPR is not due to increased stability of the mRNA (Liu and Scott 1998).

The relative transcription rate of *CYP6D1* in the LPR and a susceptible strain was measured using an in vitro run-on transcription assay. When nuclei from the susceptible strain were used in the run-on assay only a trace of CYP6D1 mRNA was detected (Liu and Scott 1998). In contrast, when nuclei from the LPR strain were used for the run-on assay abundant CYP6D1 mRNA was detected. Quantitation of the relative intensities of the CYP6D1 signals between aabys and LPR revealed approximately a 10-fold difference (Liu and Scott 1998), which is comparable to the differences in *CYP6D1* expression observed by Northern hybridization (Tomita et al. 1995; Liu and Scott 1996). This was the first direct evidence for increased transcription as an underlying cause of insecticide resistance (Liu and Scott 1998).

The increased rate of transcription of *CYP6D1* was due to factors on autosomes 1 and 2 in the LPR strain (Liu and Scott 1998). This is consistent with what is known about the linkage of monooxygenase-mediated resistance (Scott and Georghiou 1986a; Liu and Scott 1995) and overexpression of CYP6D1 mRNA and protein levels (Liu and Scott 1996). Interestingly, resistance (via increased transcription of *CYP6D1*) is mediated both by *cis-* and *trans-*regulatory factors (Liu et al. 1995).

Recently, the *CYP6D1* 5′ flanking sequences from the LPR and five pyrethroid-susceptible strains have been reported (Scott et al. 1999) and some differences have been observed that could play a role in the differential expression of *CYP6D1* in these strains. However, much more study will be needed before the mechanisms underlying the increased transcription of *CYP6D1* in LPR are fully understood.

In addition to the LPR strain, *CYP6D1* was sequenced from pyrethroid-susceptible strains of houseflies. Comparison of the four *CYP6D1* alleles reveals that the deduced protein sequence from the LPR allele differs from that of the CS, aabys and ISK alleles by 8, 11 and 7 amino acids, respectively (Tomita et al. 1995). Among them, six amino acids are the same in CS, aabys and ISK, but are different from LPR: Asp to Ala (150), Ile to Leu (153), Thr to Ser (165), Glu to Gln (218), Thr to Asn (225) and Met to Ile (227) (Tomita et al. 1995). The observed amino acid substitutions occur at two highly variable regions among cytochromes P450 in family 6, and the changes at residues 218, 225 and 227 are close to a proposed substrate binding region (Gotoh and Fujii-Kuriyama 1989). Whether or not the amino acid differences among the CYP6D1 proteins results in different catalytic activity toward pyrethroid insecticides remains to be elucidated.

CYP6D1 mRNA expression is developmentally regulated with no CYP6D1 mRNA detectable in eggs, larvae, or pupae (Scott et al. 1996). High levels of mRNA were found in adults from 1 to 6 days old. The low levels of expression observed in day-3 pupae may be attributed to pharate adults in this sample. This pattern matches that observed for the CYP6D1 protein (Wheelock et al. 1991). Results from studies on individual tissues indicate that the CYP6D1 protein is found in many tissues throughout the housefly abdomen with the relative abundance being fat bodies > proximal intestine > reproductive system > rectum (Lee and Scott 1992; Scott and Lee 1993b). CYP6D1 is also expressed in all tagmata and in thoracic ganglia of the housefly (Korytko and Scott 1998). The level of expression of CYP6D1 at each of these tissues and/or sites was higher in LPR compared with susceptible strains (Lee and Scott 1992; Scott and Lee 1993b). This suggests there is no single tissue or site within the LPR housefly that is responsible for pyrethroid resistance. Furthermore, monooxygenase-mediated detoxification at the level of the target tissue may help to explain the high levels of resistance to pyrethroids in the LPR strain.

Elucidation of the overall constraints that determine substrate turn over or inhibitor potency for a specific P450 could be particularly useful in defining cross-resistance patterns or developing insect P450-specific inhibitors (i.e. synergists). Known CYP6D1 substrates include methoxyresorufin, benzo[*a*]pyrene (Wheelock and Scott 1992a), chlorpyrifos (Hatano and Scott 1993), phenanthrene (Korytko et al. 2000) and phenoxybenzyl pyrethroids (Wheelock and Scott 1992b; Zhang and Scott 1996; Korytko and Scott 1998). Thus, CYP6D1 is capable of carrying out a wide variety of metabolic reactions (aromatic hydroxylation, *O*-dealkylation, etc.). However, small changes in some of these structures result in compounds that are not significantly metabolized, such as ethoxyresorufin, pentoxyresorufin or ethoxycoumarin (Wheelock and Scott 1992a). The discovery that CYP6D1 detoxified pyrethroids and also activated an organophosphate insecticide suggests that the rotational use of pyrethroids (selection for increased levels of CYP6D1) and organophosphates (selection for decreased levels of CYP6D1) could, at least theoretically, slow the evolution of this resistance. The recent report that a novel insecticide (chlorfenapyr) was more toxic to pyrethroid-resistant hornflies compared with a susceptible strain (enhanced monooxygenase-mediated activation was implicated as the cause) (Sheppard and Joyce 1998) suggests this strategy may extend beyond organophosphates. CYP6D1 was found to be strongly inhibited by xanthotoxin, chlorpyrifos, β-naphthoflavone, PBO, substituted alkynylpyrenes and substituted phenanthrenes (Scott 1996; Scott et al. 2000). These studies demonstrated that identification of isoform-selective inhibitors of P450s within an insect, and between species, are possible. In addition, isosafrole and verbutin were shown to be a potent synergists of pyrethroid insecticides in adult houseflies (Scott 1996; Scott et al. 2000).

5
Summary of the Lessons Learned from *CYP6D1*

Studies of *CYP6D1* have shown that resistance can occur by increased transcription of a P450 leading to increased expression of the protein and, thus, increased detoxification of the insecticide (Liu and Scott 1998). Interestingly, this can occur by *cis* and *trans* regulation of the transcription process (Liu and Scott 1998). It is also noteworthy that the factor(s) responsible for *trans* regulation of the P450(s) involved in resistance may also be capable of regulating the expression of other P450s that are not necessarily involved in resistance (at least in houseflies). This adds another level of complexity to the study of insect P450s, because not only is the substrate specificity variable, but regulation of P450 genes are also subject to controls that will vary. Knowing which subsets of P450s are elevated by the same regulatory factor(s) may help to understand cross-resistance patterns.

Studies of CYP6D1 have shown that resistance can occur due to detoxification via a single P450 and that the metabolic attack can be limited to a single site on the insecticide (Zhang and Scott 1994). It is also clear that b_5 is needed for CYP6D1-mediated detoxification of pyrethroids, and is involved in this case of insecticide resistance (Zhang and Scott 1996).

Terriere has proposed that "the same regulatory genes may be involved in both induction and biochemical resistance" (Terriere 1983, 1984), and Plapp (1984) proposed the existence of a "master gene on chromosome 2" (housefly) that was involved in metabolism-mediated insecticide resistance. Recent results hint that both of these hypotheses may be correct. PB induction is linked to autosome 2 (Liu and Scott 1997b) and *CYP6D1* is on autosome 1.

Increased transcription of *CYP6D1* in LPR flies (i.e. monooxygenase-mediated resistance to pyrethroid insecticides) is controlled by factors on autosomes 1 and 2 (Liu and Scott 1996, 1998). *CYP6D1* is not inducible with PB in the LPR strain, but it is inducible in susceptible strains (Lee and Scott 1989; Scott et al. 1996). Therefore, one plausible idea is that there is a factor on autosome 2 of susceptible houseflies which is elevated in response to PB treatment and acts to increase transcription of other genes. If this factor was under the control of a regulatory element (e.g. a repressor), and if this regulatory element was mutated, increased constitutive expression of several P450s could result. Provided at least one of these P450s could carry out insecticide detoxification, the strain would then be resistant to that insecticide.

This scenario is consistent with what is observed in LPR. Based on this model, we might expect other P450s (not necessarily involved in insecticide resistance) to be elevated in the LPR strain. This has been observed. For example, *CYP6A1,* which is on autosome 5 in the housefly (Carino et al. 1994), is also expressed at elevated levels in the LPR relative to susceptible strains (Carino et al. 1992). If such a model is correct, understanding cross-resistance patterns could become much more difficult. For example, it would be possible that several different P450s, all of which could be increased by the factor on autosome 2, could be responsible for detoxification of different insecticides. Obviously, the above model is not entirely consistent with what is known about *CYP6A1.* In Rutgers houseflies, *CYP6A1* is over-expressed due to a factor on autosome 2, yet *CYP6A1* expression in Rutgers is inducible with PB (although to a lesser extent than susceptible strains) (Carino et al. 1992, 1994; Cohen and Feyereisen 1995). Clearly more work is needed to identify the PB-mediated factor on autosome 2, its regulatory elements, and the regulatory elements involved in expression of housefly P450s, before we will know if the ideas proposed by Terriere (1983, 1984) and Plapp (1984) are correct.

As has been pointed out previously (NRC 1986; Jutsum et al. 1998), monitoring is an important component of resistance management. Identification of specific probes for P450s involved in resistance will provide a means for rapid and sensitive detection of this mechanism, and will also be useful to study the population dynamics and evolution of resistance. However, it appears unlikely that identical P450 isoforms will be responsible for resistance to a given insecticide in two different species. For example, probes for CYP6D1 (nucleotide or antiserum) do not cross-react with P450s in other species (Wheelock et al. 1991; Scott et al., unpubl. data). There is, however, the potential that the mechanism responsible for the increased transcription of *CYP6D1* (leading to over-expression and resistance) could be similar in many species. If this were the case, the goal of developing a molecular-based assay for monooxygenase-mediated resistance could be possible.

Insecticide resistance is a compelling model for studies of evolutionary biology (Guillemaud et al. 1998), as exemplified by the work on organophosphate resistance in *Culex* mosquitoes (Chevillon et al. 1999). One of the major mechanisms of organophosphate resistance in these mosquitoes is gene amplification (Mouches et al. 1986; Field and Devonshire 1991; Hemingway and Karunarantne 1998). Given that CYP6D1-mediated pyrethroid resistance is not due to gene amplification it will be intriguing to compare the forces shaping monooxygenase-mediated (i.e. increased gene transcription) resistance in comparison with esterase-mediated (i.e. gene amplification) resistance.

Acknowledgements. This work was supported by a grant from the National Institutes of Health (GM 47831).

References

Agosin M (1985) Role of microsomal oxidations in insecticide degradation. In: Kerkut GA, Gilbert LI (eds) Comprehensive insect physiology, biochemistry, and pharmacology, vol 12. Pergamon Press, New York, pp 647–712

Ahmad S (1986) Enzymatic adaptations of herbivorous insects and mites to phytochemicals. J Chem Ecol 12:533–539

Brattsten LB, Wilkinson CF (1973) Induction of microsomal enzymes in the southern armyworm (*Prodenia eridania*). Pestic Biochem Physiol 3:393–407

Brattsten LB, Holyoke CW Jr, Leeper JR, Raffa KF (1986) Insecticide resistance: challenge to pest management and basic research. Science 231:1255–1260

Brown AWA (1958) Insecticide resistance in arthropods. World Health Organization, Geneva

Carino F, Koener JF, Plapp FW Jr, Feyereisen R (1992) Expression of the cytochrome P450 gene *CYP6A1* in the housefly, *Musca domestica*. In: Mullin CA, Scott JG (eds) Molecular mechanisms of insecticide resistance: diversity among insects. ACS Symposium series 505. American Chemical Society, Washington, DC, pp 31–40

Carino FA, Koener JF, Plapp FW Jr, Feyereisen R (1994) Constitutive overexpression of the cytochrome P450 gene *CYP6A1* in a housefly strain with metabolic resistance to insecticides. Insect Biochem Mol Biol 24:411–418

Chevillon C, Raymond M, Guillemaud T, Lenormand T, Pasteur N (1999) Population genetics of insecticide resistance in the mosquito *Culex pipiens*. Biol J Linnean Soc 68:147–157

Cohen MB, Feyereisen R (1995) A cluster of cytochrome P450 genes of the CYP6 family in the housefly. DNA Cell Biol 14:73–82

Conney AH (1967) Pharmacological implications of microsomal enzyme induction. Pharmacol Rev 19:317–366

Conney AH (1982) Induction of microsomal enzymes by foreign chemicals and carcinogenesis by polycyclic aromatic hydrocarbons. Cancer Res 42:4875–4917

Epstein PM, Curti M, Jansson I, Huang C-K, Schenkman JB (1989) Phosphorylation of cytochrome P450: regulation by cytochrome b_5. Arch Biochem Biophys 271:424–432

ffrench-Constant RH (1999) Target site mediated insecticide resistance: what questions remain? Insect Biochem Mol Biol 29:397–403

Field LM, Devonshire AL (1991) Insecticide resistance by gene amplification in *Myzus persicae*. In: Denholm I, Devonshire A, Holloman DW (eds) Resistance 91. Elsevier, New York, pp 240–250

Georghiou GP (1980) Insecticide resistance and prospects for its management. Resid Rev 76:131–145

Georghiou GP, Lagunes-Tejeda A (1991) The occurrence of resistance to pesticides in arthropods. Food and Agriculture Organization of the United Nations, Rome

Georghiou GP, Saito T (eds) (1983) Pest resistance to pesticides. Plenum Press, New York

Gotoh O, Fujii-Kuriyama Y (1989) Evolution, structure, and gene regulation of cytochrome P-450. In: Ruckpaul K, Rein H (eds) Frontiers in biotransformation, vol 1. Taylor and Francis, New York, pp 195–243

Guengerich FP (1995) Human cytochrome P450 enzymes. In: Ortiz de Montellano PR (ed) Cytochrome P450: structure, mechanism, and biochemistry. Plenum Press, San Francisco, pp 473–535

Guengerich FP (1997) Comparisons of catalytic selectivity of cytochrome P450 subfamily enzymes from different species. Chemico-Biol Interact 106:161–182

Guillemaud T, Lenormand T, Bourguet D, Chevillon C, Pasteur N, Raymond M (1998) Evolution of resistance in *Culex pipiens*: allele replacement and changing environment. Evolution 52:443–453

Hatano R, Scott JG (1993) Anti-P450$_{lpr}$ antiserum inhibits the activation of chlorpyrifos to chlor-pyrifos-oxon in housefly microsomes. Pestic Biochem Physiol 45:228–233

Hemingway J, Karunarantne SHPP (1998) Mosquito carboxylesterases: a review of the molecular biology and biochemistry of a major insecticide resistance mechanism. Med Vet Entomol 12:1–12

Hodgson E (1983) The significance of cytochrome P-450 in insects. Insect Biochem 13:237–246

Hodgson E (1985) Microsomal mono-oxygenases. In: Kerkut GA, Gilbert LC (eds) Comprehensive insect physiology, biochemistry and pharmacology, vol 11. Pergamon Press, Oxford, pp 647–712

Hodgson E, Kulkarni AP (1983) Characterization of cytochrome P-450 in studies of insecticide resistance. In: Georghiou GP, Saito T (eds) Pest resistance to pesticides. Plenum Press, New York

Hodgson E, Silver IS, Butler LE, Lawton MP, Levi PE (1991) Metabolism. In: Hayes WJ Jr, Laws ER Jr (eds) Handbook of pesticide toxicology, vol 1. General principles. Academic Press, New York, pp 107–168

Jutsum AR, Heaney SP, Perrin BM, Wege PJ (1998) Pesticide resistance: assessment of risk and the development and implementation of effective management strategies. Pestic Sci 54:435–446

Konno T, Hodgson E, Dauterman WC (1989) Studies on methyl parathion resistance in *Heliothis virescens*. Pestic Biochem Physiol 33:189–199

Korytko PJ, Quimby FW, Scott JG (2000) Metabolism of phenanthrene by housefly CYP6D1 and dog liver cytochrome P450. J Biochem Molec Toxicol 14:20–25

Korytko PJ, Scott JG (1998) CYP6D1 protects thoracic ganglia of houseflies from the neurotoxic insecticide cypermethrin. Arch Insect Biochem Physiol 37:57–63

Kotze AC (1993) Cytochrome P450 monooxygenases in larvae of insecticide-susceptible and – resistant strains of the Australian sheep blowfly, *Lucilia cuprina*. Pestic Biochem Physiol 46:65–72

Kotze AC, Wallbank BE (1996) Esterase and monooxygenase activities in organophosphate-resistant strains of *Oryzaephilus surinamensis* (Coleoptera: Cucujidae). J Econ Entomol 89:571–576

Kulkarni AP, Hodgson E (1980) Metabolism of insecticides by mixed function oxidase systems. Pharmacol Ther 8:379–475

Lee SST, Scott JG (1989) Microsomal cytochrome P450 monooxygenases in the housefly (*Musca domestica* L): biochemical changes associated with insecticide resistance and phenobarbital induction. Pestic Biochem Physiol 35:1–10

Lee SST, Scott JG (1992) Tissue distribution of microsomal cytochrome P-450 monooxygenases and their inducibility by phenobarbital in the housefly, *Musca domestica* L. Insect Biochem Mol Biol 22:699–711

Lindberg RLP, Negishi M (1989) Alteration of mouse cytochrome P450$_{coh}$ substrate specificity by mutation of a single amino-acid residue. Nature 339:632–634

Liu N, Scott JG (1995) Genetics of resistance to pyrethroid insecticides in the housefly, *Musca domestica*. Pestic Biochem Physiol 52:116–124

Liu N, Scott JG (1996) Genetic analysis of factors controlling elevated cytochrome P450, CYP6D1, cytochrome b$_5$, P450 reductase and monooxygenase activities in LPR houseflies, *Musca domestica*. Biochem Genet 34:133–148

Liu N, Scott JG (1997a) Inheritance of CYP6D1-mediated pyrethroid resistance in housefly (Diptera: Muscidae). J Econ Entomol 90:1478–1481

Liu N, Scott JG (1997b) Phenobarbital induction of CYP6D1 is due to a *trans* acting factor on autosome 2 in houseflies, *Musca domestica*. Insect Mol Biol 6:77–81

Liu N, Scott JG (1998) Increased transcription of CYP6D1 causes cytochrome P450-mediated insecticide resistance in housefly. Insect Biochem Mol Biol 28:531–535

Liu N, Tomita T, Scott JG (1995) Allele-specific PCR reveals that the cytochrome P450$_{lpr}$ gene is on chromosome 1 in the housefly, *Musca domestica*. Experientia 51:164–167

Lu AYH, Coon MJ (1968) Role of hemoprotein P-450 in fatty acid omega-hydroxylation in a soluble enzyme system for liver microsomes. J Biol Chem 243:1331–1332

Mansuy D (1998) The great diversity of reactions catalyzed by cytochromes P450. Comp Biochem Physiol C 121:5–14

Mouches C, Pasteur N, Berge JB, Hyrien O, Raymond M, De Saint Vincent BR, De Silvestri M, Georghiou GP (1986) Amplification of an esterase gene is responsible for insecticide resistance in a California *Culex* mosquito. Science 233:778–780

Nelson DR (1998) http://drnelson.utmem.edu/nelsonhomepage.html

Nelson DR, Koymans L, Kamataki T, Stegeman JJ, Feyereisen R, Waxman DJ, Waterman MR, Gotoh O, Coon MJ, Estabrook RW, Gunsalus IC, Nebert DW (1996) P450 superfamily: update on new sequences, gene mapping, accession numbers and nomenclature. Pharmacogenetics 6:1–42

National Research Council (1986) Executive summary, pesticide resistance strategies and tactics for management. In: National Research Council (ed) Pesticide resistance strategies and tactics for management. National Academy Press, Washington, DC, pp 1–9

Oppenoorth FJ (1985) Biochemistry and genetics of insecticide resistance. In: Kerkut GA, Gilbert LI (eds) Comprehensive insect physiology, biochemistry, and pharmacology. Pergamon Press, Oxford, pp 731–774

Peterson JA, Prough RA (1986) Cytochrome P-450 reductase and cytochrome b$_5$ in cytochrome P-450 catalysis. In: Ortiz de Montellano PR (ed) Cytochrome P-450 structure, mechanism, and biochemistry. Plenum Press, New York, 89 pp

Plapp FW Jr (1984) The genetic basis of insecticide resistance in the housefly: evidence that single locus plays a major role in metabolic resistance to insecticides. Pestic Biochem Physiol 22:194–201

Pompon D (1987) Rabbit liver cytochrome P-450 LM$_2$: roles of substrates, inhibitors and cytochrome b$_5$ in modulating the partition between productive and abortive mechanisms. Biochemistry 26:6429–6435

Rendic S, Di Carlo FJ (1997) Human cytochrome P450 enzymes: a status report summarizing their reactions, substrates, inducers, and inhibitors. Drug Metab Rev 29:413–580

Roush RT, Tabashnik BE (eds) (1990) Pesticide resistance in arthropods. Chapman and Hall, New York

Sawicki RM (1962) Insecticidal activity of pyrethrum extract and its four insecticidal constituents against houseflies. III. Knock-down and recovery of flies treated with pyrethrum extract with and without piperonyl butoxide. J Sci Food Agric 13:283–291

Scott JG (1991) Insecticide resistance in insects. In: Pimentel D (ed) Handbook of pest management in agriculture, vol 2. CRC Press, Boca Raton, p 663

Scott JG (1996) Inhibitors of CYP6D1 in housefly microsomes. Insect Biochem Mol Biol 26:645–649

Scott JG (1999) Molecular basis of insecticide resistance: cytochromes P450. Insect Biochem Mol Biol 29:757–777

Scott JG, Georghiou GP (1985) Rapid development of high-level permethrin resistance in a field-collected strain of housefly (Diptera: Muscidae) under laboratory selection. J Econ Entomol 78:316–319

Scott JG, Georghiou GP (1986a) The biochemical genetics of permethrin resistance in the Learn-PyR strain of housefly. Biochem Genet 24:25–37

Scott JG, Georghiou GP (1986b) Mechanisms responsible for high levels of permethrin resistance in the housefly. Pestic Sci 17:195–206

Scott JG, Lee SST (1993a) Purification and characterization of a cytochrome P-450 from insecticide susceptible and resistant strains of housefly, *Musca domestica* L. Arch Insect Biochem Physiol 24:1–19

Scott JG, Lee SST (1993b) Tissue distribution of microsomal cytochrome P-450 monooxygenases and their inducibility by phenobarbital in the insecticide resistant LPR strain of housefly, *Musca domestica* L. Insect Biochem Mol Biol 23:729–738

Scott JG, Sridhar P, Liu N (1996) Adult specific expression and induction of cytochrome P450$_{lpr}$ in house flies. Arch Insect Biochem Physiol 31:313–323

Scott JG, Liu N, Wen Z (1998) Insect cytochromes P450: diversity, insecticide resistance and tolerance to plant toxins. Comp Biochem Physiol C 121:147–155

Scott JG, Liu N, Wen Z, Smith FF, Kasai S, Horak CE (1999) House fly cytochrome P450 *CYP6D1*:5 prime flanking sequences and comparison of alleles. Gene 226:347–353

Scott JG, Foroozesh M, Hopkins NE, Alefantis TG, Alworth WL (2000) Inhibition of cytochrome P450 6D1 by alkynylarenes, methylenedioxyarenes and other substituted aromatics. Pestic Biochem Physiol 67:63–71

Sheppard DC, Joyce JA (1998) Increased susceptibility of pyrethroid-resistant hornflies (Diptera: Muscidae) to chlorfenapyr. J Econ Entomol 91:398–400

Smith FF, Scott JG (1997) Functional expression of housefly (*Musca domestica*) cytochrome P450 *CYP6D1* in yeast (*Saccharomyces cerevisiae*). Insect Biochem Mol Biol 27:999–1006

Sun CN, Tsai YC, Chiang FM (1992) Resistance in the diamondback moth to pyrethroids and benzoylphenylureas. In: Mullin CA, Scott JG (eds) Molecular mechanisms of insecticide resistance: diversity among insects. ACS Symposium Series 505. American Chemical Society, Washington, DC, pp 149–167

Terriere LC (1983) Enzyme induction, gene amplification, and insect resistance to insecticides. In: Georghiou GP, Saito T (eds) Pest resistance to pesticides. Plenum Press, New York, pp 265–298

Terriere LC (1984) Induction of detoxication enzymes in insects. Annu Rev Entomol 29:71–88

Terriere LC, Yu SJ (1974) The induction of detoxifying enzymes in insects. J Agric Food Chem 22:366–376

Tomita T, Scott JG (1995) cDNA and deduced protein sequence of *CYP6D1*: the putative gene for a cytochrome P450 responsible for pyrethroid resistance in the housefly. Insect Biochem Mol Biol 25:275–283

Tomita T, Liu N, Smith FF, Sridhar P, Scott JG (1995) Molecular mechanisms involved in increased expression of a cytochrome P450 responsible for pyrethroid resistance in the housefly, *Musca domestica*. Insect Mol Biol 4:135–140

Valles SM, Yu SJ (1996) Detection and biochemical characterization of insecticide resistance in the German cockroach (Dictyoptera: Blattelidae). J Econ Entomol 89:21–26

Vatsis KP, Gurka DP, Hollenberg PF (1980) Involvement of cytochrome b$_5$ in the NADPH-dependent regioselective hydroxylation of *N*-methylcarbazole by cytochrome P-450LM2 and P-450LM4 in a reconstituted liver microsomal enzyme system. In: Gustafsson J, Carlstedt-Duke J, Mode A, Rafter J (eds) Biochemistry, biophysics, and regulation of cytochrome P-450. Elsevier, New York, pp 347–350

Vincent DR, Moldenke AF, Farnsworth DE, Terriere LC (1985) Cytochrome P-450 in insects. 6. Age dependency and phenobarbital induction of cytochrome P-450, P-450 reductase, and monooxygenase activities in susceptible and resistant strains of *Musca domestica*. Pestic Biochem Physiol 23:171–181

Wen Z, Scott JG (1997) Cross-resistance to imidacloprid in strains of German cockroach (*Blattella germanica*) and housefly (*Musca domestica*). Pestic Sci 49:367–371

Wheelock GD, Scott JG (1989) Simultaneous purification of a cytochrome P-450 and cytochrome b$_5$ from the housefly *Musca domestica* L. Insect Biochem 19:481–488

Wheelock GD, Scott JG (1990) Immunological detection of cytochrome P450 from insecticide resistant and susceptible houseflies (*Musca domestica*). Pestic Biochem Physiol 38:130–139

Wheelock GD, Scott JG (1992a) Anti-P450$_{lpr}$ antiserum inhibits specific monooxygenase activities in LPR housefly microsomes. J Exp Zool 264:153–158

Wheelock GD, Scott JG (1992b) The role of cytochrome P450$_{lpr}$ in deltamethrin metabolism by pyrethroid resistant and susceptible strains of houseflies. Pestic Biochem Physiol 43:67–77

Wheelock GD, Konno Y, Scott JG (1991) Expression of P450$_{lpr}$ is developmentally regulated and limited to housefly. J Biochem Toxicol 6:239–246

Wilkinson CF (1967) Penetration, metabolism, and synergistic activity with carbaryl of some simple derivatives of 1,3-benzodioxole in the housefly. J Agric Food Chem 15:139–147

Wilkinson CF (1983) Role of mixed-function oxidases in insecticide resistance. In: Georghiou GP, Saito T (eds) Pest resistance to pesticides. Plenum Press, New York, pp 175–206

Wilkinson CF (1985) Role of mixed-function oxidases in insect growth and development. In: Hedin PA (ed) Bioregulators for pest control. ACS Symp Ser 276. American Chemical Society, Washington, DC, pp 161–176

World Health Organization (1957) Expert committee on insecticides. WHO Tech Rep Ser 7th Rep, 125 pp

World Health Organization (1976) 22nd Report by the Expert Committee on Insecticides. WHO Tech Rep Ser 585:77

Yu SJ (1996) Insect glutathione S-transferases. Zool Stud 35:9–19

Zhang M, Scott JG (1994) Cytochrome b_5 involvement in cytochrome P450 monooxygenase activities in housefly microsomes. Arch Insect Biochem Physiol 27:205–216

Zhang M, Scott JG (1996) Cytochrome b_5 is essential for cytochrome P450 6D1-mediated cypermethrin resistance in LPR houseflies. Pestic Biochem Physiol 55:150–156

Mechanisms of Organophosphate Resistance in Insects

BLAIR D. SIEGFRIED and MICHAEL E. SCHARF[1]

1
Introduction

The organophosphate (OP) insecticides were first introduced as pest control agents over 50 years ago. These compounds still account for approximately 30% of the registered synthetic insecticides and acaricides in the US, and are used in a variety of agriculture, public hygiene and medical settings (Ware 1994). Organophosphate insecticides cause toxicity through inhibition of acetyl-cholinesterase (AChE), which is responsible for the degradation of the excitatory neurotransmitter, acetylcholine, thereby terminating transmission of nerve impulses at cholinergic synapses. Inhibition of this enzyme prolongs the residence time of acetylcholine at synapses resulting in hyper-excitation and eventual death.

The organophosphates are one of the largest groups of insecticides in use today and exhibit a wide range of structural diversity. They exist as neutral ester or amide derivatives of phosphoric acid that carry a phosphoryl or thio-phosphoryl group (Fig. 1). The basic structural skeleton of the organophosphates allows a very wide variety of constituents to be attached without loss of toxicity. Importantly, many organophosphates, specifically those possessing a thiophosphate linkage, require oxidative activation because the true phosphate is a more potent inhibitor of acetylcholinesterase.

Organophosphates were first introduced as insecticides in 1944, although early compounds such as TEPP (tetraethyl pyrophosphate) were soon discontinued because of very high mammalian toxicity, a consequence of a shared target site between insects and mammals (Eto 1974). Parathion was the first successfully introduced OP insecticide and it is still in use today. Later developments in organophosphate chemistry, such as the phosphoester methyl substituted compounds and malathion, increased selectivity and resulted in widespread acceptance of these compounds for commercial field application. However, the future of organophosphate insecticides is uncertain because of high mammalian toxicity and concern for worker safety and environmental protection. Of all the organophosphate insecticides in use worldwide, 30 are

[1] Department of Entomology, 202 Plant Industry Bldg., University of Nebraska, Lincoln, NE 68583-0816, USA

Isaac Ishaaya (Ed.): Biochemical Sites of
Insecticide Action and Resistance
© Springer-Verlag Berlin, Heidelberg 2001

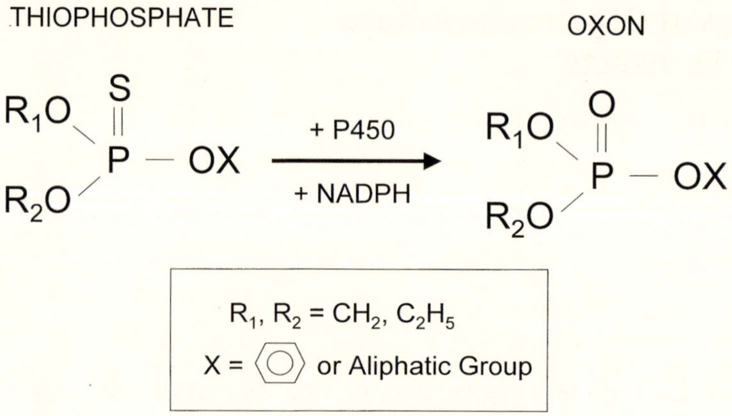

THIOPHOSPHATE   OXON

$$\underset{R_2O}{\overset{R_1O}{\diagdown}} \overset{S}{\underset{\diagup}{\overset{\|}{P}}} - OX \quad \xrightarrow[+ \ NADPH]{+ \ P450} \quad \underset{R_2O}{\overset{R_1O}{\diagdown}} \overset{O}{\underset{\diagup}{\overset{\|}{P}}} - OX$$

$R_1, R_2 = CH_2, C_2H_5$

$X = \langle\bigcirc\rangle$ or Aliphatic Group

Fig. 1. General structure of organophosphate insecticides

classed as extremely hazardous, 29 are highly hazardous, 24 are moderately hazardous, and 8 are slightly hazardous (Perry et al. 1998). With legislation such as the Food Quality Protection Act in the US, increased scrutiny with regard to existing and future registrations is likely because of the potential for high mammalian toxicity.

Organophosphate insecticides had largely replaced the organochlorines by the mid-1960s because of the widespread occurrence of resistance to both DDT and cyclodiene insecticides (Georghiou 1986). However, resistance to organophosphates was soon reported in a variety of agricultural and medically important insects. By 1959, only 15 years after their introduction, organophosphate resistance had been reported in 20 agricultural species and 9 species of importance to public health (Brown 1961). By 1991, organophosphate resistance was present in 241 species of arthropods and accounted for nearly 50% of the reported instances of resistance (Georghiou and Lagunes-Tejeda 1991).

Organophosphate resistance is concentrated in the medically important dipterans, but is also common in agriculturally important orders such as Lepidoptera, Coleoptera and Homoptera. The taxonomic distribution of OP resistance parallels that observed for insecticide resistance in general, and reflects the widespread use of these compounds for controlling pest species and general selection across taxonomic levels (Fig. 2).

2
Physiological Mechanisms of Resistance

The intoxication of an insect by an insecticide encompasses three levels of pharmacological interaction: barriers to penetration, increased metabolic detoxification, and altered molecular interaction of the insecticide with its ulti-

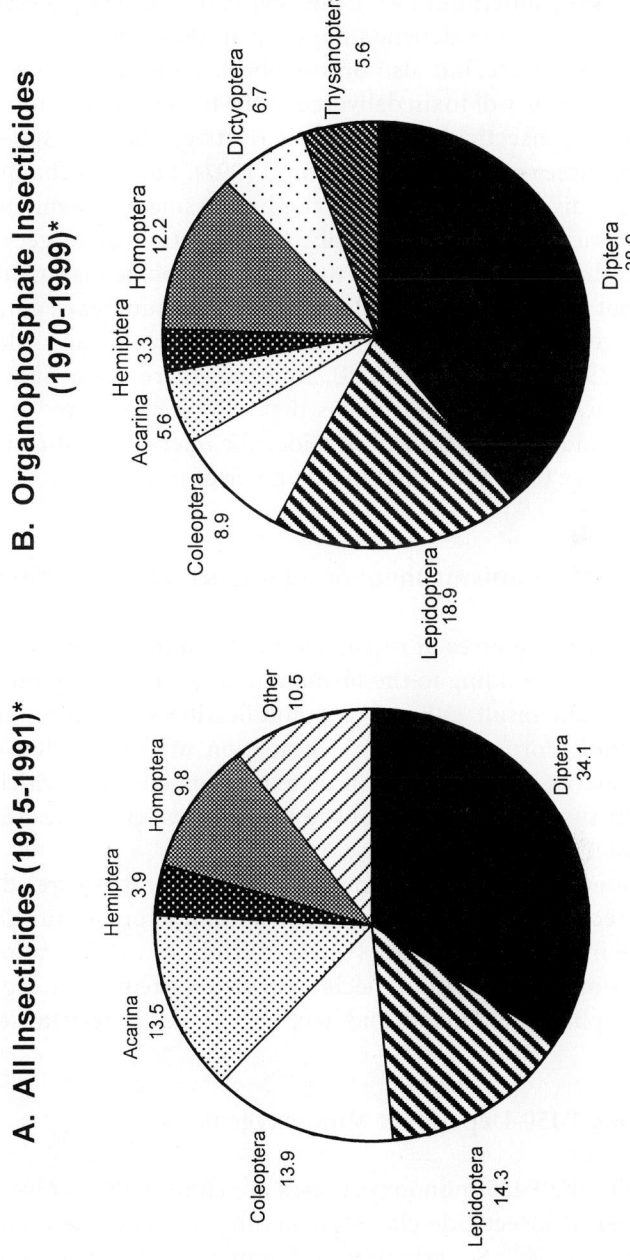

Fig. 2. Distribution of **A** insecticide- and **B** organophosphate-resistant species by order. (**A** adapted from Georghiou and Lagunes-Tejeda 1991; **B** adapted from a search of the National Agricultural Library AGRICOLA citation database and includes 168 citations related to organophosphate resistance in insects)

mate target site (Soderlund 1997). The ability of insects to tolerate exposure to an insecticide therefore depends not only on the interaction of a toxic molecule with a target site, but also on the pharmacokinetic processes that affect the rate and quantity of toxin delivered to the target site. The number of known mechanisms of insecticide resistance is relatively small despite the chemical diversity of these compounds (Soderlund 1997). These mechanisms have been variously subdivided, but generally can be classified as involving either: penetration barriers, enhanced metabolic detoxification, or target site insensitivity. The mechanisms of resistance to organophosphate insecticides are similar to those that have been extensively reviewed by a number of authors for insecticides in general (e.g., Oppenoorth 1985; Soderlund and Blomquist 1991; Soderlund 1997; Perry et al. 1998). However, there are a number of unique characteristics of organophosphates that will be considered in the following discussion. Additionally, we will consider the interaction of resistance mechanisms that have been identified in a number of pest species.

2.1
Resistance Mechanisms Involving Enhanced Biotransformation

There are many instances of resistance due to an increased capacity to metabolize insecticides, leading to the formation of less toxic metabolites. Increased metabolism can result either from modification of existing enzyme forms, making them more suitable for degradation of insecticides, or via factors leading to increased production of detoxification enzymes which were already available in susceptible insects but in much lower quantities (Soderlund and Bloomquist 1991).

Organophosphates possess a number of structural features that are unique among insecticides and provide important sites for biotransformation that result both in increased toxicity (i.e., activation) or detoxification. The following discussion will focus on specific enzyme systems that are important to organophosphate metabolism and have been linked to resistance mechanisms.

2.1.1
Cytochrome-P450-Dependent Monooxygenases

The cytochrome P450 monooxygenases are commonly involved in resistance to a number of insecticide classes, including the OPs (see reviews by Agosin 1985; Hodgson 1985; Soderlund and Bloomquist 1990; Soderlund 1997; Scott 1999). Cytochrome P450s are heme-containing proteins which are present in all cellular organisms, and are microsomal (endoplasmic reticulum-bound) in eukaryotes. The P450s require the cofactor NADPH for catalysis of oxidative reactions, which includes metabolism of endogenous substances such as hormones and deleterious exogenous materials such as secondary plant compounds and insecticides. The net result of these oxidations is the conversion of apolar substrates to more polar products that either perform some biological function (as with hormones) or are more readily acted upon by secondary conjugation enzymes and excreted from the body (as with xenobiotics). In addition, increased P450 expression is often inducible by apolar substances. In many instances, P450 inducers are specific substrates, while, in others, there does not appear to be a direct relationship with catalysis (Terriere 1984; Hodgson 1985).

Experimental evidence now suggests that P450s exist in multiple forms within any organism, including insects, with each form having a potentially narrow substrate specificity (Feyereisen 1999; Scott 1999). P450s are typically over-expressed in insecticide-resistant populations, and multiple resistance-associated forms have been identified based on protein- (Ronis et al. 1988) and nucleic-acid-based experimental approaches (Scott et al. 1994; Pittendrigh et al. 1997). As it relates to insecticide resistance, however, the molecular basis for this overexpression has until recently remained largely unknown. In *Musca domestica*, both P450-based pyrethroid resistance and phenobarbital inducibility (in susceptible populations) map to an autosomal locus separate from the structural gene which encodes a well-characterized P450 protein (CYP6D1; Liu and Scott 1996, 1997). This "trans" regulation of constitutive overexpression in pyrethroid resistant populations is thought to result from a mutation to a gene regulating the induction process. To date, no parallel understanding exists for any other P450s involved in OP resistance.

Relative to other OP resistance mechanisms, there have been substantially fewer incidences where cytochrome P450s have been conclusively linked with OP resistance. To make such identifications, researchers commonly rely on synergism studies of OP toxicity (e.g., Siegfried et al. 1990; Berrada et al. 1994; Scharf et al. 1996; Miota et al. 1998) or metabolism studies employing radiolabeled OP insecticides (e.g., Siegfried et al. 1990; Hatano and Scott 1993; Zhao et al. 1994). Throughout the literature, however, speculation of the involvement of cytochrome P450s in resistance to the OP (and other insecticide) classes is based solely on results of biochemical determinations (i.e., P450 content by carbon monoxide difference spectra or NADPH-dependent metabolism of model substrates). Unless in combination with synergism, OP metabolism and/or selection-based studies, biochemical results are considered circumstantial and should be interpreted with great care (Scott 1991). This is especially true when dealing with field-collected insect populations which are resistant to a number of insecticide classes, or when the model substrates used in metabolism studies do not share structural features with the OP insecticide in question.

OP insecticides possess multiple structural features which can be potentially acted upon by cytochrome P450s. Of primary importance is the activation of OP insecticides (e.g., Siegfried et al. 1990; Hatano and Scott 1993; Dunkov et al. 1997). The phosphorothioate OPs require sulfoxidation to the corresponding oxon before they effectively inhibit acetylcholinesterases. With regard to resistance, however, the documentation of reduced oxon formation in OP-resistant insects as specifically mediated by reduced P450 sulfoxidation is extremely uncommon. In fact, the only reported case of reduced sulfoxidation occurred in an OP-resistant *Heliothis virescens* strain (Konno et al. 1989) which exhibited significantly lower rates of methyl-parathion sulfoxidation in vitro.

Other structural features of organophosphate insecticides are potentially susceptible to P450-based oxidative metabolism, and are therefore most likely to contribute to OP resistance. These include deethylation and demethylation of the phosphate ester, aromatic ring hydroxylation, and ester cleavage. The five relevant structural features (and examples of OP insecticides which possess these structures) which can potentially be subject to modification by P450 are: (1) *O*-ethyl substitutions (e.g., chlorpyrifos, parathion); (2) *O*-methyl substitutions (e.g., chlorpyrifos-methyl, methyl parathion), (3) *N*-methyl substitutions (e.g., dimethoate), (4) aromatic rings (e.g., parathion and methyl parathion), and (5) ester linkages (e.g., parathion, chlorpyrifos, diazinon). Of the relatively few insect P450s isolated and studied to date, only two cases exist where OP metabolism has been documented by specific P450 isoforms: (1) sulfoxidation of chlorpyrifos by *Musca domestica* CYP6D1 (Hatano and Scott 1993) and, (2) sulfoxidation and ester cleavage of diazinon by *Drosophila melanogaster* CYP6A2 (Dunkov et al. 1997).

2.1.2
Glutathione *S*-Transferases

The glutathione *S*-transferases (GSTs) are a detoxification enzyme family which typically act upon electrophilic metabolites of various insecticides, catalyzing a conjugation to the tripeptide glutathione and thus facilitating more rapid excretion (Dauterman 1985; Yu 1996). As discussed below, however, GSTs can play a major role in OP resistance, especially those which possess *O*-methyl substitutions. GSTs are relatively stable, soluble (cytosolic) enzymes, that exist in multiple forms within most insects examined to date (Yu 1996). Like P450, increased GST expression is often inducible by xenobiotic exposure at levels inversely proportional to baselines (Yu 1996), suggesting similar gene regulation factors may be responsible for induction and constitutive overexpression of the P450s and GSTs (Soderlund 1997).

Several approaches have been employed by researchers to document GST involvement in resistance and to biochemically characterize their activity in association with resistance. As mentioned previously, insecticide synergism bioassays are the most straight-forward approach for identifying the involvement of various resistance mechanisms. Synergists which inhibit GSTs include diethyl maleate, methyl iodide, *tert*-phenylbutenone (Raffa and Priester 1985), and saligenin cyclic phosphate (Shiotsuki et al. 1992). Conclusive documentation of the involvement of GST in OP metabolism/resistance is made by studying the metabolism or radiolabeled insecticide substrates. Elevated in vivo formation of immobile polar metabolites provides favorable evidence; however, glutathione-dependence of polar metabolite formation by cytosolic fractions in vitro is the most conclusive experimental result to implicate GSTs in resistance (Motoyama and Dauterman 1980). Another interesting and very informative technique reported in the literature for identifying GST involvement in OP resistance is the spectrophotometric quantification of S-4-nitrophenylglutathione formation following incubation of methyl parathion with purified GST protein (Chiang and Sun 1993).

The two model substrates commonly used to identify elevated GST activity in association with resistance are chlorodinitro- and dichloronitrobenzene (CDNB and DCNB, respectively). Typically, pH optima for metabolism of these substrates is in the 8–9 range, with CDNB being conjugated at much higher rates than DCNB. It is also important to note that elevation of GST activity alone should not be considered as diagnostic for resistance unless accompanied by supporting data from metabolism studies. Because multiple detoxification enzymes can act upon the same functional groups and multiple GST forms can be present, it may also be necessary to conduct model substrate and insecticide metabolism studies with GST proteins that have been purified to homogeneity.

Regarding OP resistance, GSTs have a historically well-documented involvement in the primary metabolism of O-methyl-substituted thiophosphate- and -oxon analogs of OP insecticides. However, the metabolic products commonly identified are not only O-alkyl-, but include O-aryl- and phosphonate-conjugates (Motoyama and Dauterman 1980). With respect to more recent literature, molecular characterizations of individual GST proteins involved in OP resistance are less common. One GST isoform (termed GST-3) purified from *Plutella xylostella* (L.) was found to be elevated in a methyl-parathion-resistant strain and to rapidly metabolize and conjugate methyl parathion to form S-4-nitrophenylglutathione (Chiang and Sun 1993). Subsequent molecular efforts led to the cloning and heterologous expression of the *P. xylostella* GST-3 isozyme (Huang et al. 1998) and indicated that factors other than gene amplification are responsible for elevated expression of the protein. Multiple GST genes from OP-resistant *M. domestica* strains have recently been isolated, cloned and/or characterized by Wang et al. (1991), Syvanen et al. (1994, 1996), and Zhou and Syvanen (1997). At the time of this review, however, the role of only one of the expression products of these genes with respect to OP metabolism is understood (Syvanen et al. 1996).

2.1.3
Hydrolytic Enzymes

Hydrolytic enzymes (esterases, carboxylesterases) are particularly important in the biotransformation of organophosphate insecticides because the phosphate ester is potentially susceptible to hydrolysis, and its integrity is essential to the ability of these compounds to inhibit the target enzyme, acetylcholinesterase. Additionally, the carboxylester moiety associated with the side chain of certain OP molecules such as malathion is a potential site of enzymatic hydrolysis and detoxification. Oxon analogs of organophosphate insecticides may also act as potent inhibitors of hydrolytic enzymes which act to sequester the phosphate ester and prevent interaction with the target site.

The hydrolase enzymes constitute a large family of enzymes that are distinguished based on electrophoretic mobility and substrate and inhibitor specificity. Insect hydrolases exist in multiple forms, which can be resolved by electrophoresis and detected using simple spectrophotometric assays with substrates such as the acyl esters of α- and β-naphthol or p-nitrophenol.

The involvement of hydrolytic enzymes in resistance has been inferred based on synergism bioassays employing such compounds as DEF (S,S,S-tributyl phosphorotrithioate) and TEPP (O,O,O-triphenyl phosphate) which enhance toxicity by inhibition of hydrolytic metabolism. More commonly, elevated total esterase activity or elevated activity of a specific isozyme in crude homogenates of insect tissue is cited as evidence of involvement of this enzyme system. While such data provide a fast and relatively inexpensive means to assess resistance status, results must be interpreted cautiously as the enzymes involved with hydrolysis of model substrates may be different from those responsible for insecticide metabolism. Additionally, elevated activity is not always correlated with resistance status (Siegfried and Scott 1992). Unfortunately, the corroborating data that documents the activity of a specific esterase isozyme in the hydrolysis of an organophosphate insecticide using radiolabeled insecticide substrates is often lacking.

Although enhanced esterase activity has been associated with organophosphate resistance in many insect pests, the biochemical and molecular mechanisms of esterase-mediated resistance have been elucidated in only a few species. Two different functions have been established thus far in terms of the ability of these enzymes to confer resistance to organophosphate insecticides: (1) overproduction of esterases that act to sequester the phosphate ester and (2) qualitative changes in enzyme structure resulting in increased catalytic activity. The following discussion will focus on species that have been extensively studied and for which the role of a detoxifying esterase has been well characterized.

2.1.3.1
Quantitative Changes (Gene Amplification)

The most common of the mechanisms involving hydrolytic enzymes in organophosphate resistance is related to esterases that detoxify organophosphate insecticides by sequestering them through phosphorylation of the enzyme's catalytic center (Devonshire and Field 1991). The over-produced esterases act primarily as "high affinity sponges" that are preferentially phosphorylated by molar equivalents of organophosphate (Newcomb et al. 1997). Perhaps the most extensive characterization of a resistance mechanism for any insecticides involves the biochemical and molecular investigations documenting the overexpression of these esterase enzymes in certain aphid species and in *Culex* mosquitoes.

Populations of the peach-potato aphid, *Myzus persicae* have developed resistance to organophosphate insecticides around the world in both field and glasshouse populations (Devonshire and Field 1991). In all cases studied, the primary resistance mechanism identified involves the synthesis of high levels of one of two closely related, insecticide-degrading esterases (Field and Devonshire 1992). Two forms of the esterase can confer resistance, E4 which is associated with an A1,3 chromosomal translocation and FE4, which is associated with aphids of normal karyotype (Devonshire and Sawiki 1979). The FE4 esterase is a variant of E4 with a slightly higher electrophoretic mobility. Devonshire (1977) found that the total carboxylesterase activity was 60-fold higher in organophosphate-resistant aphids than in susceptible aphids, and there was a positive correlation between resistance level and carboxylesterase activity. The observed doubling in esterase content through a series of seven progressively more resistant clones led to the conclusion that gene amplification was the basis of elevated enzyme synthesis and insecticide resistance (Devonshire and Sawicki 1979).

Purification and characterization of the E4 isozyme were conducted by Devonshire and Moores (1982). The purified E4 isozyme has an estimated M_r of 65,000 and exhibited only weak phosphatase activity. More specifically, recovery of activity following inhibition was observed only after 10 h suggesting that the main role of this enzyme was to sequester the organophosphate molecule thereby preventing it from interacting with the target molecule, acetylcholinesterase. Polyclonal antiserum raised against the E4 protein recognized up to 3% of the total protein present in resistant aphids as E4 protein, thereby providing resistant aphids with a large pool of available sequestration protein.

Confirmation that increased E4 content and organophosphate resistance was the result of gene amplification was provided by Field et al. (1988), when E4 cDNA was used as a probe to document increased esterase gene copy number in resistant aphids with either the E4 or the FE4 enzyme. Binding of the cDNA probe to EcoRI-digested aphid DNA revealed that amplified esterase sequences were present on an 8-kb fragment in translocated aphids with E4 and a 4-kb fragment in aphids with FE4. The correlation between restriction patterns and karyotypes indicated close linkage between the translocation and the events leading to esterase overproduction.

Some resistant variants of *M. persicae* have been shown to lose resistance quickly (within several generations) in the absence of exposure to insecticide (ffrench-Constant et al. 1988). When these variants were examined at a molecular level, the translocation was found to be associated with instability of resistance. Although loss of resistance occurred at varying rates and response to selection also varied, loss and recovery of resistance occurred only in clones with the translocated karyotype. The revertant susceptible aphids retained the translocation, and exhibited a slightly higher level of E4-encoding mRNA relative to susceptible clones lacking the chromosomal translocation (Field and Devonshire 1992). Interestingly, reversion of resistance involves a decline in E4 enzyme activity and transcript abundance to levels near to that of fully susceptible clones but does not involve a loss of amplified DNA sequences (Field et al. 1989a). The loss of resistance in these clones is associated with a reduction or loss of DNA methylation, but re-selection for resistance does result in a concomitant increase in DNA methylation (Field et al. 1989b; Hick et al. 1996).

The greenbug, *Schizaphis graminum* (Homoptera: Aphididae), is a worldwide pest of small grains and sorghum, and has been repeatedly shown to develop resistance to organophosphate insecticides (reviewed by Ono et al. 1994). Organophosphate resistance in *S. graminum* is associated with elevated esterase activity, and two distinct patterns of enhanced esterase isozymes have been detected by native polyacrylamide gel electrophoresis (Ono et al. 1994; Shufran et al. 1996). The type I strain of resistant *S. graminum* produces a single esterase band that is undetectable in susceptible strains. Type II resistant aphids produce multiple, darkly staining esterase bands although one band, different from type I, predominates. A combination of the two esterase banding patterns has not been observed in field populations.

Based on in vivo and in vitro metabolism experiments, resistance apparently results from the increased levels of esterases that bind more insecticide molecules relative to the susceptible strain (Siegfried and Ono 1993a,b; Ono et al. 1994) and is similar to the mechanisms previously described for *M. persicae*. The kinetic properties of partially purified esterase from type II aphids suggest its overproduction based on a threefold elevated V_{max} and similar K_M in the resistant vs. susceptible strains with α-naphthyl butyrate as substrate (Siegfried and Zera 1994). However, a polyclonal antibody raised against the purified type II esterase resulted in signals of similar intensity in Western blots of total homogenates of susceptible type I and type II *S. graminum* (Siegfried et al. 1997a), indicating that the amount of the type II esterase was similar in all strains. In contrast, an antiserum to esterase E4 of *M. persicae* reacted with the esterase from the type I *S. graminum* to give a more intense signal than with either the susceptible or type II strain (Siegfried et al. 1997a). These results suggest that the type I greenbug esterase and esterase E4 of *M. persicae* have shared epitopes, and may be similar in structure. However, the type II esterase did not cross-react with the E4 antiserum suggesting the two esterases are not as similar. Type I and II esterases share certain characteristics such as high sensitivity to inhibition by paraoxon and $HgCl_2$ and partial inhibition with eserine while lacking reactivity toward acetylthiocholine (Siegfried et al. 1997b; Ono et al. 1999).

Gene amplification has been shown to be the molecular mechanism of organophosphate resistance in the type I strain of *S. graminum* but not in the type II strain (Ono et al. 1999). Repeated Southern hybridization using a type I esterase DNA fragment obtained by degenerate polymerase chain reaction (PCR) showed no evidence for amplification of similar esterase sequences in the genome of the type II strain. In contrast, the type I greenbugs exhibited a four- to eightfold higher number of copies of the gene than the susceptible strain, and the amplified esterase sequences in *S. graminum* were methylated as in *M. persicae*. Sequences of the gene encoding type I *S. graminum* confirm that this gene is very similar to the amplified esterase E4/FE4 genes of *M. persicae*, with 73% identity shared between inferred amino acid sequences (Ono et al. 1999).

Because of the similarities between esterase E4/FE4 of *M. persicae* and the type I esterase of *S. graminum*, it is possible that genes may have evolved from a common ancestor and have since diverged. Base changes and size differences in the second intron between the two species support this hypothesis. The change in the sequence arrangement that resulted in amplification may be a relatively recent event that occurred independently in the two species. Because of the shared ancestral origin, similarities between genes could have resulted in similar amplification processes in response to selective pressures from exposure to organophosphate insecticides in both insects.

In *Culex* mosquitoes, resistance to organophosphate insecticides has been widely reported and has also been correlated with highly active carboxylesterases, although the number of esterases that have been identified is much greater than in *M. persicae* or *S. graminum* (Devonshire and Field 1991; Soderlund 1997). Esterase isozymes in *Culex* mosquitoes are numbered according to the order of discovery and named A or B depending on the specificity of the enzyme for α- or β-naphthyl ester substrates. World-wide surveys of resistance to organophosphate insecticides have revealed that only two loci involving increased hydrolytic activity have developed as major resistance alleles. These two loci (*Est-2* or esterase B and *Est-3* or esterase A) code for carboxyl ester hydrolases that are over-expressed in resistant strains and result in the detoxification of organophosphate esters first by sequestration and then by hydrolysis following esterase reactivation. Six distinct esterase isozymes have been described for esterase B and four have been described at the esterase A locus. In some situations, both esterase A and B loci have been identified to co-occur in the same resistant strains (Wirth et al. 1990; Rooker et al. 1996), and the most common resistance-associated phenotype is co-elevation of esterases A2 and B2 which occur in complete linkage disequilibrium (i.e., the two esterases are always co-elevated) (Vaughan et al. 1995).

The esterases A and B apparently comprise two gene families that have homology within, but not between, the groups. Antiserum to esterase A1 cross-reacts with other A-esterases, but not with B-esterases, and B1 antiserum similarly reacts only with other B-esterases (Mouchés et al. 1987). These results suggest that the A- and B-esterases constitute distinct groups of enzymes that are products of separate but closely linked genetic loci (Soderlund 1997). However, Karunaratne et al. (1995a,b) have shown that antiserum raised to nondenatured A2-esterase does cross-react with the B2-esterase, although the specificity was approximately 50-fold lower for B2 than A2, suggesting that the two enzymes share common epitopes and amino acid sequences.

Immunological experiments have also demonstrated quantitative increases in esterase protein with estimates of esterase B1 in a Californian strain of C. quinquefasciatus having 500-fold more esterase B1, comprising 6–12% of its total protein (Mouchés et al. 1987). Although elevated esterase activity in resistant mosquitoes was assumed for many years, direct inhibition of isozymes by oxon analogs of OP insecticides provided the first documentation that resistance resulted from a quantitative change in the production of a specific esterase isozyme (Cuany et al. 1993; Jayawardena et al. 1994).

The characterization of gene amplification in Culex as a molecular basis of resistance has been widely reported and reviewed (see Devonshire and Field 1991; Soderlund and Bloomquist 1991; Soderlund 1997). The molecular cloning of cDNAs corresponding to amplified esterases A2 and B1 has facilitated a number of studies of the molecular basis of resistance in mosquitoes. A partial cDNA for esterase B1 was employed as a probe to identify amplified esterase B1 sequences in a resistant strain of C. quinquefasciatus from California which was shown to have approximately 250-fold amplification of the esterase B1 locus (Mouchés et al. 1986). The B1 probe has also been used to examine DNA of Culex strains derived from field and laboratory populations (Raymond et al. 1991a) and was shown to recognize sequences encoding esterase B1 and B2. The coding sequences of esterases A2 and B2 have also been determined, and probes derived from these sequences have been used to demonstrate conclusively that the over-expression of these esterases also results from gene amplification (Vaughan and Hemingway 1995; Vaughan et al. 1995).

Restriction-fragment-length-polymorphism (RFLP) analysis of the B1 and B2 genes and their flanking regions suggests that the two genes are alleles which have been independently amplified (Raymond et al. 1991; Vaughan et al.

1995). In contrast, flanking sequences of esterase B2 from different continents were identical, and led to the hypothesis that the initial amplification event occurred once and has rapidly spread world-wide (Raymond et al. 1991). Similar results have been observed for the B1 and A2 electromorph as mosquitoes from various localities exhibited similar restriction maps and nucleotide sequences (Qiao and Raymond 1995; Guillemaud et al. 1996). However, based on differences in kinetic properties of purified esterases A2 and B2 from a number of resistant *C. quinquefasciatus*, Ketterman et al. (1993) concluded that multiple amplification events have occurred. Further support for the multiple origin hypothesis was provided by Vaughan et al. (1995) who reported two electrophoretically identical B1-esterase isozymes from California and Cuba likely having different origins based on differences in RFLP patterns from their respective genomic DNA.

The occurrence of gene amplification as a resistance mechanism in *Culex* and *M. persicae* represent some of the most well-characterized biochemical and molecular resistance mechanisms to date. Documentation of gene amplification in *Schizaphis graminum* provides further evidence that enhanced carboxylesterase activity as a result of gene amplification is responsible for resistance. There are many examples of increased carboxylesterase activity associated with organophosphate resistance, and, in some species, the esterases have been purified and characterized biochemically. However, in many cases, there is only correlative data linking elevated activity with resistance.

2.1.3.2
Qualitative Changes

The presence of qualitatively altered esterases associated with organophosphate resistance has been suggested in a number of OP-resistant species (Whyard et al. 1995). The basis of such qualitative changes has been proposed to involve possible alterations to the structure of an esterase to increase its ability to degrade insecticides (Parker et al. 1991). Such a mechanism has been proposed for four dipterans, *Musca domestica* (Oppenoorth and van Asperen 1960), *Lucilia cuprina* (Hughes and Raftos 1985), *Chrysomya putoria* (Townsend and Busvine 1969) and *Culex tarsalis* (Whyard et al. 1995) and for one Lepidopteran, *Plodia interpunctella* (Beeman and Schmidt 1982).

The proposed structural changes have apparently resulted in the loss of activity for model substrates such as α-naphtholic esters and has led to the "mutant ali-esterase hypothesis" (Oppenoorth and van Asperen 1960). This hypothesis states that a carboxylesterase is modified so that it acquires OP hydrolytic activity at the expense of activity for artificial substrates (Parker et al. 1996).

Such an alteration was initially proposed for insect strains that had developed resistance to malathion. Malathion possesses two carboxyl ester moieties in the leaving group of the insecticide that are potential sites of hydrolytic detoxification. Specifically, the cleavage of the β-ethyl ester group has been shown to be an important pathway of detoxification (Oppennoorth 1985). Purification of malathion carboxylesterase (MCE) in both *Culex tarsalis* and *Lucilia cuprina* has provided strong evidence that mutant enzymes exist which are absent in susceptible strains (Parker et al. 1991; Whyard and Walker 1994; Whyard et al. 1995). In *C. tarsalis*, purification studies revealed that malathion-susceptible mosquitoes contain a single malathion-hydrolyzing enzyme, whereas malathion-resistant strains have a second highly active enzyme that is not detectable in the susceptible insects (Whyard et al. 1995). Initial studies of MCE activity in resistant *L. cuprina* indicated that resistance was associated with overproduction of the enzyme because of similarities in kinetic constants between resistant and susceptible strains (Whyard and Walker 1994). However, subsequent analysis of esterases among strains with varying resistance levels suggested that the esterase expressed in highly resistant insects is qualitatively different from the malathion-hydrolyzing carboxylesterase in susceptible insects (Smyth et al. 1996).

Two types of resistance to organophosphate insecticides have been associated with changes in ali-esterase activity in the higher Dipterans (Campbell et al. 1998a). One type is referred to as "diazinon" resistance which confers broad cross-resistance among OPs, especially those with two ethoxy groups attached to the central phosphorus atom, but little or no resistance to malathion. The second type is referred to as the malathion type, and confers broad cross-resistance among OPs, especially dimethyl-substituted OPs, coupled with exceptionally high level resistance to malathion itself (Bell and Busvine 1967; Townsend and Busvine 1969; Campbell et al. 1998b). Activity toward naphtholic ester substrates is greatly reduced in both resistant phenotypes, although increased phosphotriesterase activity, defined as hydrolysis of the phosphoester linkage of the oxon form, is increased in both resistance types. Malathion carboxylesterase activity is also elevated in the malathion-resistant strains, but reduced in the diazinon-resistant strains (Campbell et al. 1997).

Until recently, a molecular basis for the ali-esterase theory has been lacking. However, a major resistance gene from *L. curpina* known as *Rop-1* has been shown to correspond to the cloned *LcαE7* gene. This gene encodes the major ali-esterase of *L. curpina* which is also known as the esterase isozyme, E3. Alleles isolated from OP-susceptible and diazinon-resistant strains have revealed a novel amino acid substitution in the enzyme's active site (Gly137→Asp). This substitution appears to simultaneously increase the rate of dephosphorylation yielding OP turnover and abolishes ali-esterase activity (Newcomb et al. 1997). A second mutation in the *LcαE7* (Trp→Leu) gene was reported by Campbell et al. 1998a) and, like the Gly137→Asp, is predicted to be close to the catalytic residues of the enzyme. Both amino acid substitutions within the active site of the enzyme cause all of the ali-esterase, OP hydrolase and MCE activity changes associated with OP-resistance status.

2.2
Target Site Insensitivity

Reduced sensitivity of acetylcholinesterase to inhibition by OP insecticides has been documented in a variety of agricultural, medical and veterinary pest arthropods. The target site for organophosphate insecticides involves inhibition of acetylcholinesterase (AChE) which is responsible for the degradation of the excitatory neurotransmitter, acetylcholine, thereby ending transmission of nerve impulses at cholinergic synapses. The determination of AChE insensitivity in insects relies on in vitro measurement of AChE activity from nervous tissue preparations (usually the head). The assay is conducted in the presence of various concentrations of insecticide inhibitors or for various inhibition times in the presence of a single inhibitor concentration from both resistant and susceptible strains. Differences in the rate of inhibition between resistant and susceptible strains is usually indicative of AChE insensitivity. However, such results must be interpreted with caution since AChE insensitivity is often accompanied by other resistance mechanisms.

AChEs that are less susceptible to inhibition by OPs are known to confer resistance in a variety of arthropods. The molecular basis of the altered forms of AChE, however, is known only for *D. melanogaster* and Colorado potato beetle, *Leptinotarsa decemlineata*. The molecular understanding of AChE insensitivity has been facilitated by sequencing and identification of the genomic structure of the *Ace* locus, which encodes acetylcholinesterase in *D. melanogaster* (Fournier et al. 1989) and the detection of OP-resistant strains (Fournier et al. 1992). The *Ace* gene of a malathion-resistant *D. melanogaster* strain was found to contain a single amino acid substitution, a conversion of a conserved phenylalanine residue (Phe386) to tyrosine. Introduction of this mutation into susceptible flies through P-element transformation resulted in both increased tolerance to malathion as well as reduced sensitivity to malaoxon inhibition.

Three additional point mutations have been identified from field strains of OP-resistant *D. melanogaster* that confer resistance to organophosphates: Phe115→Ser, Ile199→Thr/Val, and Gly303→Ala. Each of the three strains examined contained at least two of the four resistance-associated mutations, and one strain contained all four. All four mutations lie within the domains of the *Ace* gene that have been associated with binding and hydrolysis of acetylcholinesterase substrates and inhibitors. To assess the effect of single and multiple mutations, single mutations and various combinations were introduced into the wild-type *Ace* sequence and expressed in *Xenopus laevis* oocytes. Assays of the inhibition of the expressed acetylcholinesterases by various OPs showed that sequences with multiple mutations gave generally higher levels of insensitivity to inhibition relative to single mutations.

In *L. decemlineata*, altered AChE is a major mechanism resulting in azinphos-methyl resistance (Argentine et al. 1994). Purification and kinetic analysis of the AChE from resistant strains indicate that the alteration appears to affect both the esteratic subsite and peripheral anionic site of the enzyme (Zhu and Clark 1994, 1995a). The molecular cloning and sequencing of the AChE complementary cDNA from a nearly isogenic strain has allowed identification of an azinphos-methyl-resistance associated mutation (Zhu and Clark 1995b; Zhu et al. 1996). Mutational analysis of AChE coupled to an enzyme inhibition assay using azinphos-methyl-oxon in individual insects has indicated that a serine to glycine point mutation is most likely responsible for reduced sensitivity of AChE to inhibition. This point mutation was found in all individuals from the azinphos-methyl-resistant strain but was not found in a susceptible strain and was also found in 80% of AChE cDNA sequences amplified by polymerase chain reaction from the nearly isogenic strain. Interestingly, the Ser→Gly mutation that was associated with resistance does not correspond to any of the resistance-conferring mutations identified in *D. melanogaster* and is outside the binding domain implicated in acetylcholine hydrolysis and inhibitor binding. The Ser→Gly mutation was proposed to alter the secondary structure of the AChE molecule resulting in a conformational change of the molecule (Zhu and Clark 1997), leading to alterations in the multiple binding sites and is consistent with biochemical and kinetic analyses.

2.3
Interactions Between Resistance Mechanisms

There is increasing evidence that different resistance genes provide resistance to the same insecticide and are present at the same time in populations of resistant insects (Raymond et al. 1989b), and a common theme among many organophosphate resistant insects is the co-occurrence of resistance mechanisms in the same species. The presence of multiple resistance mechanisms acting at different levels has the potential to produce highly synergistic interactions that can result in virtual immunity in some populations (Soderlund and Blomquist 1991). Numerous examples exist where multiple detoxification enzymes appear to act in concert or where insensitive acetylcholinesterase co-occurs with elevated metabolism, but few of these studies have documented the interaction and relative contribution of multiple resistance mechanisms. Most commonly, elevated esterase activity has been identified in combination with insensitive acetylcholinesterases (e.g., Zhu and Brindley 1992; Bryne and Devonshire 1993; Siegfried and Ono 1993a; Sakata and Miyata 1994; Moores et al. 1994) or increased oxidative metabolism (Siegfried and Scott 1992; Scharf et al. 1999).

The co-occurrence of hydrolytic and oxidative metabolism is intriguing because the oxidation can result in increased rates of activation, especially for phosphorothioate OPs, while hydrolysis represents detoxification. In the German cockroach, *Blattella germanica*, results of independent studies have identified a common occurrence of elevated hydrolytic (i.e., esterase) and oxidative (i.e., cytochrome P450) detoxification-based capabilities in OP-resistant German cockroach strains (Siegfried et al. 1990; Scharf et al. 1998). The co-occurrence of these mechanisms was first noted using synergism and metabolism studies conducted on a field-collected, chlorpyrifos-resistant strain (Dursban-R; 22-fold resistant; Siegfried et al. 1990). NADPH-dependent microsomal metabolism was enhanced in the resistant strain and, although this oxidative activity resulted in increased levels of non-toxic metabolites, formation of chlorpyrifos-oxon was also enhanced. The increased rate of oxidation and insecticide activation was apparently offset by subsequent hydrolytic metabolism of chlorpyrifos-oxon. In assays of cytosolic preparations with both chlorpyrifos and chlorpyrifos-oxon as substrates, hydrolytic activity was detected only when chlorpyrifos-oxon was used as substrate and was elevated in the resistant strain.

Insensitivity of acetylcholinesterase has also been shown to exist in combination with increased metabolic detoxification. By itself, AChE insensitivity may be of minor importance but, in combination with increased metabolic detoxification, acts to prolong the time that metabolic degradation can take place. In *Lygus hesperus* (Hemiptera: Lygaeidae), both AChE insensitivity and increased hydrolytic activity were detected in a number of strains with varying levels of resistance (Zhu and Brindley 1992). However, the level of resistance was independent of the degree of AChE insensitivity, because strains that exhibited high AChE insensitivity were still susceptible in the absence of high carboxylesterase activity.

Raymond et al. (1989b) proposed a pharmacokinetic model to explain the potential interaction of resistant genes in which the joint effect of different mechanisms of resistance can be expressed in terms of epistasis at the physiological level based on data collected from OP-resistant *Culex pipiens*. The type of epistasis predicted by this model was dependent on the particular mechanisms involved in resistance. Resistance due to a reduced rate of penetration would combine multiplicatively with other factors, while resistance due to detoxification and insensitivity of the target would combine additively. Large-scale testing of this model has yet to be conducted, although it may provide a framework for examining interactions when multiple mechanisms have been identified.

3
Summary

Organophosphate insecticides have provided effective pest control for nearly 50 years, although their efficacy has been challenged by widespread adaptation among pest populations. Resistance to these compounds among pest species has been widely investigated, but a thorough understanding of resistance mechanisms at a biochemical and molecular level is still emerging. Organophosphate insecticides are represented by a diversity of structures that are susceptible to metabolic detoxification. Additionally, minor alterations in the structure of the target molecule, acetylcholinesterase, has been shown to confer insensitivity to organophosphate inhibition.

With respect to cytochrome P450, a number of different metabolic reactions that result in both activation and detoxification of organophosphate molecules have been associated with resistance. Determination of the relative importance of the different pathways is necessary to fully understand the role of oxidative metabolism in resistance mechanisms. Hydrolytic metabolism has also been commonly reported to be involved in organophosphate resistance, although the function of these enzymes is dependent on whether there is a quantitative or qualitative change in activity. Determination of gene amplification as a molecular basis of organophosphate resistance represents an important step in developing a molecular understanding of resistance evolution. However, there are many examples of correlative data linking enhanced carboxylesterase activity with resistance that have yet to be thoroughly examined at a biochemical and molecular level. Therefore, it would be premature to assume that gene amplification is always associated with increased esterase activity and OP resistance. The recent discovery and elucidation of molecular changes in enzyme structure associated with malathion carboxylesterase and ali-esterases also deserve consideration in examining increased hydrolytic metabolism as a basis of resistance. With regard to target site sensitivity, specific mutations to the acetylcholinesterase gene that reduce affinity or produce conformational changes that affect inhibitor binding have recently been identified. However, the number of possible mutations and their potential interactions have not been thoroughly investigated.

The future of organophosphate insecticides in pest management is uncertain due to regulatory efforts to reduce environmental and human health risks that have been associated with this class of insecticides. However, given the widespread use of these compounds and relative lack of alternatives in many cropping systems, it is likely that resistance management and an understanding of resistance mechanisms will take on increasing importance in maintaining organophosphates as viable pest management alternatives.

References

Agosin M (1985) Role of microsomal oxidations in insecticide degradation. In: Kerkut GA, Gilbert LI (eds) Comprehensive insect physiology, biochemistry and pharmacology, vol 12. Pergamon Press, Oxford, pp 647–712

Argentine JA, Zhu KY, Lee SH, Clark JM (1994) Biochemical mechanisms of azinphosmethyl resistance in isogenic strains of Colorado potato beetle. Pestic Biochem Physiol 48:63–78

Beeman RW, Schmidt BA (1982) Biochemical and genetic aspects of malathion-specific resistance in the Indian meal moth (Lepidoptera: Pyralidae). J Econ Entomol 75:945–949

Bell JD, Busvine JR (1967) Synergism of organophosphates in *Musca domestica* and *Chrysomyia putoria*. Entomol Exp Appl 10:263–269

Berrada S, Fournier D, Cuany A, Nguyen TX (1994) Identification of resistance mechanisms in a selected laboratory strain of *Cacopsylla pyri* (Homoptera: Psyllidae): altered acetylcholinesterase and detoxifying oxidases. Pestic Biochem Physiol 48:41–47

Brown (1961) The challenge of insecticide resistance. Bull Entomol Soc Am 7:6–19

Byrne FJ, Devonshire AL (1993) Insensitive acetylcholinesterase and esterase polymorphism in susceptible and resistant populations of the tobacco whitefly *Bemesia tabaci* (Genn.). Pestic Biochem Physiol 45:34–42

Campbell PM, Trott JT, Claudioanos C, Smyth K-A, Russell RJ, Oakeshott JG (1997) Biochemistry of esterases associated with organophosphate resistance with comparisons to putative orthologues in other Diptera. Biochem Genet 35:17–40

Campbell PM, Newcomb RD, Russell RJ, Oakeshott JG (1998a) Two different amino acid substitutions in the ali-esterase, E3, confer alternative types of organophosphorus insecticide resistance in the sheep blowfly, *Lucilia cuprina*. Insect Biochem Mol Biol 28:139–150

Campbell PM, Yen JL, Mausoumi A, Russell RJ, Batterham P, McKenzie JA, Oakeshott JG (1998b) Cross-resistance patterns among Australian sheep blowfly, *Lucilia cuprina* (Diptera: Calliphoridae), resistant to organophosphorus insecticides. J Econ Entomol 91:367–375

Cuany A, Jandani J, Bergè J, Fournier D, Raymond M, Georgiou GP, Pasteur N (1993) Action of esterase B1 on chlorpyrifos in organophosphate-resistant *Culex* mosquitoes. Pestic Biochem Physiol 45:1–6

Chiang FM, Sun CN (1993) Glutathione transferase isozymes of diamondback moth larvae and their role in the degradation of some organophosphorus insecticides. Pestic Biochem Physiol 45:7–14

Dauterman WC (1985) Insect metabolism: extramicrosomal. In: Kerkut GA, Gilbert LI (eds) Comprehensive insect physiology, biochemistry and pharmacology, vol 12. Pergamon Press, Oxford, pp 713–730

Devonshire AL (1977) The properties of a carboxylesterase from the peach-potato aphid, *Myzus persicae* (Sulz.) and its role in conferring insecticide resistance. Biochem J 167:675–683

Devonshire AL, Field LM (1991) Gene amplifications and insecticide resistance. Annu Rev Entomol 36:1–24

Devonshire AL, Moores GD (1982) A carboxylesterase with broad substrate specificity causes organophosphorus, carbamate and pyrethroid resistance in peach-potato aphids (*Myzus persicae*). Pestic Biochem Physiol 18:253–246

Devonshire AL, Sawicki RM (1979) Insecticide-resistant *Myzus persicae* as an example of evolution by gene amplification. Nature 280:140–141

Dunkov BC, Guzov VM, Mocelin G, Shotkoski F, Brun A, Amichot M, ffrench-Constant RH, Feyereisen R (1997) The *Drosophila* cytochrome P450 gene *Cyp6a2*: structure, localization, heterologous expression, and induction by phenobarbital. DNA Cell Biol 16:1345–1356

Eto M (1974) Organophosphorus pesticides: organic and biological chemistry. CRC Press, Boca Raton

Feyereisen R (1999) Insect P450 enzymes. Annu Rev Entomol 44:507–533

Field LM, Devonshire AL (1992) Esterase gene conferring insecticide resistance in aphids. In: Mullen CA, Scott JG (eds) Molecular mechanisms of insecticide resistance. ACS symposium series 505. American Chemical Society, Washington, DC, pp 209–217

Field LM, Devonshire AL, Ford BG (1988) Molecular evidence that insecticide resistance in peach-potato aphids (*Myzus persicae* Sulz.) results from amplification of an esterase gene. Biochem J 251:309–312

Field LM, Devonshire AL, ffrench-Constant RH (1989a) The combined use of immunoassay and a DNA diagnostic technique to identify insecticide-resistant genotypes in the peach-potato aphid *Myzus persicae* (Sulz.). Pestic Biochem Physiol 34:174–178

Field LM, Devonshire AL, ffrench-Constant RH, Forde BG (1989b) Positive correlation between methylation and expression of amplified insecticide-resistance genes. FEBS Lett 243:323–327

ffrench-Constant RH, Devonshire AL, White RP (1988) Spontaneous loss and reselection of resistance in extremely resistant *Myzus persicae* (Sulzer). Pestic Biochem Physiol 30:1–10

Fournier D, Karch F, Bride J-M, Hall LMC, Bergè J-B, Speirer P (1989) *Drosophila melanogaster* acetylcholinesterase gene: structure, evolution and mutation. J Mol Biol 210:15–22

Fournier D, Bride J-M, Hoffmann F, Karch F (1992) Acetylcholinesterase: two types of modifications confer resistance to insecticide. J Biol Chem 267:14270–14274

Georghiou GP (1986) The magnitude of the resistance problem. In: Pesticide resistance: strategies and tactics for management. National Academy of Sciences Press, Washington, DC

Georghiou GP, Lagunes-Tejeda A (1991) The occurrence of resistance to pesticides in arthropods. Food and Agriculture Organization of the United Nations, Rome

Guillemaud T, Rooker S, Pasteur N, Raymond M (1996) Testing the unique amplification event and the worldwide migration hypothesis of insecticide resistance genes with sequence data. Heredity 77:535–543

Hatano R, Scott JG (1993) Anti-P450$_{lpr}$ antiserum inhibits the activation of chlorpyrifos to chlorpyrifos oxon in housefly microsomes. Pestic Biochem Physiol 45:228–233

Hick CA, Field LM, Devonshire AL (1996) Changes in the methylation of amplified esterase DNA during loss and reselection of insecticide resistance in peach-potato aphids, *Myzus persicae*. Insect Biochem Mol Biol 26:41–47

Hodgson E (1985) Microsomal monooxygenases. In: Kerkut GA, Gilbert LI (eds) Comprehensive insect physiology, biochemistry and pharmacology, vol 11. Pergamon Press, New York, pp 225–322

Huang HS, Hu NT, Yao YE, Wu CY, Chiang SW, Sun CN (1998) Molecular cloning and heterologous expression of a glutathione S-transferase involved in insecticide resistance from the diamondback moth, *Plutella xylostella*. Insect Biochem Mol Biol 28:651–658

Hughes PB, Raftos DA (1985) Genetics of an esterase associated with resistance to organophosphorus insecticides in the sheep blowfly, *Lucilia cuprina* (Weidemann) (Diptera: Calliphoridae). Bull Entomol Res 75:535–544

Jayawardena KGI, Karunaratne SHPP, Ketterman AJ, Hemingway J (1994) Determination of the role of elevated B2 esterase in insecticide resistance in *Culex quinquefasciatus* (Diptera: Culicidae) from studies on the purified enzyme. Bull Entomol Res 84:39–43

Karunaratne SHPP, Hemingway J, Jayawardena KGI, Dassanayaka V, Vaughan A (1995a) Kinetic and molecular differences in the amplified and non-amplified esterases from insecticide-resistant and susceptible *Culex quinquefasciatus* mosquitoes. J Biol Chem 270:31124–31128

Karunaratne SHPP, Jayawardena KGI, Hemingway J (1995b) The cross-reactivity spectrum of a polyclonal antiserum raised against the native amplified A2 esterase involved in insecticide resistance. Pestic Biochem Physiol 53:75–83

Ketterman AJ, Karunaratne SHPP, Jayawardena KGI, Hemingway J (1993) Qualitative differences between populations of *Culex quinquefasciatus* in both the esterases A$_2$ and B which are involved in insecticide resistance. Pestic Biochem Physiol 47:142–148

Konno T, Hodgson E, Dauterman WC (1989) Studies on methyl parathion resistance in *Heliothis virescens*. Pestic Biochem Physiol 33:189

Liu N, Scott JG (1996) Genetic linkage of elevated cytochrome P450, cytochrome b$_5$, P450 reductase, CYP6D1 and monooxygenase activities in pyrethroid resistant houseflies, *Musca domestica*. Biochem Genet 34:133–148

Liu N, Scott JG (1997) Phenobarbital induction of CYP6D1 is due to a *trans* acting factor on autosome 2 in houseflies, *Musca domestica*. Insect Mol Biol 6:77–81

Miota F, Scharf ME, Ono M, Marcon P, Meinke LJ, Wright RJ, Chandler LD, Siegfried BD (1998) Mechanisms of methyl and ethyl parathion resistance in the western corn rootworm (Coleoptera: Chrysomelidae). Pestic Biochem Physiol 61:39–52

Moores GD, Devine GJ, Devonshire AL (1994) Insecticide-insensitive acetylcholinesterase can enhance esterase-based resistance in *Myzus persicae* and *Myzus nicotianae*. Pestic Biochem Physiol 49:114–120

Motoyama N, Dauterman WC (1980) Glutathione S-transferases: their role in the metabolism of organophosphorus insecticides. Rev Biochem Toxicol 2:49–69

Mouchés C, Pasteur N, Bergé J-B, Hyrien O, Raymond M, De Saint Vincent B (1986) Amplification of an esterase gene is responsible for insecticide resistance in a Californian *Culex* mosquito. Science 233:778–780

Mouchés C, Magnin M, Bergé J-B, De Silvestri M, Beyssat V, Pasteur N, Georghiou GP (1987) Overproduction of detoxifying esterases in organophosphate-resistant *Culex* mosquitoes and their presence in other insects. Proc Natl Acad Sci USA 84:2113–2116

Newcomb RD, Campbell PM, Ollis DL, Cheah E, Russell RJ, Oakeshott JG (1997) A single amino acid substitution converts a carboxylesterase to an organophosphorus hydrolase and confers insecticide resistance on a blowfly. Proc Natl Acad Sci USA 94:7464–7468

Ono M, Richman JS, Siegfried BD (1994) Characterization of general esterases from susceptible and parathion-resistant strains of the greenbug (Homoptera: Aphididae). J Econ Entomol 87:1430–1436

Ono M, Swanson JJ, Field LM, Devonshire AL, Siegfried BD (1999) Amplification and methylation of an esterase gene associated with insecticide resistance in greenbugs, *Schizaphis graminum* (Rondani) (Homoptera: Aphididae). Insect Biochem Mol Biol 29:1065–1073

Oppenoorth FJ (1985) Biochemistry and genetics of insecticide resistance. In: Kerkut GA, Gilbert LI (eds) Comprehensive insect physiology, biochemistry and pharmacology, vol 12. Pergamon Press, Oxford, pp 731–773

Oppenoorth FJ, van Asperen K (1960) Allelic genes in the housefly producing modified enzymes that cause organophosphate resistance. Science 132:298–300

Parker AG, Russel RJ, Delves A, Oakshott JG (1991) Biochemistry and physiology of esterases in organophosphate-susceptible and -resistant strains of the Australian sheep blowfly, *Lucilia cuprina*. Pestic Biochem Physiol 41:305–318

Parker AG, Campbell PM, Spackman ME, Russell RJ, Oakeshott JG (1996) Comparison of an esterase associated with organophosphate resistance in *Lucilia cuprina* with an orthologue not associated with resistance in *Drosophila melanogaster*. Pestic Biochem Physiol 55:85–95

Pasteur N, Sinegre G, Gabinaud A (1981) *Est*-2 and *est*-3 polymorphisms in *Culex pipiens* L. from southern France in relation to organophosphate resistance. Biochem Genet 19:499–508

Perry AS, Yamamoto I, Ishaaya I, Perry RY (1998) Insecticides in agriculture and the environment. Springer, Berlin Heidelberg New York

Pittendrigh B, Aronstein K, Zinkovsky E, Andreev O, Campbell B, Daly J, Trowell S, ffrench-Constant RH (1997) Cytochrome P450 genes from *Helicoverpa armigera*: expression in a pyrethroid-susceptible and -resistant strain. Insect Biochem Mol Biol 27:507–512

Qiao C-L, Raymond M (1995) The same esterase B1 haplotype is amplified in insecticide-resistant mosquitoes of the *Culex pipiens* complex from the Americas and China. Heredity 74 339–345

Raffa KF, Priester TM (1985) Synergists as research tools and control agents in agriculture. J Agric Entomol 2:27–45

Raymond M, Beyssat-Arnaouty V, Sivasubramanian N, Mouchè C, Georghiou GP, Pasteur N (1989a) Diversity of the amplification of various esterases B responsible for organophosphate resistance in *Culex* mosquitoes. Biochem Genet 27:417–423

Raymond M, Heckel DG, Scott JG (1989b) Interactions between pesticide genes: model and experiment. Genetics 123:543–551

Raymond M, Callaghan A, Fort P, Pasteur N (1991) Worldwide migration of amplified insecticide resistance genes in mosquitoes. Nature 350:151–153

Ronis MJ, Hodgson E, Dauterman WC (1988) Characterization of multiple forms of cytochrome P450 from an insecticide resistant strain of housefly (*Musca domestica*). Pestic Biochem Physiol 32:74–90

Rooker S, Guillemaud T, Bergé J, Pasteur N, Raymond M (1996) Coamplification of A and B esterase gene as a single unit in the mosquito *Culex pipiens*. Heredity 77:555–561

Sakata K, Miyata T (1994) Biochemical characterization of carboxylesterase in the small brown planthopper *Laodelphax striatellus* (Fallén). Pestic Biochem Physiol 50:247–256

Scharf ME, Hemingway J, Reid BL, Small GJ, Bennett GW (1996) Toxicological and biochemical characterization of insecticide resistance in a field-collected strain of *Blattella germanica* (Dictyoptera: Blattellidae). J Econ Entomol 89:322–331

Scharf ME, Neal JJ, Bennett GW (1998) Changes of insecticide resistance levels and detoxication enzymes following insecticide selection in the German cockroach, *Blattella germanica* (L.). Pestic Biochem Physiol 59:67–79

Scharf ME, Meinke LJ, Wright RJ, Chandler LD, Siegfried BD (1999) Metabolism of carbaryl by insecticide resistant and susceptible western corn rootworm populations (Coleoptera: Chrysomelidae). Pestic Biochem Physiol 63:85–96

Scott JA, Collins FH, Feyereisen R (1994) Diversity of cytochrome P450 genes in the mosquito, *Anopheles Albimanus*. Biochem Biophys Res Commun 205:1452–1459

Scott JG (1991) Investigating mechanisms of insecticide resistance: methods, strategies and pitfalls. In: Roush RT, Tabashnik BE (eds) Pesticide resistance in arthropods. Chapman and Hall, New York, pp 39–59

Scott JG (1999) Cytochrome P450 and insecticide resistance. Insect Biochem Mol Biol 29:757–777

Shiotsuke T, Kakimoto T, Eto M (1992) Irreversible inactivation of glutathione transferases by saligenin cyclic phosphates. Pestic Biochem Physiol 42:119

Shufran RA, Wilde GE, Sloderbeck PE (1996) Description of three isozyme polymorphisms associated with insecticide resistance in greenbug (Homoptera: Aphididae) populations. J Econ Entomol 89:46–50

Siegfried BD, Scott JG (1992) Biochemical characterization of hydrolytic and oxidative enzymes in insecticide resistant and susceptible strains of the German cockroach (Dictyoptera: Blattellidae). J Econ Entomol 85:1092–1098

Siegfried BD, Ono M (1993a) Mechanisms of parathion resistance in the greenbug *Schizaphis graminum* (Rondani). Pestic Biochem Physiol 45:24–33

Siegfried BD, Ono M (1993b) Parathion toxicokinetics in resistant and susceptible strains of the greenbug (Homoptera: Aphididae). J Econ Entomol 86:1317–1323

Siegfried BD, Zera AJ (1994) Partial purification and characterization of a greenbug (Homoptera: Aphididae) esterase associated with resistance to parathion. Pestic Biochem Physiol 49: 132–137

Siegfried BD, Scott JG, Roush RT, Zeichner BC (1990) Biochemistry and genetics of chlorpyrifos resistance in the German cockroach, *Blattella germanica* (L.). Pestic Biochem Physiol 38:110–121

Siegfried BD, Swanson JJ, Devonshire AL (1997a) Immunological detection of greenbug (*Schizaphis graminum*) esterases associated with resistance to organophosphate insecticides. Pestic Biochem Physiol 57:165–170

Siegfried BD, Ono M, Swanson JJ (1997b) Purification and characterization of a carboxylesterase associated with organophosphate resistance in the greenbug, *Schizaphis graminum* (Homoptera: Aphididae). Arch Insect Biochem Physiol 36:229–240

Smyth KA (1996) Biochemical and physiological differences in malathion carboxylesterase activities of malathion-susceptible and -resistant lines of the sheep blowfly *Lucilia cuprina*. Pestic Biochem Physiol 54:48–55

Soderlund DM (1997) Molecular mechanisms of insecticide resistance. In: Sjut V (ed) Molecular mechanisms of resistance to agrochemicals, vol 13: Chemistry of plant protection. Springer, Berlin Heidelberg New York, pp 21–56

Soderlund DM, Bloomquist JR (1990) Molecular mechanisms of insecticide resistance. In: Roush RT, Tabashnik BE (eds) Pesticide resistance in arthropods. Chapman and Hall, New York, pp 58–95

Syvanen M, Zhou ZH, Wang JY (1994) Glutathione transferase gene family from the housefly *Musca domestica*. Mol Gen Genet 245:25–31

Syvanen M, Zhou ZH, Wharton J, Goldsbury C, Clark A (1996) Heterogeneity of the glutathione transferase genes encoding enzymes responsible for insecticide degradation in the housefly. J Mol Evol 43:236–240

Terriere LC (1984) Induction of detoxication enzymes in insects. Annu Rev Entomol 29:71–88

Townsend MG, Busvine JR (1969) The mechanisms of malathion-resistance in the blowfly, *Chrysomya putoria*. Entomol Exp Appl 12:243–267

Vaughan A, Hemingway J (1995) Mosquito carboxylesterase Estα21 (A2). Cloning and sequence of the full-length cDNA for a major insecticide resistance gene worldwide in the mosquito *Culex quinquefasciatus*. J Biol Chem 270:17044–17049

Vaughan A, Magdalena R, Hemingway J (1995) The independent gene amplification of electrophoretically indistinguishable B esterases from the insecticide-resistant mosquito *Culex quinquefasciatus*. Biochem J 305:651–658

Wang JY, McCommas S, Syvanen M (1991) Molecular cloning of a glutathione S-transferase overproduced in an insecticide-resistant strain of the housefly (*Musca domestica*). Mol Gen Genet 227:260–266

Ware GW (1994) The pesticide book. Thomson Publications, Fresno

Whyard S, Walker VK (1994) Characterization of malathion carboxylesterase in the sheep blowfly *Lucilia cuprina*. Pestic Biochem Physiol 50:198–206

Whyard S, Downe AER, Walker VK (1995) Characterization of a novel esterase conferring insecticide resistance in the mosquito *Culex tarsalis*. Arch Insect Biochem Physiol 29:329–342

Wirth M, Marquine M, Georghiou GP, Pasteur N (1990) Esterases A2 and B2 in *Culex quinquefasciatus* (Diptera: Culicidae): role in organophosphate resistance and linkage. J Med Entomol 27:202–206

Wu D, Scharf ME, Neal JJ, Suiter DR, Bennett GW (1998) Mechanisms of fenvalerate resistance in the German cockroach, *Blattella germanica* (L.). Pestic Biochem Physiol 61:53–62

Yu SJ (1996) Insect glutathione S-transferases. Zool Studies 35:9–19

Zhao G, Liu W, Knowles CO (1994) Mechanisms associated with diazinon resistance in western flower thrips. Pestic Biochem Physiol 49:13–23

Zhou ZH, Syvanen M (1997) A complex glutathione transferase gene family in the housefly *Musca domestica*. Mol Gen Genet 256:187–194

Zhu KY, Brindley WA (1992) Significance of carboxylesterase and insensitive acetylcholinesterase in conferring organophosphate resistance in *Lygus hesperus* populations. Pestic Biochem Physiol 43:223–231

Zhu KY, Clark JM (1994) Purification and characterization of acetylcholinesterase from the Colorado potato beetle, *Leptinotarsa decemlineata* (Say). Insect Biochem Mol Biol 24:453–461

Zhu KY, Clark JM (1995a) Cloning and sequencing of a cDNA encoding acetylcholinesterase in Colorado potato beetle, *Leptinotarsa decemlineata* (Say). Insect Biochem Mol Biol 25:1129–1138

Zhu KY, Clark JM (1995b) Comparisons of kinetic properties of acetylcholinesterase purified from azinphosmethyl-susceptible and -resistant strains of Colorado potato beetle. Pestic Biochem Physiol 51:57–67

Zhu KY, Clark JM (1997) Validation of a point mutation of acetylcholinesterase in Colorado potato beetle by polymerase chain reaction coupled in enzyme inhibition assay. Pestic Biochem Physiol 57:28–35

Zhu KY, Lee SH, Clark JM (1996) A point mutation of acetylcholinesterase associated with azinphosmethyl resistance and reduced fitness in Colorado potato beetle. Pestic Biochem Physiol 55:100–108

Insect Midgut as a Site for Insecticide Detoxification and Resistance

Guy Smagghe and Luc Tirry[1]

1
Introduction

Pesticide resistance is a severe and important problem in situations where mainly chemicals are used to kill pests. However, apart from the economic, social and environmental costs associated with this problem, resistant insects and mites are a physiological marvel. Some strains have become so resistant to a given insecticide that they can survive exposure to virtually any dose. So there are numerous reasons for studying the underlying mechanisms by which insects become resistant to insecticides. Such studies are important for both the applied and basic aspects of insecticide resistance, as well as providing valuable information for workers in allied fields. For instance, if the biochemical basis of the resistance can be determined, then it may be possible to design a highly sensitive monitoring technique, which is one of the key factors in developing successful resistance management programs.

In order to exert activity, insecticide molecules have to enter the body and cross several barriers before arriving at their target site(s) of action. In the insect body, they can reversibly be absorbed by all kinds of tissue components, detoxified, and finally excreted via the feces, either unchanged or after metabolic conversion. The toxic effect exerted by a certain quantity of a chemical depends then on its intrinsic reactivity with the target site(s) of action, and on the time-course of its concentration at the target site.

For the purpose of this chapter, we focus on the insect midgut since it plays a major role in the buildup of toxicity for insecticides, it can serve as a target tissue for novel insecticidal modes of action, and modified midgut activities may imply resistance development.

The midgut epithelium is the first physical barrier after oral intake and it is a tissue that contains an armory of digestive enzymes that is needed for food conversion in nutrients but that also detoxifies insecticides. After a brief introduction on the midgut structure and its enzymatic capacity, we will discuss in more detail the significance of metabolic enzyme systems in the insect gut that

[1] Laboratory of Agrozoology, Faculty of Agricultural and Applied Biological Sciences, Ghent University, Coupure Links 563, B-9000 Ghent, Belgium

Isaac Ishaaya (Ed.): Biochemical Sites of
Insecticide Action and Resistance
© Springer-Verlag Berlin, Heidelberg 2001

lead to resistance, particularly for different groups of insect growth regulators (IGRs).

Then, we survey and discuss the penetration process through the gut epithelium for several insecticides of various groups, followed by their translocation in the insect body and then the impact of excretion via the feces on their toxicity. Subsequently, we attempt to describe these processes as far as possible in quantitative terms for developing a theoretical model of the digestive-absorption architecture of the insect midgut.

Finally, we present some current and potential contributions of physiological and biochemical techniques using insect midgut to test and screen new insecticide activities and resistance. Interference by insecticide components in the midgut morphology or activities can be the basis of new insecticide actions, leading to inhibition of feeding, thus reducing crop damage, or to lethality. Good examples in this field are *Bacillus thuringiensis* insecticidal crystal proteins and digestive enzyme inhibitors. Together, these assays offer valuable tools for research in many areas: basic biochemical and physiological processes in the insect midgut, mode of action of insecticides, evolution, metabolism, pharmacokinetics and molecular genetics of insecticide resistance.

2
The Insect Gut: a Natural Digestive-Absorption Architecture

Structurally, the insect gut or the alimentary food canal is composed of three separate elements: the stomodaeum, or foregut, and proctodaeum, or hindgut, arise as invaginations of the embryonic ectoderm; the midgut (mesenteron) is endodermal in origin (Fig. 1). However, it should be noted that a great diversity of form and function among insects and developmental stages exists (Grassé 1949, 1951; Chapman 1985b; Lehane and Billingsley 1996). These three elements join together to form a continuous tube late in the embryonic development in nearly all insects. The epithelium which forms the alimentary canal consists of a single layer of cells irrespective of its embryonic origin. For fore- and hindgut, these ectodermal epithelial cells produce a cuticular lining that is shed and renewed at each molt. Midgut cells are specialized in the production and secretion of digestive enzymes and in the absorption of nutrients. In the hindgut resorption of water and salts, amino acids and sugars is performed before excretion of the feces. The foregut is concerned with ingestion of food and passing it back to the midgut. The different elements are connected with each other with sphincter-like structures. The Malpighian tubules open at or near the junction of the mid- and hindgut.

Typically, the midgut cells do not produce a cuticle but a delicate peritrophic membrane which acts as a lining of the midgut and which compartmentalizes the midgut into an endoperitrophic and ectoperitrophic region (Terra and Ferreira 1994; Chapman 1985b). Production of this chitin-protein matrix is not linked to the molting cycle and may vary in relation to feeding. Typically, endoproteinases are secreted and pass through the peritrophic membrane into the endoperitrophic space and are responsible for the initial stages of digestion.

The principal cells of the midgut are columnar with apically microvilli lining a brush border (Fig. 1). Production and secretion of midgut enzymes are associated with abundant Golgi bodies containing electron-dense granules, larger membrane-bound vesicles or abundant rough endoplasmic reticulum. Midgut cells have a limited life: they form by mitotic division and differentiation from regenerative cells, undergo a cycle of changes associated with enzyme production and absorption and finally degenerate. These regenerative cells are distributed singly or in groups, nidi, at the base of the midgut epithelium, and have relatively little cytoplasm, being basiphilic with abundant free ribosomes. Goblet cells are scattered through the midgut epithelium of larval Lepidoptera, their luminal plasma membrane is typically deeply invaginated to form a flask-shaped cavity and they are implicated in the regulation of K^+ titers in the gut lumen.

Figure 1 schematically demonstrates for herbivorous caterpillars, here exemplified by the tobacco hornworm *Manduca sexta*, the transformation of food proteins into body tissue involving a complex set of processes (Woods and Kingsolver 1999). Initially, foliage is torn or snipped into platelets that pass into the foregut and then in the midgut. The latter contains high concentrations of K^+ and is usually highly alkaline (Dow 1986), with pH > 11 in the middle sections especially. As the platelets flow posteriorly, proteins are liberated into solution and attacked by soluble proteolytic enzymes. These digestion products consist of small- to medium-size peptides that move to the brush border membrane; here, they are broken into smaller fragments and free amino acids by membrane-bound hydrolytic enzymes. These smaller units are then absorbed into the epithelium by diffusion or by amino acid: cation symporters in the membrane. Energy for this carrier-facilitated transport is derived from a large (ca. 150 mV) lumen-positive electrochemical gradient. Herewith, the permeability of the gut elements varies for different compounds (Table 1; Maddrell and Gardiner 1980). In addition, molecules can pass the cell membrane by pinocytosis or phagocytosis, though few investigations have been carried out. From the midgut epithelium, amino acids are distributed in the hemolymph throughout the insect body (Santos et al. 1983; Broadway and Duffey 1986; Dow 1986; Giordana et al. 1989; Reuveni and Dunn 1994).

The digestion process in insects is generally adapted to the diet on which the species feeds and is coordinated with the way of food intake and the nutritional requirements. In addition, there are changes in enzyme activity that are regulated by the insect's development, external factors such as temperature and enzyme inhibitors in food, and the insect's behavior (for review see Chapman

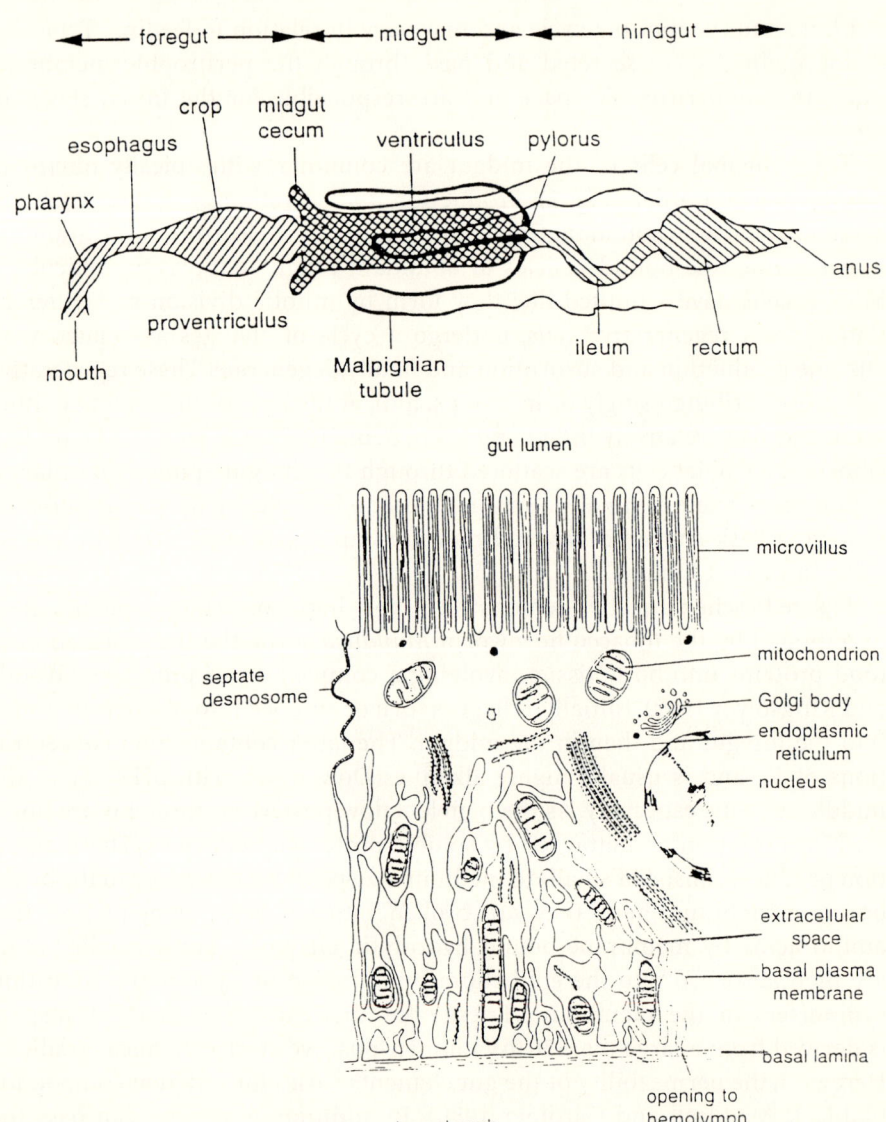

Fig. 1.

1985a; Lehane and Billingsley 1996). For instance, the polyphagous caterpillars of the cotton leafworm, *Spodoptera littoralis*, secrete high levels of proteases, amylases and invertases, which are closely related to the protein level in the diet or with the environmental temperature (Ishaaya et al. 1971). All these changes simultaneously direct the breakdown and absorption of stomach pesticides.

Basic biochemical studies on digestion in different laboratories from the 1970s on (Dadd 1970; House 1974; Applebaum 1985; Dow 1986; Lehane and Billingsley 1996) elucidated different digestive enzymes with parallel compar-

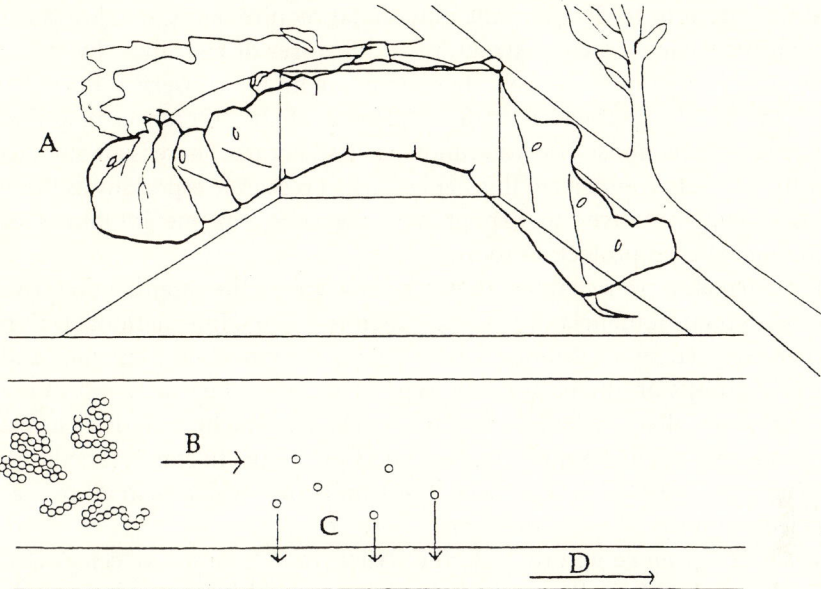

Fig. 1. *Top* (page 296) Basic structures of the insect alimentary canal. *Bottom* Diagram of a principal midgut cell in *Manduca sexta* (modified from Chapman 1985b). *Top* (page 297) A fifth-instar *M. sexta* on tomato: leaf proteins are transformed into body tissue in four steps: consumption of protein (*A*), digestion of protein into fragments (*B*), absorption of fragments across the epithelium (*C*), and construction of tissue from the absorbed fragments (*D*). (Woods and Kingsolver 1999)

Table 1. Permeability in the gut elements of *Schistocerca gregaria*. (Maddrell and Gardiner 1980)

Substance	M_r	Foregut	Ileum	Rectum
Urea	62	0.059	82.8	75.1
Glycine	75	0.016	57.6	37.3
Alanine	89	–	25.7	12.6
Serine	105	0.010	40.0	–
Proline	115	–	32.9	21.8
Glucose	180	0.006	13.9	16.8
Sucrose	342	0.004	0.5	0.4
Ouabain	585	0.022	0.96	0.25
Inulin	5200	–	0.012	0.014

Permeability $P = C/Nat$, with C: total number of radioactive counts in time t (s), N: radioactive compound concentration in bathing fluid (cpm/ml), and a: the gut sac area sac (cm^2). Results are expressed as $P \times 10^6$ cm.

isons of enzyme pH optima and pH values in the gut of various insects. Obvious correlations could be drawn between the kind of enzymes secreted and the presence of essential dietary components in macromolecular form, or between the extent of feeding and the nutritional requirements. In addition, it is firm that the basic molecular structure and features of hydrolysis by insect digestive enzymes, as in the vertebrate system, are highly conserved in evolution, with the kinetic characteristics a corollary of slight differences in amino acid sequence superimposed on a common basic structure. There is an obvious need for further achievements in this field of insect research, especially as there are forthcoming initiatives to exploit new digestive enzyme inhibitors as biotechnological crop protection tools.

For the purpose of this chapter, we will briefly survey the essential digestive enzymes with their nomenclature as recommended by the International Union of Biochemistry (Enzyme Nomenclature 1978). Enzymes are classified and divided into groups on the basis of the type of reaction they catalyze. All the digestive enzymes that we shall deal with here catalyze the hydrolytic cleavage of C-N bonds present in casein (protease) and C–O bonds in starch (amylase), sucrose (invertase) and trehalose (trehalase), an action which is in all probability their main physiological function.

Proteases are grouped according to the structure of their active center and amino acid sequence. Broadly speaking, insect endopeptidases, that are loosely defined as proteolytic enzymes cleaving internal peptide bonds and exhibiting various degrees of amino acid specificity, belong to one of the following three groups: (a) serine proteases (EC 3.4.21) like trypsin (EC3.4.21.4) and chymotrypsin (EC3.4.21.1), with an active center located at the histidine and serine moieties, (2) SH-proteases (EC 3.4.22) with a cysteine in the active center, and (3) carboxyl proteases (EC 3.4.23) in which an acidic residue is involved in the catalytic process. Proteases are the most common proteolytic enzymes found in the digestive tract of insects, some of them acting in neutral or alkaline pH as in the case of *S. littoralis*, while others act in acidic medium such as in aphids.

A-Amylases are endoglycosidases that hydrolyze 1,4-α-D-glucosidic linkages in polysaccharides containing three or more 1,4-α-D-glucose units. The insect midgut trehalase hydrolyzes the disaccharide trehalose to glucose, to assure its absorption into the hemolymph. In addition, the glucose formed is an essential energy source for generating the physiological processes needed for larval growth and development as well as adult flight. Invertase activity has been demonstrated in bacteria, fungi, higher plants and in the intestine of several animals. As early as 1964, Khan reported on invertase activity in midgut tissues of *Locusta migratoria* that followed food intake and ecdysis.

3
Enzymatic Metabolism of Pesticides Involved in Resistance

Apart from target site modifications, e.g. the most important are acetylcholinesterase (AChE) modification for organophosphorus (OPs) and carbamate insecticides, enhanced metabolism of the chemical, predominantly in the insect midgut, largely mediates resistance to insecticides. The processes by which pesticides are metabolized may be divided into two general groups referred to as phase I and phase II reactions (Williams 1959; Dauterman and Hodgson 1978; Agosin 1985; Dauterman 1985). Oxidations, reductions and hydrolysis are typical phase I reactions, and they are generally involved in metabolizing the three main insecticide groups, pyrethroids, OPs and carbamates, and IGRs. Essentially, phase I reactions introduce a hydrophilic functional group into the molecule which increases water solubility for immediate elimination via feces and/or makes it suitable for the phase II conjugating enzymes (e.g., glutathione S-transferases). Phase II reactions combine water-soluble endogenous conjugating agents derived from a carbohydrate, a protein source or a sulfur component (e.g., glutathione) with phase I metabolites. The conjugated products which are the result of this union are generally more polar, less lipid soluble and more readily eliminated via the feces than either the parent compound or its phase I metabolites as well as being, in general, less toxic.

A number of insecticides contain ester-type groups and are susceptible to hydrolysis. This includes OP, carbamate and pyrethroid insecticides and IGRs like the chitin synthesis inhibitory benzoylphenyl ureas (BPUs). The enzymes which catalyze these reactions are referred to as hydrolases. Hydrolytic reactions are the only phase I reactions that do not utilize high-energy intermediates. Hydrolysis splits ester compounds by the addition of water to yield acid and alcohol (Dauterman and Hodgson 1978; Dauterman 1985):

$$R^1COOR + H_2O \xrightarrow{\text{esterases}} R^1COOH + ROH$$

These enzymes are distributed throughout the plant and animal kingdom and have been found in a variety of tissues. In the early 1950s, Aldridge (1953) introduced a still useful classification of esterases that was based on the sensitivity towards OPs, such as diethyl 4-nitrophenylphosphate (Paraoxon, E600). A-Esterases [EC 3.1.1.2] are referred to as those which are not inhibited by OPs and even hydrolyze them. Since α-esterases preferentially cleave aromatic esters, e.g. phenyl acetate, they have also been designated as arylesterases or aromatic esterases. In contrast, β-esterases (EC 3.1.1.1) are stoichiometrically inhibited by OPs through irreversible phosphorylation of the active site serine. This makes them, formerly also called ali-esterases or non-specific esterases, serine hydrolases. Based on their sensitivity towards physostigmine, β-esterases can be further differentiated into cholinesterases and carboxylesterases (Augustinsson 1958).

The study of carboxylesterases is relatively simple as compared to several other detoxification mechanisms. There are no cofactors required and good inhibitors are available which act in vivo as more or less specific synergists, e.g. DEF (S,S,S-tributylphosphorotrithioate) and profenofos. Many of the hydrolytic detoxification enzymes show high activity towards several car-boxylesters: in particular α- and β-naphthyl acetate are excellent tools since they offer a sensitive quantitative method as well as simple staining method after electrophoretic purification. These carboxylesterases have been shown to be involved in resistance and high levels of enzyme activity have been associated with resistance to OP, carbamate and pyrethroid insecticides and acaricides.

Another group of phase I reactions involve the cytochrome-P_{450}-linked monooxygenase reactions by mixed function oxidases of the endoplasmic reticulum that are found in a variety of organs and tissues of several species, including mammals, bacteria, plants, fungi, arthropods and fish (Williams 1959; Agosin 1985). In most cases, this results in oxidative detoxification of pesticides; while in others metabolism results in activation of substances to produce toxic compounds. Chlorfenapyr is a pyrole insecticide/acaricide and a good example of such pro-insecticide chemicals that are activated by oxida-tive metabolism, suggesting that they may selectively target resistant insects with increased enzyme activities (Black et al. 1994).

The insect microsomal system, which is in many respects similar to that of mammalian liver (for review see Agosin 1985), corresponds to a multi-enzyme system requiring molecular oxygen and usually NADPH:

$$RH + O_2 + NADPH + H^+ \rightarrow R - OH + NADP^+ + H_2O$$

This mechanism implies a theoretical stoichiometry of substrate/electron reductant/molecular oxygen of $1:1:1$, and the key enzymes in the total reac-tion are cytochrome P_{450} and NADPH-cytochrome P_{450} reductase located in the endoplasmic reticulum. Mayer and Durrant (1979) reported in housefly on the stoichiometry of 12 membrane-bound P_{450}, 4 membrane-bound b_5 and 1 mem-brane-bound reductase which was very similar to that reported for mam-malian liver microsomes (Estabrook et al. 1972). Substrates metabolized include not only a variety of insecticides such as DDT-type compounds, car-bamates, phosphorothioates, cyclodienes, pyrethrins and synergists of the 1,3-benzodioxole type, but also polycyclic aromatic hydrocarbons, strong mutagens (e.g. nitrofuazone and vinyl chloride), endogenous and synthetic insect steroids (for instance ecdysone 20-hydroxylase forms 20-hydroxyecdysone from ecdysone), endogenous and synthetic juvenile hor-mones (e.g., methoprene) and insect pheromones.

For the chitin synthesis inhibitory BPUs, several studies on metabolism in vertebrates and insects have been published (Hammok and Quistad 1981; Retnakaran et al. 1985). The reports reflect a diversity of metabolic capacity depending on species and stage; however, with two major types of degrada-tion. The prototype diflubenzuron (DFB) may be hydrolyzed by cleavage of the amide bonds to give 4-chlorophenylurea, 2,6-difluorobenzamide, 2,6-difluorobenzoic acid and 4-chloroaniline; or it may be oxidized. The oxidation is seen as a ring hydroxylation on the aniline or the benzoic ring to form polar materials, followed by a conjugation. These conjugated products are then rapidly eliminated from the body.

On resistance development to DFB, Pimprikar and Georghiou (1979) were the first to report very high levels of resistance (>1000-fold) in a housefly *Musca domestica* strain stressed with DFB.

As documented by Moffit et al. (1988) and Sauphanor et al. (1998), the codling moth *Cydia pomonella* shows very high levels of resistance to DFB in the US and France, respectively. When testing for in vivo mixed function oxidases using the fluorescent substrate 7-ethoxycoumarin-*O*-deethylase in individual neonate larvae, these enzymes were demonstrated to be involved in the 45,000-fold larval resistance to DFB. In addition, no significant difference was noted between the non-specific esterase activity in resistant and susceptible strains.

In some strains of the cotton leafworm, *S. littoralis*, resistance to DFB was also increased to a factor of about 300 (El-Guindy et al. 1983). El Saidy et al. (1989) reported that DFB and teflubenzuron (TFB) were hydrolyzed rapidly by all tissues tested, and the gut wall was the most active tissue reaching 61% hydrolytic breakdown for DFB and 16% for TFB, followed by the gut plus content and the carcass. Interestingly, profenofos and DEF could inhibit DFB and TFB degradation under optimal conditions. Prior to this, Ishaaya and Casida (1980) had reported that these OP-compounds can inhibit insect esterases in the larval gut. So the strong synergism activity of DEF and profenofos indicated that the major route of detoxification in *S. littoralis* was through hydrolysis. Oxidative metabolism was found to be of minor importance for resistance.

For a multi-resistant (MR) strain of *S. littoralis*, El Saidy et al. (1990) measured esterase activities in vitro with α- and β-naphthyl acetate as substrate and showed that the optimum enzyme assay involved about 1.5 µg protein (equivalent to about 0.02 mg midgut weight), phosphate buffer with pH of 7.5, and incubation at 30 °C for 30 min (Fig. 2). Under these conditions, the Michaelis constant K_m was 13.8 mM for a susceptible (S) strain and 35.9 mM for the MR-strain. The esterase activity in the MR strain (28 µmol/mg protein min) was also 1.6-fold higher than in the S-strain. About 50% of the esterase activity in the midgut homogenate of the MR-strain was due to cholinesterase compared with 30% in the S-strain. These results indicated that enhanced levels of esterases contribute, at least in part, to the resistance of *S. littoralis*. In addition, modification of the esterases as the mechanism of resistance is also suggested. This agrees with Auda (1986) who reported that modified AChE is involved as resistance mechanism in *M. domestica*.

In greenhouse strains of the beet armyworm, *Spodoptera exigua*, that showed reduced susceptibility for DFB and TFB, Van Laecke et al. (1995) dissected midguts and explained resistance by a significant increase in mixed

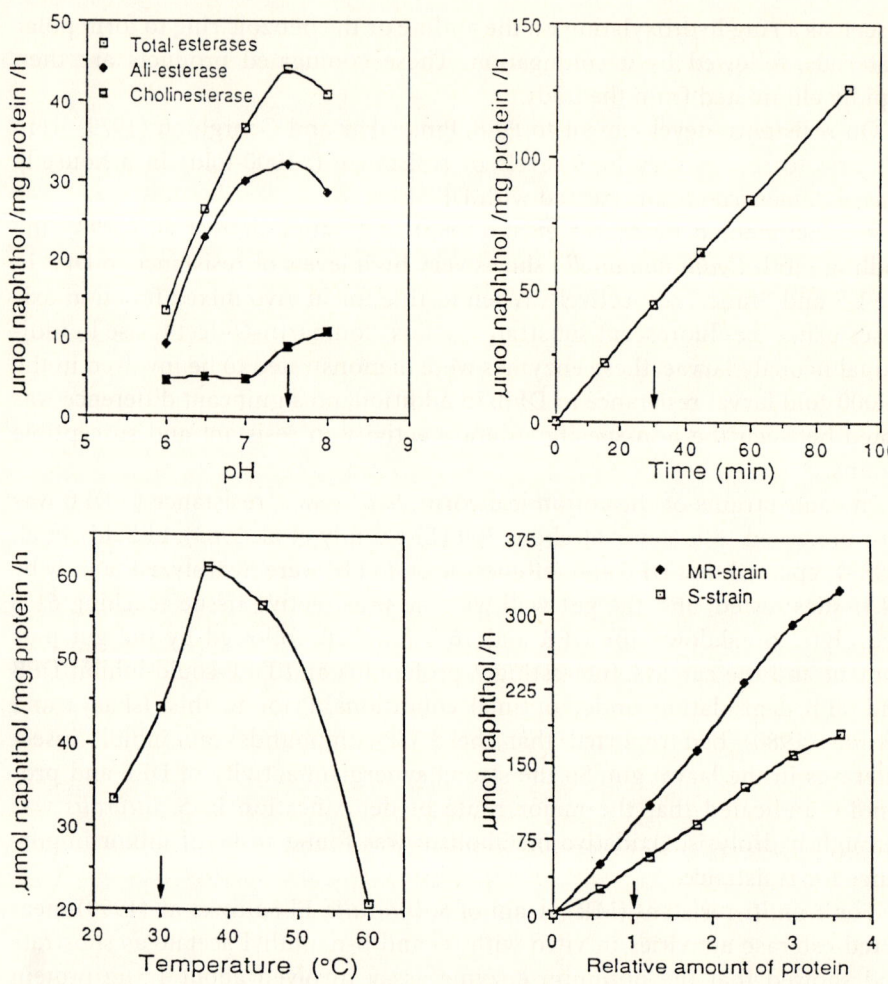

Fig. 2. Effect of pH, incubation time, reaction temperature and enzyme level on esterase activity in midgut homogenates of multi-resistant (*MR*) and susceptible (*S*) cotton leafworm *Spodoptera littoralis*. The *arrow* designates the standard assay conditions. (El Saidy 1991)

function oxidases (MFOs). These MFO-activities were measured spectrophotometrically with substrates *p*-nitroanisole and *p*-chloro-*N*-methylaniline. They also found an enhanced activity of carboxylamidases and glutathione *S*-transferases (phase II reactions) as compared with a susceptible strain. Typically, no significant differences were observed between the rates of general ester hydrolysis using α- and β-naphthyl acetate. So, from the above investigations, it is clear that the responsible mechanisms of resistance for BPUs can depend on the insect species studied. Whereas in some resistant pests hydrolysis was the most important, in others mixed function oxidations were predominant.

For the dibenzoylhydrazines, another group of IGRs represented by tebufenozide, the first commercial non-steroidal ecdysone agonist, Smagghe et al. (1998a,b, 1999b,c) reported on its toxicity for *S. exigua* and selected insects artificially by continuous exposure to sublethal doses. They revealed that tebufenozide molecules show a high stability, and high performance liquid chromatograpy (HPLC) profiles in different body tissues also demonstrated the production of alcoholic, ketone and aldehyde metabolites as a result of oxidation of the alkyl substituents of the two benzoyl rings. The majority of oxygenases in last-instar caterpillars was present in their midgut proving the gut to be the primary source of degradation enzymes, and the Michaelis constant, K_m, was 0.31 mM and V_{max} 5.35 nmol/mg protein min. In addition, it was demonstrated that piperonyl butoxide, an inhibitor of enzymes for oxidative metabolism, acted as a synergist with tebufenozide. This concurs with the results of Darvas et al. (1998) in the mosquito *Aedes aegypti*. As such, it is hypothesized that the major first-phase route for tebufenozide's detoxification is via oxidation in the gut. Further, inactive polar and very polar metabolites were found. Cleavage between the carbonyl and the amide moiety resulting in hydrolytic products could not be identified; however, this process cannot be excluded. The option that hydrolysis is of minor importance though agrees with assays in which addition of DEF could not really enhance the toxicity of ecdysone agonist.

When testing strains of *S. exigua* from different locations in Europe that resisted current chemicals, Smagghe et al. (1999c) were able to report a relationship between toxicity of tebufenozide and oxidative enzyme activity in the insect body. To quantify the enzyme activity of monooxygenases, a spectrophotometric technique with *p*-nitroanisole as substrate was used. A close correlation with $R^2 = 0.89$ was found between LC_{50} values and oxidative activities suggesting that the lower toxicity of tebufenozide was the result of an increase in oxidative breakdown of the molecules. Based on the results so far, it is considered that 12–15 nmol/mg protein.min can be used as a critical threshold level for total body oxidative enzyme activity. This corresponds with a critical threshold level for LC_{50} of about 3 mg/l. However, it should be noted that further evaluation with a wide range of field strains, especially from areas with high levels of multi-resistance, is required before firm conclusions can be drawn. Besides, optimization of the technique as a monitoring tool for resistance is needed to make it more time efficient and user friendly, so that this information in turn should help in using the appropriate tactics to reduce resistance risks. For instance, in each well of a 96-well microtiter plate, the enzyme source of one individual last-instar or ten third-instars can be tested (Schuyesmans 1998). This concept concurs with the biochemical immunoassay test for esterase diagnosis in an individual peach-potato aphid, *Myzus persicae*, which was developed in Rothamsted (Devonshire et al. 1986) to detect and monitor for resistance towards OPs, carbamates and pyrethroids.

4

Impact of Ingestion, and Penetration and Disposition in the Insect Body on Resistance to Pesticides

An understanding of the factors that govern the penetration of insecticide molecules into the body, either across the cuticle (application by contact) or through the gut wall (by ingestion), is not only fundamental to a clear insight into how their toxic action comes about, but also to their effective use. Reduced cuticular penetration of insecticides was firstly reported as a resistance process in the 1960s and the mechanism confers across a range of insecticides (Scott 1990). Most studies so far have studied and discussed the impact of penetration through the insect cuticle (Welling and Paterson 1985). However, insecticides do not enter the body by cuticular penetration only: a large group of compounds is ingested with the food and reaches the hemolymph and target and non-target organs after passage through the gut wall.

This aspect of penetration through the gut wall is much less extensively studied than cuticular penetration because most likely it requires more technical skill from the investigator. The question of how fast an insecticide can pass through the gut wall has been studied in vitro with dialysis experiments as an in vitro model for the in vivo process. Parts of insect gut were transformed into a small sac and filled with an insecticide solution, as shown in Fig. 3. These sacs were bathed in a suitable buffer solution and the time-course of the concentration of insecticide appearing in the bathing fluid was measured.

In this manner, in the early 1970s, Shah and Guthrie (1970) evaluated the penetration rate of five insecticides (nicotine, carbaryl, malathion, dieldrin and DDT) through the larval midgut of tobacco hornworm, *M. sexta*, and the cockroach, *Blaberus discoidalis*. Midguts were suspended in a stoppered, 40-ml glass tube and submerged in 30 ml of buffer. The tube was fitted with devices for introducing 0.1 µg [^{14}C]-radiolabeled insecticide into the gut lumen and for bubbling air into the buffer. The buffer was physiological saline (pH 6.6) containing 6% propylene glycol and 0.033% Tween 80. Constant temperature of incubation was maintained at 30 °C in a water bath and, 8 min after the introduction of insecticide, samples were taken of the serosal fluid, luminal fluid and midgut tissue (Fig. 3). Their results showed that transport through the gut wall was found to be a passive diffusion process because neither the presence of inhibitors of carbohydrate or encouplers of oxidative phosphorylation, nor the absence of oxygen had any effect. However, there were large differences between the two insect species. Whereas the *Manduca* midgut released the five insecticides rather very quickly, the insecticides penetrated very slowly through the gut of the cockroach in the surrounding medium. In addition, they documented that the higher the polarity of the insecticide compound the

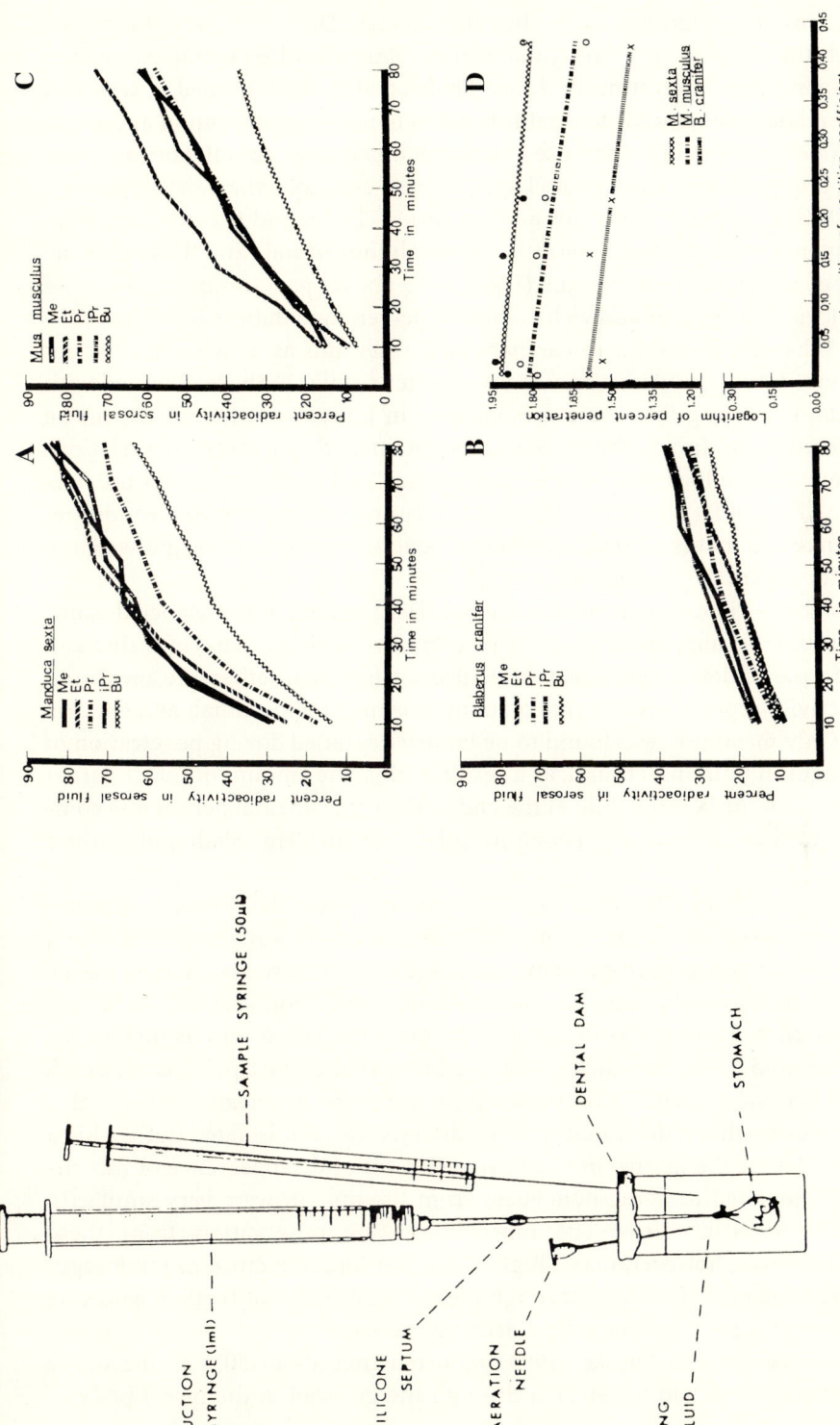

Fig. 3. *Left* Schematic picture of the bioassay to evaluate penetration rates of [¹⁴C]-labeled insecticides through isolated insect midgut (Connor et al. 1978). *Right* Comparative rates of penetration of dialkoxy analogs of dimethoate through the isolated gut of *Manduca sexta* (**A**), *Blaberus cranifer* (**B**) and *Mus musculus* (**C**), and rate of penetration plotted as logarithm of partition coefficient against logarithm of percentage of penetration in 80 min (**D**). (Shah et al. 1972)

higher was its penetration rate. In these assays, DDT and dieldrin passed through the gut wall very slowly and were greatly absorbed by the gut wall.

Besides these interesting findings, Shah et al. (1972) studied a series of closely related compounds to establish the dynamics of penetration and physical properties. For this, they used dialkoxy analogs of dimethoate, with the alkoxy group ranging from methyl to butyl, and evaluated the penetration rate through the midgut of *Manduca* and *Blaberus*. They found that the larger the substituent the slower the penetration through the gut wall, and the higher the extent of absorption by the gut (Fig. 3). As above, penetration in *M. sexta* midgut was much faster and yielded higher percentages than in *Blaberus* cockroach. When these pioneer researchers performed this assay with isolated gut of the mouse *Mus musculus*, results were rather similar to those obtained with *M. sexta*, indicating that these mechanisms in insects and mammals can be considered to be similar. From the results obtained, they quantified a relationship between the logarithm of penetration rate and the logarithm of the partition coefficient P (*n*-octanol/water) as a linear regression (Fig. 3). In addition, they remarked that the slower the analog penetrated through the gut wall, the larger the extent of metabolism.

The above-cited experiments were carried out with radiolabeled compounds and, because the midgut of insects is very rich in enzymes that can degrade pesticides, these researchers also evaluated whether the measured radioactivity represented parent compound or metabolites (Shah and Guthrie 1971). Only malathion was found to be largely degraded during penetration of the midgut in both insects and, as a result of this, the amount of intact parent compound in the bathing fluid at the end of the experimental period was comparable to that of the slowly penetrating DDT or dieldrin (Shah and Guthrie 1971).

A few years later, Conner et al. (1978) circumvented the problem of metabolic degradation during gut penetration by taking the foregut of honeybees, *Apis mellifera*, that contains no enzymatic activity. In these assays, the passage of compounds from different insecticide classes (parathion, carbaryl and dieldrin) was studied and it was found that the rate of penetration was highest for parathion and lowest for dieldrin. As carbaryl is the most polar compound, there is no direct relationship between polarity and penetration rate in this case. It was further shown that in vitro dialysis through isolated gut wall is a good model for the in vivo process because rates of disappearance of insecticide solutions fed to living honeybees from the midgut were very similar to those observed in vitro. The most interesting observation from these researchers was, however, that a high concentration of sucrose in the foregut decreased the rate of passage through the gut wall, and that further, with carbaryl, this effect was reflected by a decrease in toxicity.

More recently, Van Laecke (1993) reported that about 50% of the orally administered DFB had penetrated through the gut wall in the insect body of

S. exigua after 3 h. For TFB, another BPU, this percentage was reached already 2 h after ingestion. Due to this rapid transport, maximal amounts of BPU were retained at 2–3 h after oral uptake in the integument representing the target site tissue for this group of IGRs. El Saidy (1991) studied the penetration in a multi-resistant strain on *S. littoralis* and found that DFB was absorbed more slowly than in a susceptible strain, reaching at its maximum 17 and 21% in their respective body tissues. For TFB, penetration yielded about 25 and 32% in resistant and susceptible caterpillars, respectively, in a 24-h period after application. So, reduced penetration can be considered as a mechanism of different toxicities and resistance for BPUs.

For four dibenzoyl hydrazine analogs, Smagghe and Degheele (1993, 1994b) and Smagghe et al. (1999a) determined their penetration through the gut and absorption in the hemolymph after oral uptake of [^{14}C]-labeled compound in different insects (Table 2). In both *S. exigua* and *S. littoralis* about 20% of the applied amount had penetrated through the gut wall of last instars after 6 h of ingestion. In contrast, penetration and absorption in the body tissues of the Colorado potato beetle, *Leptinotarsa decemlineata*, varied widely (from 60 to only 10%) depending on the compound, mainly because of rapid elimination via the excrements without being taken up in the body hemocoel. Interestingly, halofenozide (RH-0345), that was highly toxic for beetle larvae, was accumulated rapidly in the body and yielded about 40% at 12 h after oral intake.

So, for this section on penetration it is clear that there is still a long way to go. Evidently, the heterogeneity of the gut structures and properties within and between species are major stumbling blocks for quantifying the penetration process and for the straightforward interpretation and extrapolation of

Table 2. Log of partition coefficient (log *P*), half-time ($t_{0.5}$ in min)[a], rate constant of excretion (k in min^{-1})[b], and oral toxicity (LC$_{50}$ in mg/liter) of four dibenzoylhydrazines in two lepidopteran species: *Spodoptera exigua* and *Spodoptera littoralis*, and in the coleopteran *Leptinotarsa decemlineata*. (Data from Smagghe 1995; Smagghe et al. 1998a,b, 1999a–c; Nakagawa et al. 1999)

Product	log *P*	*S. exigua*			*S. littoralis*			*L. decemlineata*		
		$t_{0.5}$	$k \times 10^3$	LC$_{50}$	$t_{0.5}$	$k \times 10^3$	LC$_{50}$	$t_{0.5}$	$k \times 10^3$	LC$_{50}$
RH–5849	2.45	115	8.681	110	–	–	–	367	1.887	1.8
Tebufenozide (RH–5992)	4.39	181	3.818	0.60	233	2.969	8.49	90	22.44	54
Methoxyfenozide (RH–2485)	3.93	147	4.709	0.38	187	3.705	1.14	–	–	–
Halofenozide (RH–0345)	3.37	–	–	–	230	3.015	44.8	404	1.716	1.4

[a] Half–time ($t_{0.5}$) (min) was determined from the slope of the regression curves of absorption–excretion assays with [^{14}C]–labeled ecdysone agonist.
[b] Rate constant k (min^{-1}) was determined by using the formula $k = 0.693/t_{0.5}$ (O'Brien 1967).

penetration data among different insects. However, we feel that the relationship between penetration rate and physicochemical properties of insecticides needs further study, preferably to be carried out with a homologous series of compounds.

After passing the gut wall, the pesticide molecules arrive in the circulating hemolymph, and are distributed throughout the insect body, either free in solution or absorbed by hemolymph proteins or hemocytes. Two distinct phases can be more or less clearly distinguished here: at first, distribution predominates with rising internal concentrations until approaching equilibrium in all organs and tissues. Afterwards, elimination is the most important aspect in which the time-course of concentrations in the various organs is characterized by more or less parallel curves with downward slopes.

Hence, an important pharmacokinetic parameter is the volume of distribution that relates the amount of pesticide in the body to its concentration in the hemolymph. This has no physiological identity but it indicates the extent to which a compound is absorbed by, or partitioned into, various organs other than the hemolymph. An apparent distribution volume much larger than the body volume, on a milliliter per gram basis, points to an affinity of the compound for various tissues which is much larger than that for the hemolymph. Welling and Paterson (1985) re-calculated from the very few literature data a distribution volume of 2–3 ml for pyrethroids in the American cockroach *Periplaneta americana*. This is surprisingly high and indicates that it is possible that these distribution volumes can be larger than the volume of the insect, and thus that adsorption to tissue constituents, such as binding to proteins and partitioning into lipids, is a significant aspect of insecticide disposition in insects. However, so far, explicit studies on the precise role that sorption plays in the insecticide intoxication process are lacking in the literature.

Distribution of ecdysone agonists using [^{14}C]isotopes was studied in *Spodoptera* caterpillars and *L. decemlineata* following oral uptake, and a fast distribution from the food canal into the hemolymph and integument was demonstrated (Smagghe and Degheele 1993, 1994b), although there were some smaller differences for RH-5849 and tebufenozide. For methoxyfenozide (RH-2485), Smagghe et al. (1999b) reported in *Spodoptera* relatively better translocation from the gut into the hemolymph than for tebufenozide, though the rate constant of elimination via the feces appeared somewhat higher for methoxyfenozide. This suggests that the net effect of absorption in the hemolymph is about similar for both compounds.

The distribution of a pesticide in the insect body also directly influences the toxicity of pesticides towards beneficial insects. This was documented firstly by De Clercq et al. (1995) for *Podisus maculiventris*. This predatory bug was fed *S. exigua*, which is penetrated by its stylet. The bugs did suck the caterpillar's hemolymph almost completely, but they did not use the gut content as food. The results obtained indicated the importance of food chain toxicity and, although this theme is rarely studied at present, it is considered important in the compatible use of pesticides together with natural enemies in integrated pest management programs.

Concerning the role of the excretory system via the feces in the elimination of insecticides from the body, it can be noted that, from a toxicological viewpoint, this is only a pharmacokinetically significant process when it contributes to the elimination of the parent compound or its activation product (if any) from the internal tissues, since metabolites usually are not toxic.

For DFB, Chang (1978) and Chang and Stokes (1979) studied the fate of the compound after injection in houseflies, *Musca domestica*, and boll weevils, *Anthonomus grandis*, and they could not detect excretion of parent compound or to negligible levels. On the other hand, Still and Leopold (1978) reported a substantial amount of parent compound in the feces. More recently, Van Laecke (1993) documented the importance of excretion via the feces for both DFB and TFB reaching about 50% of insecticide applied at 4 h after treatment.

In a susceptible strain of *S. littoralis*, DFB was metabolized more rapidly than TFB: in the body tissues about 58% of extracted radioactivity was parent TFB whereas this was only 38% for DFB. In the excrements, parent TFB (42%) was excreted more slowly than the metabolites. This contrasts to DFB in which 79% of the total extract was present as parent product in the feces. When tested in a multi-resistant strain of *S. littoralis*, the same trend was obtained but the rate of metabolism and excretion was faster as compared to the susceptible strain (El Saidy et al. 1989; El Saidy 1991).

For the ecdysone agonist RH-5849 and tebufenozide, Smagghe and Degheele (1993, 1994b) reported that excretion of absorbed compound was rapid before being accumulated relatively fast in the insect body. For tebufenozide, 50% of ingested compound was recovered in the excrements of both *Spodoptera* armyworm species between 2 and 6 h after oral uptake, and the rate constant of excretion for tebufenozide was estimated as $3-4 \times 10^{-3} \, min^{-1}$ (Table 2). For methoxyfenozide, the data so far (Smagghe et al. 1999b) suggest a faster excretion in the excrements via the gut resulting in a lower absorption in the body tissues; however, the compound was more toxic for the two armyworms. In contrast, the rate constant of excretion in *L. decemlineata* for RH-0345, being the most toxic compound for beetle larvae, was estimated at $1.716 \times 10^{-3} \, min^{-1}$ while this was $22.44 \times 10^{-3} \, min^{-1}$ for tebufenozide that was less toxic (Table 2). Though this excretion was rapid, the levels of original tebufenozide scored in the beetle's integument should be high enough to exert a toxic action. This suggests that factors other than uptake, transport and metabolism in the insect body mediate the selective toxicity: most likely, we are dealing here with a selective affinity to bind the target sites, i.e., ecdysteroid receptor (Smagghe et al. 1996; Dhadialla et al. 1998).

In multi-resistant last instars from *S. littoralis* originating from Israel, no marked differences in absorption profile after ingestion were recorded

Table 3. Time course of parent tebufenozide (RH-5992), expressed as nanograms of original product per mg fresh body weight (ng/mg wt), after oral ingestion in the larval body of one individual last-instar, rate constant of excretion $(k)^a$ and oral toxicity (mg/l) in a multi-resistant field and susceptible laboratory strain of the cotton leafworm *Spodoptera littoralis* (data from Smagghe 1995; Smagghe et al. 1995)

Time (h after ingestion)	Susceptible (ng/mg wt)	Multi-resistant (ng/mg wt)
2	13	7
6	3	0.5
12	0.3	0.1
24	0.3	0.1
$k \times 10^3$ (min^{-1})	2.221	2.511
LC$_{50}$ (mg/l)	16.61	20.81

[a] Rate constant k determined by using the formula $k = 0.693/t_{0.5}$ (min^{-1}) (O'Brien 1967). Half-time $(t_{0.5})$ (min) was determined from the slope of the regression curves of absorption-excretion assays with [^{14}C]-labeled ecdysone agonist.

for four ecdysone agonists (RH-5849, tebufenozide, methoxyfenozide and halofenozide) when tested in comparison with laboratory susceptible larvae (Smagghe et al. 1995, 1999a). So, the excretion pattern could not explain the observed differences in toxicity. It should be remarked, however, that the LC$_{50}$ ratio between field and susceptible strain only varied between 1.1 and 3.0. When Smagghe (1995) re-calculated the amounts of original tebufenozide in the larval body of an individual last-instar of a laboratory and multi-resistant field strain from Egypt at increasing intervals following ingestion of tebufenozide (Table 3), he was able to demonstrate that the parent compound was excreted more rapidly and metabolized to a higher extent in the field strain, which may elucidate a lower toxicity. In the body of the laboratory strain, about 90% of the amount of radioactivity consisted of parent tebufenozide, whereas this was only 55% in the field strain. However, more comparative biochemical absorption-excretion studies are required before a firm conclusion can be made on the impact of pharmacokinetics in resistance development for this novel group of IGRs, especially with populations collected from regions where growers have encountered problems difficult to control.

5
Attempts for Chemical Modeling of Digestion and Absorption in Insect Midgut

From the previous sections, it is clear that the contribution of the various processes to the final toxicity of a compound not only depends on their inherent properties of absorption and digestion but also on the interaction between them. Interaction here is that the rate of the detoxification (or intoxication) reaction is affected by a foregoing, parallel, or subsequent process. Kinetic interactions are important particularly if the pharmacokinetics of a toxic compound deviate from linearity, i.e. when the rate of one or more of the processes is not proportional to the concentration of the substance taking part in that process. This kind of interaction can best be explained by the dependence of the efficacy of an insecticide on the rate of absorption. This idea can be developed more quantitatively, as was firstly demonstrated to explain toxicity differences for cuticular penetration in mustard beetles by Ford and coworkers (1981a,b), in a theoretical model describing the insecticide fate in the insect body for different populations. Such modeling can then later be used to predict absorption, digestion and elimination, and can moreover quantitatively describe the essential process(es) involved in resistance.

In such a model, an ingested amount of insecticide (Q_i) is generally absorbed from outside (Q_o) into the body at a rate that is proportional to the amount ingested: rate constant k_a. After having penetrated through the gut wall, internal insecticide is distributed between two compartments according to the rate constants k_{12} and k_{21}. In this first compartment, the compound is metabolically degraded, and it can be assumed that the enzyme activity obeys the Michaelis-Menten equation. This equation is characterized by a maximal velocity V_m and a Michaelis constant K_m. The second compartment represents insecticide elimination. The following three differential equations then describe this two-compartment open model with first-order absorption and saturable elimination in the second compartment from the first:

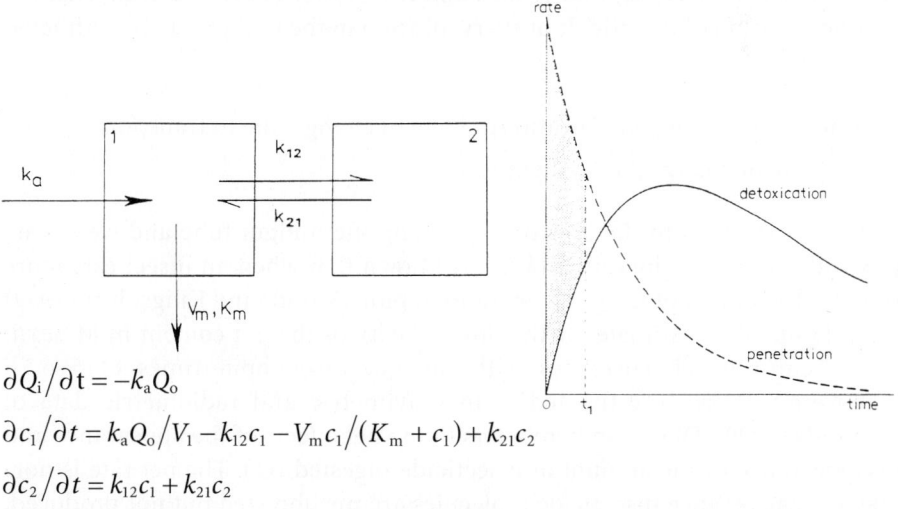

$$\partial Q_i / \partial t = -k_a Q_o$$

$$\partial c_1 / \partial t = k_a Q_o / V_1 - k_{12} c_1 - V_m c_1 / (K_m + c_1) + k_{21} c_2$$

$$\partial c_2 / \partial t = k_{12} c_1 + k_{21} c_2$$

with Q_o the amount of insecticide outside the insect body that is to be ingested orally, V_1 and V_2 the respective volumes of compartments 1 and 2, and c_1 and c_2 the concentrations in compartments 1 and 2, respectively. The amount accumulated in compartment 1 ($\partial c_1/\partial t$) depends on the absorption rate constant k_a and the amount in compartment 1, indicating that the relative elimination rate (digestion and excretion), that is expressed as a fraction of the concentration c_1, lags behind the absorption rate. So, there is a strong interaction between absorption and digestion. Integration of the differential equations elucidates the hatched area representing the amount of original insecticide absorbed up to time t_1. To illustrate how such an interaction influences the efficacy of the insecticide, the toxic effect can in this model simply be assumed to be proportional to the amount in compartment 1. When the amount Q_o is low, then the toxic effect is low and this is independent of the absorption rate constant from very low values of k_a on. With high doses of Q_o, there is at first a steep rise in the amount in compartment 1 (Q_i) with increasing absorption rates that level off to k_a-independent values reaching the plateau concentration of insecticide in compartment 1. At very high k_a-values, a situation comparable to injection directly into the hemolymph is expected. If then there is a difference in effect between absorption via oral ingestion and injection at constant dose Q_o, it is only possible if metabolism deviates from the first-order kinetics of Michaelis-Menten. In relation to compartment 2, it is to be noted that the elimination rate is proportional to the concentration c_1; degradation rate K_m is also involved.

When establishing this concept for modeling of the insect gut more concretely, we can likely describe the midgut architecture as a cylindrical tube with the insecticide flowing from the entrance through the exit. This means that the alimentary canal can be modeled for ingested insecticide molecules according to a plug-flow system, as was previously done for proteins by Penry and Jumans (1987), Dade et al. (1990) and Woods and Kingsolver (1999). As such, changes in concentration of insecticide at every location in the midgut can be estimated as follows:

[rate of change of concentration] = [rate of change due to transport]
    + [rate of change due to digestion]

Transport refers to the flow of food along the midgut tube and we equate this with the rate of ingestion (k_i). It is known that when an insect eats more rapidly, the content of the gut flows more rapidly. Woods and Kingsolver (1999) recently made an estimate for the flow velocity of the gut content in *M. sexta* of 0.93 cm/h which correlates with the gut throughput times of 5–6.5 h that were previously estimated from gravimetric and radiometric data of Reynolds (1990). Digestion here refers to the net effects of the appearance and disappearance of the amount of insecticide ingested (Q_i). The net rate is normally negative since insecticide molecules are metabolized but not produced. The net rate of production of metabolites (Q_m) locally can be negative or positive, depending on the balance between rates of metabolite production and absorption.

The gut can thus be considered as a small cylindrical volume with elements that have length Δx and radius r. The rate of change in time of Q_i in the volume element depends on Q_i entering on the left-hand side, the amount leaving on the right-hand side and the amount which is metabolized in the volume. Mathematically, the concentration of Q_i and of Q_m can be described in the following two partial differential equations, respectively:

$$\partial Q_i(x,t)/\partial t = -k_i[\partial Q_i(x,t)/\partial x] - [V_{max}Q_i(x,t)]/[K_m + Q_i(x,t)]$$

$$\partial Q_m(x,t)/\partial t = -k_i[\partial Q_m(x,t)/\partial x] - (2/r)[W_{max}Q_m(x,t)]/[L_m + Q_m(x,t)]$$
$$+ \alpha[V_{max}Q_i(x,t)]/[K_m + Q_i(x,t)]$$

with V_{max} the maximum rate of digestion, W_{max} the maximum rate of absorption, K_m the Michaelis constant of digestion, L_m the Michaelis constant of absorption, k_i the rate of ingestion/consumption (equivalent to rate of flow in the midgut) and the gut morphology (length l and radius r). Woods and Kingsolver (1999) estimated the midgut length in *M. sexta* to be 4.31 ± 0.04 cm and its radius 0.30 ± 0.004 cm. The factor $2/r$ is necessary to account for the scaling of surface area to volume of the midgut. If a fixed density of transporters on the gut surface can be assumed, then the number of transporters in any length Δx is proportional to $2\Pi r\Delta x$ and the number of molecules of Q_m to $\Pi r^2\Delta x$. Therefore, the ratio of transporters to metabolite molecules is proportional to $2/r$. So with increased radius, fewer transporters are present in the gut wall per number of metabolites in the corresponding volume.

The first term in both differential equations describes transport of insecticide molecules along the gut. The second term describes the digestion of insecticide and the absorption of metabolites, which are both processes following Michaelis-Menten kinetics. The third term in the second differential equation accounts for the appearance of metabolites due to digestion of insecticide molecules. When molecules disappear from the pool of insecticide ingested (first equation), they appear in the pool of metabolites (second equation) with α as the metric conversion factor between the two pools.

The concentration of metabolites, Q_m, which is zero at the mouth, initially rises with distance due to its production from insecticide ingested, Q_i. Further along the gut, the concentration of metabolites begins to fall as the rate of production of metabolites falls lower than the rate of absorption of metabolites across the epithelium. The overall rate of absorption of metabolites (A) in the gut can thus be defined as follows:

$$A = 2\Pi r \int_0^l [W_{max}Q_m(x)]/[L_m + Q_m(x)]$$

It should be pointed out that the absorption rate at each location is a non-linear function of Q_m since it is weighted by the Michaelis-Menten equation. The overall rate of absorption obviously depends on the values of the various parameters and also on the rate of consumption. For instance, with increasing consumption rate, absorption initially rises until it reaches a maximum at an intermediate rate of consumption and then drops. It is clear that the rate of feeding affects performance by altering the profiles of insecticide and metabolites.

Using these models of absorption and digestion in the insect midgut, it is our aim to test several insecticides, particularly IGRs, and estimate a chemical modeling of their fate in the midgut, for instance in *Spodoptera* caterpillars. We are investigating the in vivo concentration profiles of original insecticide and metabolites in susceptible and resistant populations, and will then use these data to estimate differential equations for absorption and digestion for several insecticides. As such, we can achieve fundamental insights into the midgut processes that can then be quantitatively measured and eventually extrapolated between different populations and insect species. From this viewpoint, we believe that this modeling of midgut processes can also provide extra information to help choosing the appropriate tactics for reducing resistance risks.

6
In Vitro Gut Cultures for Insecticidal Activity Studies

Next to the fact that insect midgut cultures are increasingly considered for use in virus and recombinant protein productions, such insect cell systems could also be an alternative to the mass rearing of insects for the bio-testing of viruses, toxins, growth regulators and different chemicals (Bellocnik et al. 1997). In general, a midgut culture system in vitro is simpler and free from other cells and tissues that may differentially affect their response. Cloning of cDNAs for various genes from a cell line is also easier than using whole animals because the RNA is less complex.

For a number of years, naturally occurring protease inhibitors have been explored since they interact and block the active center of the digestive enzymes, both proteases and amylases, in the gut system. Many proteinase inhibitors are rather small, very stable, abundant proteins and are therefore rather easy to purify. Those commonly used include: crystalline soybean trypsin inhibitor of Kunitz, Bownman-Birk soybean trypsin inhibitor, lima bean inhibitor, chickpea trypsin and chymotrypsin inhibitor, bovine pancreatic trypsin inhibitor and ovomucoid (Applebaum 1985; Reeck et al. 1997). These proteins bind tightly to the active site of the enzyme in the midgut and so prevent access to normal substrates. Failure by the pest insect to overcome this inhibition of digestion results in death by starvation. Nowadays, this principle of digestive enzyme inhibitors can be exploited with the current recombinant DNA-technology in transgenic plants for crop protection approaches in order to, amongst others, overcome resistance towards neurotoxic pesticides (Singh 1986; Christou et al. 1991; Gatehouse et al. 1996; Reeck et al. 1997). On the other hand, Jongsma et al. (1995) reported adaptation of *S. exigua* caterpillars by induction of gut proteinase activity in the midgut insensitive to inhibition. Nevertheless, it is believed that there is great promise in the use of genes that encode digestive enzyme inhibitors for insect control in transgenic plants. Inhibitors should, however, be selected carefully, with consideration of the physiology and biochemistry of the protein digestion in the insect midgut, and, in addition, combinations of inhibitors with other stress inducing or growth suppressing proteins should be exploited (Reeck et al. 1997). In vitro midgut cultures can in this way be employed efficiently to screen combinations of possible candidates under controlled conditions.

A related aspect of digestive enzyme inhibitors is their fate once they are ingested. Castañera and Ortega (pers. comm. 1999) found some evidence to suggest that proteinase inhibitors, in addition to their action on the digestive proteases, are able to penetrate through the gut wall and obtain access to other target organs outside the gut. Billings et al. (1992) reported that, following oral administration in mice, soybean Bowman-Birk trypsin inhibitor passed across the gut barrier and could be found in almost every organ, the blood and urine. In this context insect midgut cultures can also be of use to assess the possibility of enzyme inhibitors for absorption through the gut wall.

Next to tissue cultures, primary cell cultures of *M. sexta* midgut epithelia, as well as from *Heliothis virescens, Lymantria dispar* and *Spodoptera* sp., can routinely be prepared from cells collected from the pharate fourth-instar midgut (Loeb and Hakim 1996; Hakim et al. 1997; M. Loeb, pers. comm.). These cells include rapidly proliferating populations of morphologically undifferentiated granular and smooth appearing spherical stem cells that differentiate in culture to morphologically distinct goblet and columnar cells. After optimization in 4×6 well culture plates using about 1000 cells in 450 µl per well, this primary cell culture provides the ability to test routinely the potency of *Bacillus thuringiensis* preparations (M. Loeb, pers. comm.) and the physiological interaction of viruses (Dougherty et al. 1998).

Similarly, Belloncik et al. (1997) developed four cell lines from the Colorado potato beetle and used them to test different bacterial and mycotoxin activities. The four lines were demonstrated to be sensitive to the solubilized crystal delta-endotoxin from different *B. thuringiensis* strains. However, more investigations must be made towards improvements of *B. thuringiensis* toxin dissolutions. Addition of midgut trypsin had proven to have no effect on the amount of protein recovered following the Na_2CO_3 treatment; but, on the other hand, addition of gut juice combined with skipping the neutralization step of the dissolved toxin improved the rate of this amount.

In 1993, Sohi and coworkers established a continuous cell line (CF-203) from the midgut of the spruce budworm, *Choristoneura fumiferana*, in an attempt to replicate the multicapsid nuclear polyhedrosis virus (CfMNPC). However, when CF-203 cells were infected with CfMNPC, the cells died prematurely and no virus was produced. However, interestingly, the observed phenomenon appeared to be similar to apoptosis (Palli et al. 1997). For biological insect control, it is interesting that apoptosis may also influence the pathogenicity of a number of insect pathogens including viral infections (e.g. baculovirus). Experiments with such midgut cell lines and testing cells undergoing apoptosis with a combination of morphological observations and TUNEL-staining, to test for DNA-fragmentation at 8 days after subculturing, is a powerful technique to screen for new insecticides.

In the field of medical, veterinary and agriculturally important protozoan parasites, Cunningham (1986) and Lawrence (1997) described, for instance, the use of organ cultures of the alimentary canal of *Glossina* tsetse fly for inducing the transformation of *Trypanosoma* promastigotes to procyclic trypanosomoses. They pointed out a strong argument for using these midgut cultures, that they maintained in a medium formulation that was based on the amino acid composition of tsetse fly hemolymph plus 20% FCS at 28°C, for further investigations that could complement and expedite ongoing efforts to suppress deleterious parasites and enhance the effectiveness of beneficial ones. Another example is the culturing of the insect vector stages of the malaria parasite *Plasmodium* in *Drosophila* cells. It is evident that midgut cultures in controlled in vitro circumstances are also useful to identify factors and test conditions that may stimulate and/or inhibit growth of such important parasites transmitted by insects to man and animals.

Acknowledgements. Dr. Guy Smagghe acknowledges a post-doctoral grant from the National Fund for Scientific Research, Brussels.

References

Agosin M (1985) Role of microsomal oxidations in insecticide degradation. In: Kerkut GA, Gilbert LI (eds) Comprehensive insect physiology, biochemistry and pharmacology, vol 12. Pergamon Press, Oxford, pp 647–712

Aldridge WN (1953) Serum seterases. I. Two types of esterases (A and B) hydrolyzing p-nitrophenyl acetate, propionate and butyrate, and a method for their determination. Biochem J 53:110–117

Agosin M (1985) Role of microsomal oxidations in insecticide degradation. In: Kerkut GA, Gilbert LI (eds) Comprehensive insect physiology, biochemistry and pharmacology, vol 12. Pergamon Press, Oxford, pp 647–712

Applebaum SW (1985) Biochemistry of digestion. In: Kerkut GA, Gilbert LI (eds) Comprehensive insect physiology, biochemistry and pharmacology, vol 4. Pergamon Press, Oxford, pp 279–311

Applebaum SW, Birk Y (1972) Natural mechanisms of resistance to insects in legume seeds. In: Rodriguez JG (ed) Insects and mite nutrition. North-Holland Publishing, Amsterdam, pp 629–636

Auda M (1986) Critical approach of the problem of insecticide resistance in *Spodoptera littoralis* Boisd. and *Musca domestica* L. PhD dissertation, State University Ghent, Ghent, Belgium

Augustinsson KB (1958) Electrophoretic separation and classification of blood plasma esterases. Nature 181:1786–1789

Belloncik S, Charpentier G, Tian L (1997) Development of four cell line from the Colorado potato beetle (*Leptinotarsa decemlineata*). In: Maramorosch K, Mitsuhashi J (eds) Invertebrate cell culture – novel directions and biotechnology applications. Science Publishers, New Hampshire, pp 3–10

Billings PC, St Clair WH, Maki PA, Kennedy AR (1992) Distribution of the Bowman-Birk protease inhibitor in mice following oral administration. Cancer Lett 62:191–197

Black BC, Hollingworth RM, Ahammadsahib KI, Kukel CD, Donovan S (1994) Insecticidal action and mitochondrial uncoupling activity of AC-303,630 and related halogenated pyrroles. Pestic Biochem Physiol 50:115–128

Broadway RM, Duffey SS (1986) The effect of dietary protein on the growth and digestive physiology of larval *Heliothis zea* and *Spodoptera exigua*. J Insect Physiol 32:673–680

Chang SC (1978) Conjugation: the major metabolic pathway of C14diflubenzuron in the housefly. J Econ Entomol 71:31–39

Chang SC, Stokes JB (1979) Conjugation: the major metabolic pathway of C14diflubenzuron in the boll weevil (*Anthonomus grandis grandis* Boheman). J Econ Entomol 72:15–19

Chapman RF (1985a) Coordination of digestion. In: Kerkut GA, Gilbert LI (eds) Comprehensive insect physiology, biochemistry and pharmacology, vol 4. Pergamon Press, Oxford, pp 213–240

Chapman RF (1985b) Structure of the digestive system. In: Kerkut GA, Gilbert LI (eds) Comprehensive insect physiology, biochemistry and pharmacology, vol 4. Pergamon Press, Oxford, pp 165–212

Christou P, Ford TL, Kofron M (1991) Production of transgenic rice (*Oryzae sativa* L.) plants from agronomically important indica and japonica varieties via electric discharge particle acceleration of exogenous DNA into immature zygotic embryos. Bio/Technology 9:957–962

Conner WE, Wilkinson CF, Morse RA (1978) Penetration of insecticides through the foregut of the honeybee (*Apis mellifera* L.). Pestic Biochem Physiol 9:131–139

Cunningham I (1986) Infectivity of *Trypanosoma rhodesiense* cultivated at 28 °C with various tsetse fly tissues. J Protozool 33:226–231

Dadd RH (1970) Digestion in insects. In: Florkin M, Scheer BT (eds) Chemical zoology, vol 5. Academic Press, New York, pp 117–145

Dade WB, Jumars PA, Penry DL (1990) Supply-side optimization: maximizing absorptive rates. In: Hughes RN (ed) Behavioral mechanisms of food selection. Springer, Berlin Heidelberg New York, pp 531–556

Darvas B, Pap L, Kelemen M, Polgar LA (1998) Synergistic effects of verbutin with dibenzoylhydrazine-type ecdysteroid agonists on larvae of *Aedes aegypti* (Diptera: Culicidae). J Econ Entomol 91:1260–1264

Dauterman WC (1985) Insect metabolism: extramicrosomal. In: Kerkut GA, Gilbert LI (eds) Comprehensive insect physiology, biochemistry and pharmacology, vol 12. Pergamon Press, Oxford, pp 713–730

Dauterman WC, Hodgson E (1978) Detoxification mechanisms in insects. In: Rockstein M (ed) Biochemistry of insects. Academic Press, New York, pp 541–577

De Clercq P, Viñuela E, Smagghe G, Degheele D (1995) Transport and kinetics of distribution of diflubenzuron and pyriproxyfen in the beet armyworm *Spodoptera exigua* and its predator *Podisus maculiventris*. Entomol Exp Appl 76:189–194

Devonshire AL, Moores GD, ffrench-Constant RH (1986) Detection of insecticide resistance by immunological estimation of carboxylesterase activity in *Myzus persicae* (Sulzer) and cross reaction of the antiserum with *Phorodon humili* (Schrank) (Hemiptera: Aphilidae). Bull Entomol Res 76:97–107

Dhadialla TS, Carlson GR, Le DP (1998) New insecticides with ecdysteroidal and juvenile hormone activity. Annu Rev Entomol 43:545–569

Dougherty E, Loeb MJ, Narang N, Shapiro M (1998) Use of midgut cell cultures to elucidate the mode of action for fluorescent brighteners. Baculovirus Insect Cell Culture Conference

Dow JAT (1986) Insect midgut function. Adv Insect Physiol 19:187–328

El-Guindy MA, El-Rafai ARM, Abdel-Satter MM (1983) The pattern of cross-resistance to insecticides and juvenile hormone analogues in a diflubenzuron-resistant strain of the cotton leafworm, *Spodoptera littoralis*. Pestic Sci 14:235–245

El Saidy MF (1991) Biological and biochemical activities of benzoylphenylureas and conventional insecticides on *Spodoptera littoralis*. PhD Diss, State University Ghent, Ghent, Belgium, 339 pp

El Saidy MF, Auda M, Degheele D (1989) Detoxification mechanism of diflubenzuron and teflubenzuron in the larvae of *Spodoptera littoralis* (Boisd.). Pestic Biochem Physiol 35: 211–222

El Saidy MF, Auda M, Degheele D (1990) Esterases and resistance to organophosphorus insecticides in *Spodoptera littoralis*. Med Fac Landbouww Rijksuniv Gent 55:565–576

Enzyme nomenclature (1978) Recommendations of the nomenclature committee of the international union of biochemistry on the nomenclature and classification of enzymes. Academic Press, New York

Estabrook RW, Baron J, Franklin M, Mason I, Waterman M, Peterson J (1972) Cytochrome P-450: panacae or plague? In: Schultz J, Cameron BF (eds) The molecular basis of electron transport. Academic Press, New York, pp 197–230

Ford MG, Greenwood R, Thomas PJ (1981a) The kinetics of insecticide action. I. The properties of a mathematical model describing insect pharmacokinetics. Pestic Sci 12:175–198

Ford MG, Greenwood R, Thomas PJ (1981b) The kinetics of insecticide action,. II The relationship between the pharmacokinetics of substituted benzoyl (1*R,S*)-*cis,trans*-chrysanthemates and their relative toxicities to mustard beetles (*Phaedon cochleariae* Fab.). Pestic Sci 12: 265–284

Gatehouse JA, Powell K, Edmonds H (1996) Genetic engineering of rice for resistance to homopteran insect pests. Rice Genet III:189–200

Giordana B, Sacchi VF, Parenti P, Hanozet GM (1989) Amino acid transport systems in intestinal brush-border membranes from lepidopteran larvae. Am J Physiol 257:494–500

Grassé P-P (ed) (1949) Traité de zoologie, vol 9. Masson and Cie, Paris

Grassé P-P (ed) (1951) Traité de zoologie, vol 10. Masson and Cie, Paris

Gupta BL, Hall TA (1979) Quantitative electron probe X-ray microanalysis of electrolyte elements within epithelial tissue compartments. Fed Proc 38:144–153

Hakim RS, Hakim FT, Loeb MJ (1997) Growth of *Manduca sexta* epithelial cells in the establishment of a primary culture. In: Maramorosch K, Mitsuhashi J (eds) Invertebrate cell culture. Novel directions and biotechnology applications. Science Publishers, New Hampshire, pp 19–24

Hammock BD, Quistad GB (1981) Metabolism and mode of juvenile hormone, juvenoids, and other insect growth regulators. In: Hutson DH, Roberts TR (eds) Progress in pesticide biochemistry, vol 1. Wiley, New York, pp 1–83

House HL (1974) Digestion. In: Rockstein M (ed) The physiology of Insecta. Academic Press, New York, pp 63–117

Ishaaya I (1986) Nutritional and allelochemical insect plant interactions relating to digestion and food intake: some examples. In: Miller JR, Miller TA (eds) Insect-plant interactions, vol 7. Springer, Berlin Heidelberg New York, pp 291–333

Ishaaya I, Casida JE (1980) Properties and toxicological significance of esterases hydrolyzing permethrin and cypermethrin in *Trichoplusia ni* larval gut integument. Pestic Biochem Physiol 14:178–184

Ishaaya I, Degheele D (1988) Properties and toxicological significance of diflubenzuron hydrolase activity in *Spodoptera littoralis* larvae. Pestic Biochem Physiol 32:180–187

Ishaaya I, Moore I, Joseph D (1971) Protease and amylase activity in larvae of the Egyptian cotton leafworm, *Spodoptera littoralis*. J Insect Physiol 17:945–953

Jongsma M, Bakker P, Peters D, Bosch D, Stiekema W (1995) Adaptation of *Spodoptera exigua* larvae to plant proteinase inhibitors by induction of gut proteinase activity insensitive to inhibition. Proc Natl Acad Sci USA 92:8041–8045

Khan MA (1964) Studies on the secretion of digestive enzymes in *Locusta migratoria* L. II. Invertase activity. Entomol Exp Appl 7:125–130

Klocke JA (1987) Natural plant compounds useful in insect control. In: Waller GR (ed) ACS symposium series no 330. American Chemical Society, Washington, DC, pp 396–415

Lawrence PO (1997) Insect cell culture: an under-exploited resource for the study of parasite-host interactions. In: Maramorosch K, Mitsuhashi J (eds) Invertebrate cell culture. Novel directions and biotechnology applications. Science Publishers, New Hampshire, pp 279–287

Lehane MJ, Billingsley PF (1996) Biology of the insect midgut. Chapman and Hall, London, 486 pp

Loeb MJ, Hakim RS (1996) Insect midgut epithelium in vitro: an insect stem cell system. J Insect Physiol 42:1103–1111

Maddrell SHP, Gardiner BOC (1980) The permeability of the cuticular lining of the insect alimentary canal. J Exp Biol 85:227–237

Mayer RT, Durrant JL (1979) Preparation of homogeneous NADPH cytochrome c(P-450) reductase EC-1.6.2.4 from houseflies using affinity chromatography techniques. J Biol Chem 254:756–761

McKenzie JA (1996) Ecological and evolutionary aspects of insecticide resistance. Landes/Academic Press, Austin, 185 pp

McKenzie JA, Batterham P (1994) The genetic, molecular and phenotypic consequences of selection for insecticide resistance. Trends Ecol Evol 9:166–169

Mayer SE, Melmon KL, Gilman AG (1980) The dynamics of drug absorption, distribution and elimination. In: Gilman A et al. (eds) The pharmacological basis of therapeutics. Macmillan, New York, pp 1–24

Moffit HR, Westgrad PH, Mantey KD, Van de Baan HE (1988) Resistance to diflubenzuron in codling moth (Lepidoptera: Tortricidae). J Econ Entomol 81:1511–1515

Nakagawa Y, Smagghe G, Kugimiya S, Hattori K, Ueno T, Tirry L, Fujita T (1999) Quantitative structure-activity studies of insect growth regulators. XVI. Substituent effects of dibenzoyl-hydrazines on the insecticidal activity to Colorado potato beetle *Leptinotarsa decemlineata*. Pestic Sci 55:909–918

O'Brien RDO (1967) Insecticides, action and metabolism. Academic Press, New York, 258 pp

Palli SR, Sohi SS, Cook BJ, Primavera M, Retnakaran A (1997) Screening of 12 continuous cell lines for apoptosis. In: Maramorosch K, Mitsuhashi J (eds) Invertebrate cell culture. Novel directions and biotechnology applications. Science Publishers, New Hampshire, pp 52–60

Penry DL, Jumars PA (1987) Modeling animal guts as chemical reactors. Am Nat 129:69–96

Pimprikar GD, Georghiou GP (1979) Mechanisms of resistance to diflubenzuron in the housefly, *Musca domestica* L. Pestic Biochem Physiol 12:10–22

Reeck GR, Kramer KJ, Baker JE, Kanost MR, Fabrick JA, Behnke GA (1997) Proteinase inhibitors and resistance of transgenic plants to insects. In: Carozzi N, Koziel M (eds) Advances in insect control: the role of transgenic plants. Taylor and Francis, London, pp 157–183

Reynolds SE (1990) Feeding caterpillars: maximizing or optimizing food acquisition? In: Mellinger J (ed) Animal nutrition and transport processes 1. Nutrition in wild and domestic animals, vol 5. Karger, Basel, pp 106–118

Retnakaran A, Granett J, Ennis T (1985) Insect growth regulators. In: Kerkut GA, Gilbert LI (eds) Comprehensive insect physiology, biochemistry and pharmacology, vol 12. Pergamon Press, Oxford, pp 529–601

Reuveni M, Dunn PE (1994) Proline transport into brush order membrane vesicles from the midgut of *Manduca sexta* larvae. Comp Biochem Physiol 107A:685–691

Santos CD, Ferreira C, Terra WR (1983) Consumption of food and spatial organization of digestion in the casava hornworm, *Erinnyis ello*. J. Insect Physiol 29:707–714

Sauphanor B, Brosse V, Monier C, Bouvier JC (1998) Differential ovicidal and larvicidal resistance to benzoylureas in the codling moth, *Cydia pomonella*. Entomol Exp Appl 88:247–253

Schuyesmans S (1998) Relation between toxicity and enzyme activity for new insect growth regulators in the beet armyworm *Spodoptera exigua* (in Flemish). MSc dissertation, University Ghent, Ghent, Belgium, p 71

Scott (1990) Investigating mechanisms on insecticide resistance: methods, strategies, and pitfalls. In: Roush RT, Tabashnik BE (eds) Pesticide resistance in arthropods. Chapman and Hall, New York, pp 39–57

Shah AH, Guthrie FE (1970) Penetration of insecticides through the isolated gut of insects and mammals. Comp Gen Pharmacol 1:391–399

Shah AH, Guthrie FE (1971) In vitro metabolism of insecticides during midgut penetration. Pestic Biochem Physiol 1:1–10

Shah PV, Dauterman WC, Gunthrie FE (1972) Penetration of a series of dialkoxy analogs of dimethoate through the isolated gut of insects and mammals. Pestic Biochem Physiol 2:324–330

Singh DP (1986) Breeding for resistance to diseases and insect pests. Springer, Berlin Heidelberg New York, pp 35–61

Still GG, Leopold RA (1978) The elimination of (N-[[4-chlorophenyl)amino] carbonyl]-2,6-difluorobenzamide) by the boll weevil. Pestic Biochem Physiol 9:304–312

Smagghe G (1995) Nonsteroidal ecdysteroid agonists: biological activity and insect specificity. PhD Diss, University Ghent, Ghent, Belgium, 118 pp

Smagghe G, Degheele D (1993) Toxicity, pharmacokinetics, and metabolism of the first nonsteroidal ecdysteroid agonist, RH 5849, on *Spodoptera exempta* (Walker), *Spodoptera exigua* (Hübner), and *Leptinotarsa decemlineata* (Say). Pestic Biochem Physiol 46:149–160

Smagghe G, Degheele D (1994a) Action of a novel nonsteroidal ecdysteroid mimic, tebufenozide (RH-5992), on insects of different orders. Pestic Sci 42:85–92

Smagghe G, Degheele D (1994b) The significance of pharmacokinetics and metabolism to the biological activity of RH-5992 (tebufenozide) in *Spodoptera exempta*, *Spodoptera exigua*, and *Leptinotarsa decemlineata*. Pestic Biochem Physiol 49:224–234

Smagghe G, Audenaert L, Degheele D (1995) Tebufenozide: is toxicity correlated with pharmacokinetics and metabolism in different strains of the Egyptian cotton leafworm? Med Fac Landbouw Toeg Biol Wet Univ Gent 60:1015–1016

Smagghe G, Eelen H, Verschelde E, Richter K, Degheele D (1996) Differential effects of nonsteroidal ecdysteroid agonists in Coleoptera and Lepidoptera: analysis on evagination and receptor binding in imaginal discs. Insect Biochem Mol Biol 26:687–695

Smagghe G, Dhadialla TS, Derycke S, Tirry L, Degheele D (1998a) Tebufenozide in susceptible and artificially tolerant beet armyworm. Pestic Sci 54:27–34

Smagghe G, Wesemael W, Carton B, Tirry L (1998b) Tebufenozide and methoxyfenozide against the beet armyworm *Spodoptera exigua*. Proc Brighton Crop Protection Conf-Pests Dis, Brighton, UK, vol 1, pp 311–312

Smagghe G, Carton B, Heirman A, Tirry L (1999a) Toxicity and impact of kinetics, metabolism and binding of nonsteroidal ecdysone agonists in a susceptible and resistant strain of the cotton leafworm. Proc XIVth Int Plant Prot Congr, Jerusalem, Israel, p 152

Smagghe G, Carton B, Wesemael W, Ishaaya I, Tirry L (1999b) Ecdysone agonists: mechanism and application on *Spodoptera* species. Pestic Sci 55:243–389

Smagghe G, Medina P, Schuyesmans S, Tirry L, Viñuela E (1999c) Insecticide resistance monitoring and potential of novel insect growth regulators for managing the beet armyworm (*Spodoptera exigua* Hübner). Proc Combating Insecticide Resistance, Thessaloniki, Greece, pp 70–78

Smagghe G, Nakagawa Y, Carton B, Mourad AK, Tirry L (1999d) Comparative ecdysteroid action of ring-substituted dibenzoylhydrazines in *Spodoptera exigua*. Arch Insect Biochem Physiol 41:42–53

Sohi SS, Lalouette W, Macdonald JA, Gringorten JL, Budae CB (1993) Establishment of continuous midgut cell lines of spruce budworm (Lepidoptera: Tortricidae). In Vitro Cell Dev Biol 29A:56A

Still GG, Leopold RA (1978) The elimination of (N-[[(4-chlorophenyl)amino] carbonyl]-2,6-diflubenzamide) by the boll weevil. Pestic Biochem Physiol 9:304–312

Terra W, Ferreira C (1994) Insect digestive enzymes: properties, compartmentalization and function. Comp Biochem Physiol 109B:1–62

Turunen S (1985) Absorption. In: Kerkut GA, Gilbert LI (eds) Comprehensive insect physiology, biochemistry and pharmacology, vol 4. Pergamon Press, Oxford, pp 241–277

Van Laecke K (1993) Insecticide-detoxification mechanisms in *Spodoptera exigua* (Hübner) (Lepidoptera: Noctuidae). PhD Diss, University of Ghent, Ghent, Belgium, p 202

Van Laecke K, Smagghe G, Degheele D (1995) Detoxifying enzymes in greenhouse and laboratory strain of beet armyworm (Lepidoptera: Noctuidae). J Econ Entomol 88:777–781

Welling W, Paterson (1985) Toxicodynamics of insecticides. In: Kerkut GA, Gilbert LI (eds) Comprehensive insect physiology, biochemistry and pharmacology, vol 12. Pergamon Press, Oxford, pp 603–645

Williams RT (1959) Detoxification mechanisms. Chapman and Hall, London

Woods HA, Kingsolver JG (1999) Feeding rate and the structure of protein digestion and absorption in lepidopteran midguts. Arch Insect Biochem Phsyiol 42:74–87

Impact of Insecticide Resistance Mechanisms on Management Strategies

A. R. Horowitz[1] and I. Denholm[2]

1
Introduction

Few areas of applied entomology have advanced as rapidly or received such widespread attention in recent years as that of insecticide resistance. This reflects both the increasingly severe impact of resistance on pest and disease management programmes, and, as is evident from several other chapters in this volume, the exciting contributions that resistance is making to fundamental knowledge of insect genetics, biochemistry and physiology. These insights should, in turn, offer greater prospects for developing or fine-tuning strategies aimed at circumventing the impact of resistance on pest management, or even preventing its appearance in the first place.

Opportunities for exploiting a knowledge of mechanisms to combat resistance were outlined very effectively by Georghiou (1983) in his landmark paper entitled "Management of resistance in arthropods". As well as presenting a novel classification and a comprehensive review of operational tactics for combating resistance to insecticides, this paper also highlighted several areas of research needed to refine such tactics and adapt them to specific resistance problems. These included detailed work on the toxicological and biochemical nature of resistance mechanisms, in order to improve the detection of resistance genes and provide means of mitigating mechanisms through the use of synergists and/or novel "resistance-defeating" toxicants.

Similar views have been expressed by a number of authors including Scott (1990) and McKenzie (1996). Holloway and Forrester (1998) also emphasized how an improved understanding of resistance mechanisms could contribute to effective recommendations for resistance management strategies. Similarly, McCaffery (1998) considered an understanding of the mechanisms underlying resistance to be central to optimising the use of existing insecticide chemistry in cases where resistance already has evolved, and to understanding cross-resistance patterns within and between chemical groups.

[1] Department of Entomology, Agricultural Research Organisation, The Volcani Center, Bet Dagan 50250, Israel
[2] Department of Biological and Ecological Chemistry, IACR-Rothamsted, Harpenden, Herts AL5 2JQ, UK

Isaac Ishaaya (Ed.): Biochemical Sites of Insecticide Action and Resistance
© Springer-Verlag Berlin, Heidelberg 2001

This chapter explores the extent to which research on resistance mechanisms is contributing or might contribute to managing resistance in practice. In keeping with the scope of this volume, we restrict our subject matter to knowledge arising from biochemical and molecular studies of genes or gene products that confer resistance. Other attributes of resistance mechanisms such as their effective dominance and phenotypic expression under field conditions also play a critical role in the development and management of resistance, but have been reviewed comprehensively elsewhere (Roush and Tabashnik 1990; Denholm and Rowland 1992; McKenzie 1996).

2
Overview of Resistance Mechanisms

Insecticide resistance is an evolutionary phenomenon, reflecting the selection of mutant genes by insecticide applications. These genes encode a variety of mechanisms based primarily on enhanced detoxification of insecticides, or modifications to their target sites within arthropods. Many such mechanisms are described in detail elsewhere in this volume as well as in several earlier books or conference proceedings (e.g. Otto and Weber 1990; Roush and Tabashnik 1990; Mullin and Scott 1992; Brown 1996; McKenzie 1996; Denholm et al. 1998b). Other recent reviews focus on specific resistance mechanisms such as ones based on cytochrome P450 monooxygenases (Scott 1999), or on cataloguing known mechanisms of resistance in important multi-resistant pests such as the heliothine bollworms (McCaffery 1998).

The most extensively used insecticide classes – organochlorines, organophosphates, carbamates and pyrethroids – have generally been the most seriously threatened by resistance, and hence the major targets of research to resolve the causal mechanisms. Resistance to cyclodienes (such as dieldrin and endosulfan) usually results from a modification of the target site for this class of compounds, the GABA(γ-aminobutyric acid)-gated chloride channel of post-synaptic nerve membranes. Resistance to organophosphates (OPs) and carbamates can arise through enhanced detoxification by cytochrome P450 monooxygenases, esterases or glutathione S-transferases (GSTs), or from structural modifications of their target enzyme, acetylcholinesterase (AChE). Pyrethroid resistance can also arise through enhanced esteratic or oxidative detoxification, as well as from target-site insensitivity at the voltage-gated sodium channel in nerve membranes (so-called knockdown or *kdr* resistance).

Mechanisms of resistance to insecticides acting outside the nervous system (e.g. insect growth regulators [IGRs] and *Bacillus thuringiensis* endotoxins), or to more novel neurotoxins (e.g. the neonicotinoids), are less clearly understood, but are also likely to prove attributable to enhanced detoxification and/or target site modification.

Perhaps the most significant recent progress in the understanding of resistance mechanisms has resulted from the application of molecular biology to resistance research. Depending on the mechanism involved, resistance has been shown to arise from structural alterations of genes encoding detoxifying enzymes (Newcomb et al. 1997), or target-site proteins (ffrench-Constant et al. 1998), or through processes (e.g. amplification or altered transcription) affecting gene expression (Hemingway et al. 1998). Despite the complexity of receptors or enzymes responsible, mutations leading to resistance frequently recur in different species (Thompson et al. 1993; Martinez-Torres et al. 1997). This is especially the case for mechanisms based on decreased sensitivity of insecticide target sites. Molecular studies of insecticide resistance have identified the point mutations associated with target site insensitivity in genes encoding the three major insecticide targets: the GABA receptor (cyclodiene resistance), the voltage-gated sodium channel (pyrethroids) and AChE (OPs and carbamates) (ffrench-Constant et al. 1993; Mutero et al. 1994; Williamson et al. 1996). These are providing exciting insights into the homology of resistance mutations between species and the frequency with which they arise (ffrench-Constant et al. 1996, 1998).

One of the most difficult challenges with managing resistance is the frequent occurrence of several resistance mechanisms in the same pest individual or population. Such cases of multiple resistance are best documented for pests that have been exposed repeatedly and for long periods to a succession of insecticide groups, leading to the accumulation of mechanisms with contrasting cross-resistance characteristics. Example of such multi-resistant pests include the diamondback moth, *Plutella xylostella* (L), cotton whitefly, *Bemisia tabaci* (Gennadius), peach-potato aphid, *Myzus persicae* (Sulzer), heliothine bollworms (e.g. *Heliothis virescens* [F] and *Helicoverpa armigera* [Hübner]), Colorado potato beetle, *Leptinotarsa decemlineata* (Say), various mosquitoes, the housefly (*Musca domestica* [L]) and the German cockroach, *Blattella germanica* (L). Some of these are covered in more detail in specific sections later in the chapter.

3
Overview of Resistance Management Tactics

Since the 1970s, various countermeasures based largely on computer models have been proposed for combating resistance. Most are based on manipulating operational factors defining the rate, timing, nature and frequency of insecticide applications, and on exploiting a knowledge of pest biology in order to anticipate the selection pressure imposed by insecticides. As noted by several authors (e.g. Sawicki 1981; Roush 1989; Denholm and Rowland 1992; Georghiou 1994; Castle et al. 1999), there is no single prescription for combating resistance under all situations. Tactics must instead be tailored as carefully as possible to individual pests or pest complexes in light of ecological and genetic factors, the diversity of chemicals available, and practical constraints on the precision with which they can be implemented.

Approaches to combating resistance can be viewed from a number of different perspectives (e.g. Georghiou 1983; Roush 1989; Denholm and Rowland 1992; McKenzie 1996). The classification proposed by Georghiou (1983) is a useful one in the context of this chapter and is summarized below.

Management by moderation aims to reduce selection for resistance by preserving susceptible insects in the population through use of low application doses, less frequent applications, short-lived residues, or the creation of untreated "refuges". This approach is often the easiest to implement and involves least risk. However, the value of lowering application rates to manage resistance remains contentious, since unless overall efficacy is compromised to a substantial extent, there is a threat of increasing the number of resistance genes that could be selected (Roush 1989; Denholm and Rowland 1992; McKenzie 1996).

Management by saturation aims to overpower any resistant individuals present by using doses sufficiently high to kill resistant insects (resistance heterozygotes especially), suppressing detoxification enzymes through the use of synergists, or identifying "resistance-defeating" toxins less affected or unaffected by known resistance mechanisms.

Management by multiple attack involves using two or more unrelated pesticides in ways that reduce the selection or impact of resistance to any one chemical. The compounds can be applied simultaneously as mixtures, alternately in rotation, or in more complex spatial patterns known as mosaics. Although mixtures offer greater theoretical benefits than alternations, they require a far greater number of assumptions to be met regarding the efficacy, persistence and complementarity of partner chemicals (Roush 1989; Tabashnik 1989; Denholm et al. 1998a). All tactics in this category rely on the absence of cross-resistance between component insecticides.

Strategies implemented to contend with resistance in practice have tended to adopt combinations of the above three approaches. As an example, measures introduced in the early 1980s to combat pyrethroid resistance in the bollworm, *H. armigera*, on cotton (reviewed by Forrester et al. 1993) involved restricting the period over which pyrethroids could be used (management by moderation) and recommending the use of non-pyrethroid alternatives outside this period (management by multiple attack). Other recommendations to target pyrethroids against neonate larvae (enabling even pyrethroid-resistant phenotypes to be killed) and to co-apply pyrethroids with the synergist piperonyl butoxide (PBO) (to suppress detoxification systems) emerged from subsequent work on the underlying mechanisms, and introduced components of management by saturation into the strategy. Unfortunately, even these measures failed to prevent a gradual increase in the frequency of pyrethroid resistance in *H.*

armigera. The Australian strategy nonetheless pioneered a number of principles relating to the design, implementation and support of large-scale resistance management, and has rightly achieved a great deal of international acclaim.

A resistance management strategy introduced in Israel in the late 1980s against the whitefly, *B. tabaci,* and co-existing cotton pests, also relies heavily on restricting the use of lead compounds (in this case to a single application per season), and on rotating insecticides in a sequence intended to protect beneficial organisms and to exploit non-chemical tactics as much as possible (Horowitz et al. 1994, 1995, 1999). Again, this has not completely prevented resistance, but it has resulted in a dramatic reduction in the number of insecticide sprays on cotton. Similar results have been obtained by extending components of the Israeli strategy to the cotton/vegetable cropping systems of the southwestern USA (Dennehy et al. 1996; Dennehy and Williams 1997).

One notable feature of these and many other resistance management strategies is that they were formulated (initially at least) with little or no knowledge of the resistance mechanisms present already or of those which might arise subsequently. Their primary objective was (and continues to be) to prevent resistance phenotypes reaching economically damaging frequencies. In principle, this objective is achievable with no biochemical or genetic input whatsoever. However, there is also no doubt that the latter can, if available, be exploited to assist with all three major approaches to resistance management. Possible applications of such knowledge include:

1. development of in vitro techniques for diagnosing and monitoring specific mechanisms and genotypes (applicable to any management tactic);
2. identification of biochemical or chemical countermeasures to particular resistance mechanisms (applicable to the saturation approach); and
3. resolution and rationalization of cross-resistance patterns (vital for the successful implementation of management by multiple attack).

These opportunities are explored in more detail below.

4
Diagnosing Resistance

The theoretical framework of population genetics predicts that resistance is far easier to circumvent when the underlying resistance genes are still rare and/or very localized in their distribution (Roush and Daly 1990). Unfortunately, the development of resistance often goes unnoticed at the critical early stage, only becoming apparent once the proportion of resistant individuals is sufficiently high to visibly impair the efficacy of pesticide applications. By then, the options for modifying control regimes to combat the threat are very limited and far less likely to be effective. In addition, scope for switching to new chemical groups once resistance becomes established is diminishing as novel chemicals become increasingly difficult to discover and register, and cross-resistance conferred by broad-spectrum detoxification systems and insensitive target proteins becomes increasingly pervasive (Denholm et al. 1998a).

Monitoring programmes to detect resistance genotypes and/or phenotypes, as early as possible, and to document their distribution, should therefore be a key component of any resistance management strategy. Whole-organism bioassays, involving topical application or exposure to pesticide residues on surfaces or in food, have long been the cornerstone of such programmes, but are limited in their application and power of resolution (Roush and Miller 1986). Comparisons of LD_{50} or LD_{90} values of samples from populations – the most widely adopted approach – may be useful for detecting a high frequency of resistant insects but are far too insensitive for detection incipient resistance. Use of a "discriminating" dose or concentration corresponding to the LD_{99} or higher of baseline susceptible populations, although a better use of resources, is still subject to important statistical constraints. Firstly, the estimation of these doses is challenging because the fitting of probit or logit models is usually imprecise at the extreme ends of dose-response relationships. Secondly, unless doses are perfectly diagnostic (i.e. killing 100% of susceptible ones but no resistance ones, which is rarely the case), sample sizes required for the reliable detection of even 1% resistance may be prohibitively large (Roush and Miller 1986).

4.1
In Vitro Assays for Diagnosing Resistance

In principle, many of these difficulties can be overcome by developing more precise tests that can be applied rapidly on a large scale and which distinguish unambiguously between susceptible and resistant individuals. Given recent progress in resolving the nature of resistance mechanisms, in vitro assays for diagnosing the genes or gene products responsible for resistance are becoming increasingly feasible as alternatives to whole-organism bioassays. Their potential advantages over bioassays include the provision of more direct information on the genetic composition of a population, and the possibility of distinguishing between resistance heterozygotes and homozygotes. The latter is especially important in the early stages of selection, when virtually all-resistant alleles are present in a heterozygous condition.

In vitro assays nonetheless suffer from their own limitations and constraints, especially when applied to track resistance in field populations. Firstly, they are obviously wholly dependent on a through understanding of the biochemical and/or molecular basis of resistance in a particular species, usually obtained through detailed studies on populations in which resistance is already well advanced. This constraint may be overcome if there is an opportunity to transfer technology from one species to another, or to apply knowledge from one part of a species' range to another. A good example of the latter relates to the early detection of pirimicarb resistance caused by modified acetyl-cholinesterase (MACE) in *M. persicae* in the UK. This mechanism was first detected in southern Europe in 1990 (Moores et al. 1994), enabling its bio-chemical properties to be established and a microplate assay for pirimicarb insensitivity to be optimised before its appearance in northern Europe. As a consequence, MACE aphids were first identified in the UK in 1995 in samples from aerial suction traps, in advance of any control problems with pirimicarb being reported. This not only alerted UK growers to threats posed by MACE, but also enabled localized control failures with pirimicarb the following year (1996) to be attributed ambiguously to the new mechanism, and management recommendations to be revised accordingly (Foster et al. 1998).

Another limitation of biochemical assays is they often depend on the use of model substrates for the enzyme under consideration, in order to obtain the required sensitivity. Such substrates are generally of broad specificity and often affected by many similar enzymes; it is essential therefore that the differences in activity measured have a proven bearing on resistance, or at least correlate fully with differences in tolerance of insecticides. Difficulties with establishing a direct causal link between biochemical enzyme markers and resistance are exemplified by work on pyrethroid resistance in the bollworm *H. armigera* in Australia. Gunning et al. (1996) demonstrated a consistent association between pyrethroid resistance and enhanced esterase activity, with the most resistant populations showing an approximately 50-fold increase in enzyme activity compared with susceptible ones. This finding, in turn, highlighted an exciting opportunity to monitor resistance using electrophoretic or microplate assays based on increased esterase hydrolysis of 1-naphthyl acetate. However, the validity of these results was challenged by the conventional view, based largely on the synergism of resistance by known oxidase inhibitors including PBO, that resistance in Australia was due primarily to enhanced activity of cytochrome P450 monooxygenases (e.g. Forrester et al. 1993). The discovery that PBO suppresses esterase as well as monooxygenase activity in heliothine bollworms (Gunning et al 1998; J. Ottea, pers. comm. 1999) supports an inter-pretation of pyrethroid resistance being attributable to enhanced esteratic hydrolysis or sequestration, and reinforces the utility of an esterase-based assay as a reliable in vitro diagnostic for resistance monitoring.

In contrast to whole-organism bioassays, which recognise any type or com-bination of resistance mechanisms, biochemical and DNA diagnostics are, by definition, specific to particular metabolizing enzymes, target proteins or genes. Hence when testing field populations, the absence of an in vitro resis-tance marker shows only that insects lack certain resistance mechanisms or genes; it does not necessarily demonstrate the absence of resistance *per se*. Reliance on one specific assay can therefore be highly misleading if populations contain other unsuspected mechanisms or even different allelic forms of the same resistance gene. Given the evidence that target site resistance conferred by altered acetylcholinesterase and *kdr* can each arise from different amino acid substitutions (e.g. Mutero et al. 1994; Williamson et al. 1996; Schuler et al. 1998), DNA diagnostics specific to one particular point mutation require very careful scrutiny and validation before being incorporated into large-scale monitoring programmes. It is also important to appreciate that the predominance of resistance mechanisms can change over time, as shown by apparently seasonal fluctuations in the occurrence of biochemical markers for esterase- and monooxygenase-mediated resistance to pyrethroids in *H. armigera* in India (Kranthi et al. 1997).

With sufficient underpinning research, in vitro assays can also contend effectively with multiple resistance mechanisms. This is demonstrated clearly by research on *M. persicae*. Until relatively recently, only one resistance mechanism had been identified in this species; the over-production of carboxylesterases that degrade and/or sequester insecticidal esters (Field et al. 1988). Although this mechanism was initially implicated in strong resistance to both organophosphates and pyrethroids, subsequent analysis of the *M. persicae* sodium channel gene for mutations associated with *kdr* resistance, coupled with supporting bioassays, has shown pyrethroid resistance to be attributable largely to this target site mechanism instead (Martinez-Torres et al. 1999). Knockdown resistance had undoubtedly been present in aphid populations for many years, but overlooked due to lack of a suitable detection method. In 1990, a further mechanism based on insensitive AChE was identified in southern Europe (Moores et al. 1994). This confers resistance very specifically to the carbamate pirimicarb, which has historically proved the most effective chemical for combating high esterase-based resistance, and to triazamate, a novel aphicide that also inhibits AChE.

Collectively, these three mechanisms provide strong resistance to virtually all available chemicals except the novel neonicotinoid imidacloprid, which is consequently becoming widely used on several *M. persicae* hosts throughout Europe. To safeguard the efficacy of imidacloprid, and to continue to make best use of older insecticides, it is proving of major benefit to be able to monitor all three mechanisms separately in individual aphids through a combination of bioassays, biochemical assays and DNA diagnostics (Field et al. 1997). These methods are now being applied widely in the UK to study the dynamics of resistance, the interactions between different mechanisms, and to provide advice on the most likely control strategies for combating their occurrence within and between seasons.

Although in vitro mechanism-based assays are by no means a universal solution to the limitations of conventional bioassays, they probably represent the most generally applicable way that knowledge of mechanisms can contribute to understanding the dynamics and development of resistance in practice. Continued progress with understanding the biochemical and molecular basis of resistance, coupled with advances in the precision and throughput of such diagnostic techniques (at the DNA level especially), is likely to result in further opportunities to exploit these assays in resistance monitoring programmes.

5
Overpowering Resistance Mechanisms

Chemical synergists have long been exploited as tools for the laboratory diagnosis of particular resistance mechanisms, based on their ability to inhibit specific detoxifying pathways. They retain an important role in this respect, although recent work questioning the binding specificity of one of the most widely used compounds, PBO (Gunning et al. 1998), has highlighted the risk of drawing categorical conclusions using synergism data alone. This finding has endorsed the importance of interpreting such data with care (Scott 1990), and, whenever possible, of validating suspicions through direct insecticide binding or metabolism studies.

The use of synergists for overpowering resistance under field conditions has been constrained by a number of factors relating to their toxicity, cost, photostability, and mechanism-specificity when faced with possible multiple resistance. An exception to this has been the co-application of PBO and pyrethroids for overcoming supposed monooxygenase-mediated resistance to pyrethroids in *H. armigera* in Australia. The recommendation to use PBO followed a comprehensive study by Forrester et al. (1993) of the degree of synergism conferred by a wide range of candidate chemicals against susceptible and resistant strains of *H. armigera*. In the laboratory, pre-treatment with PBO more-or-less eliminated the strong resistance observed when testing with pyrethroids alone. Despite restrictions placed on the use of the synergist, the effectiveness of this tactic declined rapidly, suggesting the appearance of a different, non-synergizable mechanism, or even resistance to the synergist itself (Forrester and Bird 1996; Holloway and Forrester 1998).

Another example of the difficulty with exploiting synergism to combat resistance comes from work on *B. tabaci* in the southwestern USA. During the early 1990s, one response to increasing whitefly resistance was to screen numerous combinations of products for possible synergistic effects (e.g. Wolfenbarger and Riley 1994). Some mixtures of pyrethroids and organophosphates, especially of fenpropathrin and acephate, proved remarkably effective in this respect. Despite immunity to acephate and very high resistance to fenpropathrin applied alone, co-application of a fixed concentration of 1000 ppm acephate (non-toxic in its own right) with varying concentrations of fenpropathrin increased the toxicity of the pyrethroid by over 1000-fold (Simmons and Dennehy 1996). Although the biochemical basis of this extreme synergy has not been fully resolved, it probably reflected inhibition by acephate of metabolic enzymes conferring resistance to fenpropathrin and other pyrethroids applied singly.

Based on this finding, tank mixes of pyrethroids and organophosphates became widely adopted for controlling *B. tabaci*. By the end of the 1995 season, however, such mixtures had failed to control whiteflies throughout much of central Arizona (Dennehy and Williams 1997). Laboratory bioassays confirmed that over-reliance on synergized pyrethroids for field control had led to a gradual loss of synergism of fenpropathrin by acephate. As in the case of *H. armigera*, this could reflect selection of a new, non-synergizable pyrethroid resistance mechanism (e.g. target-site *kdr* resistance) or a modification of one already present.

Several other authors have proposed the use of synergists for managing resistance. For example, Archer et al. (1994) found that resistance to chlorpyrifos in the greenbug, *Schizaphis graminum* (Rondani), could be suppressed by mixtures with carbofuran, or if insects were pre-exposed to certain synergists. Much work has been conducted on resistance mechanisms and synergism in the German cockroach, *B. germanica* (Cochran 1995; Lee et al. 1996; Scharf et al. 1996, 1997, 1999). Different levels of resistance to carbamates, organophosphates and pyrethroids have been reported, and attributed to several mechanisms (including elevated monooxygenase and esterase activities and *kdr* resistance). Furthermore, in some strains, the expression of the two types of detoxification mechanism appeared to differ between life-stages. In such situations, coordinating the use of synergists and ensuring their effectiveness might be possible (Scharf et al. 1999) but would clearly be a complex task.

The future role of synergists in resistance management is difficult to judge at present. On the one hand, advances with implicating and characterizing specific detoxification systems involved in resistance should, in principle, improve prospects for identifying synergists applicable to a wider range of enzymes, and perhaps better suited for deployment in the field. On the other hand, experience gained in the USA and Australia demonstrates that tactics optimised for and specific to particular resistance mechanisms can be rapidly compromised by the genetic plasticity of insect pests. There is certainly no doubt that synergism itself is as fragile a resource as pesticide susceptibility, in need of careful management and monitoring wherever it is exploited.

6
Resolving and Exploiting Cross-Resistance

Resistance mechanisms seldom affect just one toxin. Usually, they confer different levels of resistance to a range of insecticides, through the phenomenon of cross-resistance. Resolving cross-resistance patterns conferred by individual mechanisms can be one of the most challenging aspects of resistance research, but is often essential in order to fine-tune management recommendations based, for example, on the alternation of insecticides to avoid continuous selection for the same resistance gene or mechanism. Equally importantly, detailed studies of cross-resistance may identify compounds within an insecticide group little affected or even unaffected by resistance, which offer potential as alternatives to ones whose efficacy may already be severely compromised.

Although sometimes considered a spin-off of research into resistance mechanisms (e.g. McCaffery 1998), the resolution of cross-resistance patterns is by no means dependent on biochemical or molecular data. At a superficial level it might be sufficient to bioassay resistant strains with a variety of chemicals, and quantify the level of resistance displayed to each. While yielding information of immediate practical benefit, however, this approach is generally unable to distinguish true cross-resistance patterns (i.e. ones conferred by single mechanisms) from ones reflecting the combined expression of multiple resistance mechanisms. From a resistance management perspective, this is a very important distinction unless co-existing mechanisms are physically very closely linked or co-segregate consistently for other reasons such as parthenogenetic reproduction. Empirical approaches for distinguishing between true cross-resistance and multiple resistance include: (1) repeated back-crossing of resistant populations to fully susceptible ones to establish whether resistance to one chemical does invariably co-segregate with resistance to another; and (2) reciprocal selection experiments whereby populations selected for resistance to one chemical are examined for a correlated change in response to another.

The main advantage of combining this type of work with knowledge of mechanisms is to render cross-resistance more explicable and predictable than would otherwise be the case. Cross-resistance patterns, which usually reflect either a shared target site or a common detoxification pathway, are likely to be far easier to rationalize with such knowledge to hand, and this may in turn assist with identifying or even designing "resistance-defeating" alternatives. Two comprehensive studies of cross-resistance within the pyrethroid class have provided intriguing insights into relationships between resistance levels and chemical structure, and how these depend on the type of resistance mechanism present. Forrester et al. (1993) tested a very wide range of commercialised and experimental structures against a strain of *H. armigera* exhibiting pyrethroid resistance based on a detoxification system suppressible by PBO. Changes to the alcohol moiety of pyrethroid molecules caused a dramatic variation in resistance levels, and in some cases led to resistance being virtually eliminated. The study identified no less than seven fully resistance-breaking pyrethroids, although none of these possessed the necessary physicochemical and toxicological attributes to be candidates for commercial development.

Using strains of the housefly (*M. domestica*) homozygous for either the *kdr* or more potent *super-kdr* genes conferring target-site resistance to pyrethroids, Farnham and Khambay (1995a,b) demonstrated significant differences in cross-resistance patterns conferred by the two alleles. The expression of *super-kdr* resistance again proved highly dependent on the nature of the alcohol moiety, although structure-activity relationships differed considerably from those for metabolic resistance in *H. armigera*.

Currently, one of the most exciting challenges of all is to relate such observational data to information on the three-dimensional structure and binding characteristics of receptor molecules and detoxifying enzymes. For target site mechanisms in particular, advances in understanding pharmacological aspects of insecticide-receptor interactions, assisted by computer modeling and the functional expression of mutant receptors, is rendering the design of resistance-defeating compounds acting on these sites a feasible proposition.

The possibility of taking this approach further and engineering compounds showing negative cross-resistance, whereby individuals resistant to one compound show enhanced susceptibility to another, must in some respects represent the ultimate prize for resistance management. To date, reports of this phenomenon have been rare (e.g. Elliott et al. 1986; Yamamoto et al. 1993; Guedes et al. 1997) and not readily exploitable in practice. One exception relates to the Onchocerciasis Control Programme (OCP) of the World Health Organisation, whose aim has been the elimination of river blindness in Africa through the larvicidal treatment in rivers of its vectors, members of the *Simulium damnosum* complex (Kurtak et al. 1987). In routine screening tests, larvae resistant to the organophosphate temephos were found to show enhanced susceptibility to permethrin, which was subsequently incorporated into an insecticide alternation strategy.

The novel pyrrole insecticide chlorfenapyr has exhibited increased toxicity against pyrethroid-resistant strains of the tobacco budworm, *H. virescens* (Pimprale et al. 1997), and the horn fly, *Haematobia irritans* (L) (Sheppard and Joyce 1998). In both cases the underlying mechanism is likely to reflect increased bio-activation of chlorfenapyr (a pro-insecticide) by monooxygenases contributing to enhanced detoxification of pyrethroids. Prospects for identifying negatively cross-resisted molecules through design rather than serendipity undoubtedly exist; whether or not such compounds will be seen to justify commercial development, given threats posed by the appearance of alternative resistance mechanisms, remains open to question.

7
Conclusions

Although of relatively recent appearance, insecticide resistance is now very widespread and is being increasingly used as a model system for understanding how organisms adapt to human activity and environmental stress. In recent years, biochemists and molecular biologists have established a role at the forefront of such research. Their contribution should, in our view, be evaluated first and foremost for its fundamental significance to evolutionary biology rather than its short-term practical benefits.

Although this work does have important implications for pest management, particularly for the design of new toxophores less vulnerable to known routes of resistance, its direct relevance to managing resistance is constrained on at least two counts. Firstly, time lags between discovering and characterizing resistance mean that data on the underlying mechanisms may not available when they are most needed, i.e. at the early stages of resistance development. Secondly, even if such data were available, attempts to combat specific mechanisms are continuously threatened by the genetic plasticity of insect pests, and hence the risk of selecting for alternative mechanisms. Despite our rapidly increasing knowledge of the biochemical and molecular nature of resistance, it seems certain that management strategies will continue to rely largely on "broad-brush" tactics equally applicable to whatever mechanism may be present.

Acknowledgements. We thank Martin Williamson and Isaac Ishaaya for reviewing the manuscript.

References

Archer TL, Bynum ED, Plapp FW (1994) Chlorpyrifos resistance in greenbugs (Homoptera: Aphididae): cross-resistance and synergism. J Econ Entomol 87:1437–1440

Brown TM (ed) (1996) Molecular genetics and evolution of pesticide resistance. ACS Symp Ser 645. American Chemical Society, Washington, DC

Castle SJ, Prabhaker N, Henneberry TJ (1999) Insecticide resistance and its management in cotton insects. ICAC review article on cotton production research no 5. International Cotton Advisory Committee, Washington, DC

Cochran DG (1995) Insecticide-resistance. In: Owens JM, Rust MK, Reierson DA (eds) Understanding and controlling the German cockroach. Oxford University Press, Oxford, pp 171–196

Denholm I, Rowland MW (1992) Tactics for managing pesticide resistance in arthropods: theory and practice. Annu Rev Entomol 37:91–112

Denholm I, Horowitz AR, Cahill M, Ishaaya I (1998a) Management of resistance to novel insecticides. In: Ishaaya I, Degheele D (eds) Insecticides with novel modes of action: mechanisms and application. Springer, Berlin Heidelberg New York, pp 260–282

Denholm I, Picket JA, Devonshire AL (eds) (1998b) Insecticide resistance: from mechanisms to management. CABA, Wallingford/Royal Society, London

Dennehy TJ, Williams L (1997) Management of resistance in *Bemisia* in Arizona cotton. Pestic Sci 51:398–406

Dennehy TJ, Williams L, Russell JS, Li X, Wigert M (1996) Monitoring and management of whitefly resistance to insecticides in Arizona. Proc 1996 Beltwide Cotton Prod Conf, National Cotton Council of America, Memphis, pp 135–140

Elliott M, Farnham AW, Janes NF, Johnson DM, Pulman DA, Sawicki RM (1986) Insecticidal amides with selective potency against a resistant (*super-kdr*) strain of houseflies (*Musca domestica* L.). Agric Biol Chem 50:1347–1349

Farnham AW, Khambay BPS (1995a) The pyrethrins and related compounds. XXXIX Structure-activity relationships of pyrethroidal esters with cyclic side chains in the alcohol component against resistant strains of housefly (*Musca domestica*). Pestic Sci 44: 269–275

Farnham AW, Khambay BPS (1995b) The pyrethrins and related compounds. XL Structure-activity relationships of pyrethroidal esters with acyclic side chains in the alcohol component against resistant strains of housefly (*Musca domestica*). Pestic Sci 44:277–281

ffrench-Constant RH, Rocheleau TA, Steichen JC, Chalmers AE (1993) A point mutation in a *Drosophila* GABA receptor confers insecticide resistance. Nature 363:449–451

ffrench-Constant RH, Anthony NM, Andreev D, Aronstein K (1996) Single versus multiple origins of insecticide resistance: inferences from the cyclodiene resistance gene *Rdl*. In: Brown TM (ed) Molecular genetics and evolution of pesticide resistance. ACS Symp Ser 645. American Chemical Society, Washington, DC, pp 106–116

ffrench-Constant RH, Pittendrigh B, Vaughan A, Anthony N (1998) Why are there so few resistance-associated mutations in insecticide target genes? Philos Trans R Soc Lond B Sci 353:1685–1693

Field LM, Devonshire AL, Forde BG (1988) Molecular evidence that insecticide resistance in peach-potato aphids (*Myzus persicae* Sulz.) results from amplification of an esterase gene. Biochem J 251:309–312

Field LM, Anderson AP, Denholm I, Foster SP, Harling ZK, Javed N, Martinez-Torres D, Moores GD, Williamson MS, Devonshire AL (1997) Use of biochemical and DNA diagnostics for characterising multiple mechanisms of insecticide resistance in the peach-potato aphid, *Myzus persicae* (Sulzer). Pestic Sci 51:283–289

Forrester NW, Bird LJ (1996) The need for adaptation to change in insecticide resistance management strategies: the Australian experience. In: Brown TM (ed) Molecular genetics and evolution of pesticide resistance. ACS Symp Ser 645. American Chemical Society, Washington, DC, pp 160–168

Forrester NW, Cahill M, Bird LJ, Layland JK (1993) Management of pyrethroid and endosulfan resistance in *Helicoverpa armigera* (Lepidoptera: Noctuidae) in Australia. Bull Entomol Res Suppl 1:1–132

Foster SP, Denholm I, Harling ZK, Moores GD, Devonshire AL (1998) Intensification of insecticide resistance in UK field populations of the peach-potato aphid, *Myzus persicae* (Hemiptera: Aphididae) in 1996. Bull Entomol Res 88:127–130

Georghiou GP (1983) Management of resistance in arthropods. In: Georghiou GP, Saito T (eds) Pest resistance to pesticides. Plenum Press, New York, pp 769–792

Georghiou GP (1994) Principles of insecticide resistance management. Phytoprotection 75 [Suppl]:51–59

Guedes RNC, Zhu KY, Kambhampati S, Dover BA (1997) An altered acetylcholinesterase conferring negative cross-insensitivity to different insecticidal inhibitors in organophosphate-resistant lesser grain borer, *Rhyzopertha dominica*. Pestic Biochem Physiol 58:55–62

Gunning RV, Moores GD, Devonshire AL (1996) Esterases and esfenvalerate resistance to thiodicarb in Australian *Helicoverpa armigera* Hübner (Lepidoptera: Noctuidae). Pestic Biochem Physiol 54:12–23

Gunning RV, Moores GD, Devonshire AL (1998) Inhibition of resistance-related esterases by piperonyl butoxide in *Helicoverpa armigera* (Lepidoptera: Noctuidae) and *Aphis* gossypii (Hemiptera: Aphididae). In: Jones DG (ed) Piperonyl butoxide: the insecticide synergist. Academic Press, San Diego, pp 215–226

Hemingway J, Hawkes N, Prapanthadara LA, Jayawardenal KGI, Ranson H (1998) The role of gene splicing, gene amplification and regulation in mosquito insecticide resistance. Philos Trans R Soc Lond B 353:1695–1699

Holloway JW, Forrester NW (1998) Pyrethroid resistance in Heliothine pests of cotton: mechanisms and management. In: Zalucki MP, Drew RAI, White GG (eds) Pest management – future challenges. Proc 6th Aust Appl Entomol Res Con, Univ. Queensland, Brisbane, vol 2, pp 171–178

Horowitz AR, Forer G, Ishaaya I (1994) Managing resistance in *Bemisia tabaci* in Israel with emphasis on cotton. Pestic Sci 42:113–122

Horowitz AR, Forer G, Ishaaya I (1995) Insecticide resistance management as a part of an IPM strategy in Israeli cotton fields. In: Constable GA, Forrester NW (eds) Challenging the future. Proc World Cot Res Conf 1 CSIRO, Melbourne, pp 537–544

Horowitz AR, Mendelson Z, Cahill M, Denholm I, Ishaaya I (1999) Managing resistance to the insect growth regulator, pyriproxyfen, in *Bemisia tabaci*. Pestic Sci 55:272–276

Kranthi KR, Armes NJ, Rao NGV, Raj S, Sundaramurthy VT (1997) Seasonal dynamics of metabolic mechanisms mediating pyrethroid resistance in *Helicoverpa armigera* in central India. Pestic Sci 50:91–98

Kurtak D, Meyer R, Ocran M, Ouedrago M, Renaud P, Swadogo RO, Tele B (1987) Management of insecticide resistance in control of the *Simulium damnosum* complex by the Onchocerciasis control program, West Africa: potential use of negative correlation between organophosphate resistance and pyrethroid susceptibility. Med Vet Entomol 1:137–146

Lee CY, Yap HH, Chong NL, Lee RST (1996) Insecticide resistance and synergism in field collected German cockroaches (Dictyoptera, Blattellidae) in Peninsular Malaysia. Bull Entomol Res 86:675–682

Martinez-Torres D, Devonshire AL, Williamson MS (1997) Molecular studies of knockdown resistance to pyrethroids: cloning of domain II sodium channel gene sequences from insects. Pestic Sci 51:265–270

Martinez-Torres D, Foster SP, Field LM, Devonshire AL, Williamson MS (1999) A sodium channel point mutation is associated with resistance to DDT and pyrethroid insecticides in the peach-potato aphid, *Myzus persicae* (Sulzer) (Hemiptera: Aphididae). Insect Mol Biol 8:339–346

McCaffery AR (1998) Resistance to insecticides in heliothine Lepidoptera: a global view. Philos Trans R Soc Lond B 353:1735–1750

McKenzie JA (1996) Ecological and evolutionary aspects of insecticide resistance. Landes/Academic Press, San Diego

Moores GD, Devine GJ, Devonshire AL (1994) Insecticide-insensitive acetylcholinesterase can enhance esterase-based resistance in *Myzus persicae* and *Myzus nicotianae*. Pestic Biochem Physiol 49:114–120

Mullin CA, Scott JG (1992) Biomolecular basis for insecticide resistance: classification and comparisons. In: Mullin CA, Scott JG (eds) Molecular mechanisms of insecticide resistance. ACS Sym Ser 505:1–13

Mutero A, Pralavorio M, Bride JM, Fournier D (1994) Resistance-associated point mutations in insecticide-insensitive acetylcholinesterase. Proc Natl Acad Sci USA 91:5922–5926

Newcomb RD, Campbell PM, Ollis DL, Cheah E, Russell RJ, Oakeshott JG (1997) A single amino acid substitution converts a carboxylesterase to an organophosphorus hydrolase and confers insecticide resistance on a blowfly. Proc Natl Acad Sci USA 94:7464–7468

Otto D, Weber B (eds) (1990) Insecticides: mechanism of action and resistance. Intercept, Andover, Hants

Pimprale SS, Besco CL, Bryson PK, Brown TM (1997) Increased susceptibility of pyrethroid-resistant tobacco budworm (Lepidoptera: Noctuidae) to chlorfenapyr. J Econ Entomol 90:49–54

Roush RT (1989) Designing resistance management programs: how can you choose? Pestic Sci 26: 423–441

Roush RT, Daly JC (1990) The role of population genetics in resistance research and management. In: Roush RT, Tabashnik BE (eds) Pesticide resistance in arthropods. Chapman and Hall, New York, pp 97–152

Roush RT, Miller GL (1986) Considerations for the design of insecticide resistance monitoring programs. J Econ Entomol 79:293–298

Roush RT, Tabashnik BE (eds) (1990) Pesticide resistance in arthropods. Chapman and Hall, New York

Sawicki RM (1981) Problems in countering resistance. Philos Trans R Soc Lond B 295:143–151

Scharf ME, Hemingway J, Reid BL, Small GJ, Bennett GW (1996) Toxicological and biochemical characterization of insecticide resistance in a field-collected strain of *Blattella germanica* (Dictyoptera, Blattellidae). J Econ Entomol 89:322–331

Scharf ME, Kaakeh W, Bennett GW (1997) Changes in an insecticide-resistant field population of German cockroach (Dictyoptera: Blattellidae) after exposure to an insecticide mixture. J Econ Entomol 90:38–48

Scharf ME, Lee CY, Neal JJ, Bennett GW (1999) Cytochrome P450 MA expression in insecticide-resistant German cockroaches (Dictyoptera: Blattellidae). J Econ Entomol 92:788–793

Schuler TH, Martinez-Torres D, Thompson AJ, Denholm I, Devonshire AL, Duce IR, Williamson MS (1998) Toxicological, electrophysiological, and molecular characterisation of knockdown resistance to pyrethroid insecticides in the diamondback moth, *Plutella xylostella* (L.). Pestic Biochem Physiol 59:169–182

Scott JG (1990) Investigating mechanisms of insecticide resistance: methods, strategies, and pitfalls. In: Roush RT, Tabashnik BE (eds) Pesticide resistance in arthropods. Chapman and Hall, New York, pp 39–57

Scott JG (1999) Cytochromes P450 and insecticide resistance. Insect Biochem Mol Biol 29:757–777

Sheppard DC, Joyce JA (1998) Increased susceptibility of pyrethroid-resistant horn flies (Diptera: Muscidae) to chlorfenapyr. J Econ Entomol 91:398–400

Simmons AL, Dennehy TJ (1996) Contrasts of three insecticide resistance monitoring methods for whitefly. Proc 1996 Beltwide Cotton Prod Conf, National Cotton Council of America, Memphis, pp 748–752

Tabashnik BE (1989) Managing resistance with multiple pesticide tactics: theory, evidence, and recommendations. J Econ Entomol 82:1263–1269

Thompson M, Steichen JC, ffrench-Constant RH (1993) Conservation of cyclodiene insecticide resistance-associated mutations in insects. Insect Mol Biol 2:149–154

Williamson MS, Martinez-Torres D, Hick CA, Castells N, Devonshire AL (1996) Analysis of sodium channel gene sequences in pyrethroid- resistant houseflies – progress toward a molecular diagnostic for knockdown resistance (*kdr*). In: Brown TM (ed) Molecular genetics and evolution of pesticide resistance. ACS Symp Ser 645. American Chemical Society, Washington, DC, pp 52–61

Wolfenbarger DA, Riley DG (1994) Toxicity of mixtures of insecticides and insecticides alone against B-strain sweet-potato whitefly. Proc 1994 Beltwide Cotton Prod Conf, National Cotton Council of America, Memphis, pp 1214–1216

Yamamoto I, Kyomura N, Takahashi Y (1993) Negatively correlated cross resistance: combinations of N-methylcarbamate with N-propylcarbamate or oxadiazolone for green rice leafhopper. Arch Insect Biochem Physiol 22:277–288

Subject Index

Abamectin 2, 6, 10, 35–36
Absorption modeling 310–314
Acephate 331–332
Acetamiprid 1, 5, 10, 84, 86–87, 95–96
Acetylcholine 85, 94, 96, 99–101, 222–224, 269, 279, 283–284
Acetylcholine receptors 5–7, 10, 77–101
Acetylcholinesterases 216, 221–235, 269, 274, 276–277, 283–286, 299, 301, 324–325, 330
Adenylate cyclase 192
Aedes aegypti 113, 243, 303
Aerobic respiration 243
Agelenopsis aperta 63
Agonists 98–101, 143
Agrotis ipsilon 158
Aldicarb 225–226, 229, 231
Alkylamides 48, 50, 67
Allatostatins 121
Allatotropins 121
Allelochemicals 243–244
Allosteric antagonism 64–69
Allosteric coupling 64–69
Altered acetylcholinesterase 229–234, 298, 301, 329–330
Altered target resistance 64, 256, 270–272, 283–284, 286, 324–325, 328
Alternations 326, 333
Amplified esterases 209–217
Amylases 297, 314
Anatoxin A 98, 100–101
Anopheles albimanus 224
Anopheles gambiae 244–245, 247–248
Antagonists 98–99
Anthonomus grandis 309
Apis mellifera 306
Apolysis 110
ATP 8, 174, 176, 191
ATPase 8, 173–175, 191
Autographa californica polyhedrosis virus 63
Avermectins 6–7, 28–30, 34–36
Azadirachtin 9
Azasteroids 115
Azinphosmethyl 248, 284

Bacillus thuringiensis δ-endotoxin 167–197
Backbone cyclic neuropeptide-based antagonist 152–162
Backbone cyclization 155–157
Baculovirus 316
Bemisia tabaci 3, 6–7, 223, 325, 327, 331–332
Benzoylphenyl ureas 2–3, 138–139, 300–302, 307
Bicuculline 18
Biobit 9
Biotransformation 272
Blaberus discoidalis 304–306
Blattella germanica 31–32, 285, 325, 332
Bombesin 154
Bombyx mori 108, 113, 125, 144, 158–159, 168–169, 176–182, 184–186
Boophilus microplus 224
Brush border membrane vesicles 182–184, 187–192, 195, 197
Bt receptor binding 186–188, 193, 197
Bt umbrella model 195
Bungarotoxin 5, 83–85, 94
Buprofezin 2–3, 10
Bursicon 110
Butid scorpions 58
Butyrylcholinesterase 221

Caenorhabditis elegans 20, 80–82
Calcium channels 52–56, 61–62, 192
Calliphora vicina 143
Carbamates 210, 221–235, 298–300, 303, 324–325, 332
Carbaryl 225–227, 231, 256, 304–306
Carbofuran 227–228, 332
Carboxylamidases 302
Carboxylesterases 210, 251, 276–277, 279–281, 285–286, 299–300, 330
Carcinogenesis 241
Cell lines 117, 119–120, 135–136, 139, 168, 180, 190–192, 315
Ceratitis capitata 113, 243
Chilo supressalis 134, 145
Chironomus tentans 108, 113, 139–140, 142

Printing (Computer to Film): Saladruck, Berlin
Binding: H. Stürtz AG, Würzburg